T0137131

Intelligent Systems Reference Library

Volume 136

The aim of this series is to publish a Reference Library, including novel advances and developments in all aspects of Intelligent Systems in an easily accessible and well structured form. The series includes reference works, handbooks, compendia, textbooks, well-structured monographs, dictionaries, and encyclopedias. It contains well integrated knowledge and current information in the field of Intelligent Systems. The series covers the theory, applications, and design methods of Intelligent Systems. Virtually all disciplines such as engineering, computer science, avionics, business, e-commerce, environment, healthcare, physics and life science are included.

More information about this series at http://www.springer.com/series/8578

Margarita N. Favorskaya
Lakhmi C. Jain
Editors

Computer Vision in Control Systems-4

Real Life Applications

Editors
Margarita N. Favorskaya
Institute of Informatics and Telecommunications
Reshetnev Siberian State University of Science and Technology
Krasnoyarsk
Russia Federation

Lakhmi C. Jain
Faculty of Education, Science, Technology and Mathematics
University of Canberra
Canberra, ACT
Australia

and

Bournemouth University
Poole
UK

and

KES International
Shoreham-by-Sea
UK

ISSN 1868-4394 ISSN 1868-4408 (electronic)
Intelligent Systems Reference Library
ISBN 978-3-319-88527-8 ISBN 978-3-319-67994-5 (eBook)
https://doi.org/10.1007/978-3-319-67994-5

Softcover reprint of the hardcover 1st edition 2017

Printed on acid-free paper

This Springer imprint is published by Springer Nature
The registered company is Springer International Publishing AG
The registered company address is: Gewerbestrasse 11, 6330 Cham, Switzerland

Preface

The research book is a continuation of our previous books which are focused on the recent advances in computer vision methodologies and technical solutions using conventional and intelligent paradigms.

- Computer Vision in Control Systems-1, Mathematical Theory, ISRL Series, Volume 73, Springer-Verlag, 2015
- Computer Vision in Control Systems-2, Innovations in Practice, ISRL Series, Volume 75, Springer-Verlag, 2015
- Computer Vision in Control Systems-3, Aerial ans Satellite Image Processing, ISRL Series, Volume 135, Springer-Verlag, 2018

The research work presented in the book includes a number of real-life applications including the identification of handwritten texts, watermarking techniques, the mobile robot simultaneous localization and mapping, motion control systems of mobile robots, analysis of indoor human activity, face image quality assessment, android device controlling, medical images processing, clinical decision-making and foot progression angle detection.

The book is directed to the Ph.D. students, professors, researchers and software developers working in the areas of digital video processing and computer vision technologies.

We wish to express our gratitude to the authors and reviewers for their contribution. The assistance provided by Springer-Verlag is acknowledged.

Krasnoyarsk, Russian Federation — Margarita N. Favorskaya
Canberra, Australia — Lakhmi C. Jain

Contents

About the Editors

Dr. Margarita N. Favorskaya is a Professor and Head of Department of Informatics and Computer Techniques at Siberian State Aerospace University, Russian Federation.

Professor Favorskaya is a member of KES organization since 2010, the IPC member and the Chair of invited sessions of international conferences. She serves as a reviewer in international journals (Neurocomputing, Knowledge Engineering and Soft Data Paradigms, Pattern Recognition Letters, Engineering Applications of Artificial Intelligence), an associate editor of Intelligent Decision Technologies Journal and Computer and Information Science Journal. She is the author or the co-author of 160 publications and 20 educational manuals in computer science. She co-edited three books for Springer recently. She supervised eight Ph.D. candidates and presently supervising five Ph.D. students.

Her main research interests are digital image and videos processing, remote sensing, pattern recognition, fractal image processing, artificial intelligence and information technologies.

Dr. Lakhmi C. Jain is with the Faculty of Education, Science, Technology and Mathematics at the University of Canberra, Australia, and Bournemouth University, UK. He is a Fellow of the Institution of Engineers, Australia.

Professor Jain founded the KES International for providing a professional community the opportunities for publications, knowledge exchange, cooperation and teaming. Involving around 5000 researchers drawn from universities and companies worldwide, KES facilitates international cooperation and generates synergy in teaching and research. KES regularly provides networking opportunities for professional community through one of the largest conferences of its kind in the area of KES. www.kesinternational.org.

His interests focus on the artificial intelligence paradigms and their applications in complex systems, security, e-education, e-healthcare, unmanned air vehicles and intelligent agents.

Chapter 1
Innovative Algorithms in Computer Vision

Lakhmi C. Jain and Margarita N. Favorskaya

Abstract This chapter contains a brief description of the methods, algorithms, and implementations applied in many fields of computer vision. The graphological analysis and identification of handwritten manuscripts are discussed using the examples of Great Russian writers. A perceptually tuned watermarking using non-subsampled shearlet transform is a contribution in the development of the watermarking techniques. The mobile robot simultaneous localization and mapping, as well as the joined processing of visual and audio information in the motion control systems of the mobile robots, are directed on the robotics' development. The ambient audiovisual monitoring based on a wide set of methods for digital processing of video sequences is another useful real life application. Processing of medical images becomes more and more complicated due to the enforced current requirements of medical practitioners.

Keywords Graphological analysis · Digital watermarking · Simultaneous localization and mapping · Visual and audio decision making · Indoor human activity · Face image quality assessment · Eye detection and tracking Medical image processing · Clinical decision support system · Gait monitoring

L.C. Jain (✉)
Faculty of Education, Science, Technology and Mathematics, University of Canberra, Canberra, ACT 2601, Australia
e-mail: jainlakhmi@gmail.com

L.C. Jain
Bournemouth University, Poole, UK

M.N. Favorskaya
Institute of Informatics and Telecommunications, Reshetnev Siberian State University of Science and Technology, 31, Krasnoyarsky Rabochy ave., Krasnoyarsk 660037, Russian Federation
e-mail: favorskaya@sibsau.ru

M.N. Favorskaya and L.C. Jain (eds.), *Computer Vision in Control Systems-4*,
Intelligent Systems Reference Library 136,
https://doi.org/10.1007/978-3-319-67994-5_1

1.1 Introduction

The core of any control vision system is the algorithms for raw data processing in such manner that it will be possible to obtain the non-evidence dependences in big data volumes, essential improvement of visual data, or reliable recognition of the objects of interest. Each task requires the special approaches in creation of innovative algorithms, as well as the great experimental work before its implementation in real devices. Sometimes, visual data are not enough for decision making. In these cases, audio data can be attracted successfully. The spectrum of real life applications is very wide. Therefore, this book is an attempt to contribute in some spheres of human activity including culture, robotics, human interactions, and medicine.

1.2 Chapters Included in the Book

Chapter 2 includes the issues of graphological analysis and identification of handwritten texts on the examples of the author's calligraphy courtesy provided by the Manuscript Department of the Institute of Russian Literature (Pushkin's House) of the Russian Academy of Sciences [1]. The authors discuss the main challenges of this problem due to great variety of historical texts, when even the samples of writing of some letters can greatly vary depending on the speed of writing and the writing tool used (goose quill, pen point, or pencil). Since a textual analyst follows the own manual algorithm for handwritten text identification, it is possible to design the automated information storage and retrieval system as a useful software tool in order to reduce such long process. Additionally to the existing methods and algorithms of bitmap image transformation, two methods were developed. Vectorization of bitmap images of handwritten texts builds the orientation field of image contours using the following steps: the pre-filtering of the bitmap image, construction of the orientation field for this bitmap image, filtering and extrapolation of the direction field, and searching for and tracking of image contours using the created orientation field. The core idea of this algorithm is to trace the contours in the original image that provides information about the direction of a line or a stroke near a given point. Method of vector dynamic parameterization permits to obtain more information about letter images. This method try to restore the dynamic information about the movement of a pen, when writing a letter, including the indication of the start and end points, movement direction, and the number of cyclic letter outlines. Thus, the vector dynamic representation provides the ability to visualize each letter in the form of a 3D image with x, y and t coordinates, where t is a time. Also the reader can find the detailed description of architecture of the designed software tool based on the client-server technique [2]. The system interacts with the database (around 400 Mb memory), which is pre-filled with a collection of different versions of the author's calligraphy of the classics of Russian literature A.S. Pushkin and A.S. Griboyedov (images of individual letters and their ligaments/ligatures).

Chapter 3 investigates the issues of perceptually tuned watermarking using non-subsampled shearlet transform. The multiple classification and criteria of the watermarking techniques permitted to chose very promising transformation called as digital shearlet transform and its modifications for embedding/extraction of gray-scale or color watermarks into the host images [3]. The tampering attacks categorized as the global and local ones distort a watermark; more several types of attacks can be applied for the same watermark. This causes a necessity to develop the robust watermarking algorithms, also considering the contents of the host image and watermark [4]. Such algorithms ought to find a decision in a contradictory desire to embed the maximum data volume (an image payload) and a visibility. The multiplicative watermarking techniques support a balance of the signal magnitude (fidelity) and the quality of the watermarked image (robustness) at an acceptable level. The multi-scale decomposition analysis is provided by many wavelet-like methods, including ridgelet, curvelet, brushlet, wedgelet, beamlet, contourlet, bandelet, directionlet, and shearlet. The use of the digital shearlet transform produces a variety of sub-bands for inserting the secret data due to the shearlets' multi-resolution property. These sub-bands are correlated in the digital wavelet transform but not correlated in the digital shearlet transform that makes a watermarking more secure procedure. At the same time, the non-subsampled shearlet transform is a fully shift-invariant, multi-scale, and multi-directional expansion of the shearlet transform. The proposed algorithm of a watermark embedding and extraction using a watermarking scrambling via Arnold's transform and paradigms of a human visual system demonstrates good experimental results on 44 color and monochrome images from the database Miscellaneous of University of Southern California. The scaling, disproportionate scaling, rotation, translation, and cropping attacks were simulated, and the impact of the attacks on a watermark was estimated by bit error rate and the peak signal to noise ration metrics.

Chapter 4 conducts the issues of the Simultaneous Localization And Mapping (SLAM) algorithm improvement for the task of the mobile robot simultaneous localization and mapping, which consists of two parts: the evaluation of the trajectory of the robot movement and evaluation the locations of landmarks, which depend on the coordinates of the robot at the time of each measurement. [5]. The authors consider the FastSLAM modification using the nonlinear models through the first order Taylor series expansion at the mean of the landmarks' observations. The probabilistic properties of the SLAM algorithm were considered. The proposed algorithm computes a more accurate mean and uncertainty of the landmarks moving nonlinearly. The method of calibration of Kinect-like cameras, depth map restoration using a modified interpolation technique, and filtering the noise in the RGB images are the main preprocessing methods. Additionally, the improved resampling algorithm for the particle filtering through the adaptive thresholding based on the data of the effective particle number evolution was developed. The proposed algorithm runs in real time and shows good accuracy and robustness in comparison with other modern SLAM systems. The average errors in calculating the displacement of the camera between successive frames using the adaptive threshold values is about 24% less than in the case of the strict thresholding.

Chapter 5 implies the development of fast parallel algorithms for joined processing of visual and audio information in the motion control systems of the mobile robots [6]. The main goal is the path planning of a mobile robot in real time implementation using three types of information obtained from the video camera, microphones, and laser range finders [7]. Such algorithms ought to have the appropriate precision and effectiveness in scenes with the moving objects, people, and obstacles. Fast parallel algorithms are implemented using the multiprocessor hardware structure (the well known NVIDIA Graphics Processing Unit (GPU) processor) and software platform (Compute Unified Device Architecture (CUDA) platform). Many mathematical operations widely used in the images, audio, and other signal processing, such as the matrix multiplication, Fast Fourier transform, correlation and convolution, can be executed faster in a comparison to the traditional Central Processing Unit (CPU). During this research, three algorithms were developed. The Algorithm 1 is based on the SLAM method and the mobile robot audio visual perception and attention [8]. The SLAM method collects and keeps the information for all previous executive locations for a decision making about the following locations of a mobile robot. The Extended Kalman Filter SLAM (EKF SLAM) was modified for parallel execution. Algorithm 2 and Algorithm 3 are the developed fast parallel algorithm based on the mobile robot audio perception and attention for a decision making in motion control system managed by a speaking person [9]. They consider the current audio features defining an angle of sound source arrival or location of sound source. The Steered Response Power (SRP) with PHAse Transform (PHAT) called as SRP-PHAT was applied for this purpose. The Algorithm 1, Algorithm 2, and Algorithm 3 demonstrate clearly the advantages and real time execution of using NIVIDA GPU and CUDA platform instead of CPU in the developed fast algorithm for a decision making in motion control system of a mobile robot based on audio visual information.

Chapter 6 promotes the indoor ambient audiovisual monitoring in the intelligent meeting room with a goal to determine the time events, when the states of the participants' activities are changing [10]. The chapter contains a description of some intelligent meeting rooms equipped with audio and video recording equipment [11]. However, the main discussion deals with the methods and algorithms of image and speech analysis, such as the recognition and identification of faces, detection and tracking of participants, identification of participants' positions, voice activity detection, localization and separation of sound sources, identification of speakers, and recognition of speech and non-speech acoustic events [12]. Also the combined methods are possible, for example, the estimation of the position and orientation of the speaker's head and multimodal recognition of emotions. Such audiovisual monitoring permits to control the events (a new participant enters the meeting room, the presentation begins, and the audience is given the floor) taking place in the meeting room. The proposed image processing includes the removal of low-quality frames, illumination normalization, elimination of image blur, cleaning of digital noise, and automatic segmentation and recognition of the participants' faces. In experimental section, one can find a digital comparison of three basic techniques for face recognition, such as the principal components analysis, linear discriminant analysis, and local binary pattern analysis. The authors developed a method of audiovisual recording of participant activity based on the multifunction system of video monitoring and multichannel system of audio localization that detects the position of the sound source evaluating the phase difference of signals recorded by pairs of microphones from four arrays [13].

Chapter 7 contributes in the audience analysis systems [14] and gender recognition systems [15] using the assessment of face image quality. This is a continuation of investigations of these authors during last two decades. Several standards, such as ISO/IEC 19794-5 and ICAO 9303, contain a description of parameters that provide a decision of the image suitability in the automatic face recognition systems. All parameters are grouped into two classes the textural features (sharpness, contrast and light intensity, compression ratio, and other distortions) and face features (symmetry, position, rotation, eyes visibility, and the presence of glare or shadows on the face). For such algorithms, an assessment of face image quality is a cornerstone because a success of face verification/identification depends strongly from the selected lucky frames. Three types of the quality metrics were considered, such as the texture-based, symmetry-based, and universal metrics. The authors developed the no-reference image quality assessment algorithms based on the analysis of texture and the landmark points' symmetry in the facial images [16]. Note the original methodology of the experiments' statement. Two sets were formed called as Top1 with the single best image of a person and Top3 considering three high quality images chosen by the experts and objective metrics. The proposed algorithms were tested using standard LIVE and TID2013 image database. The performance results show that the proposed algorithms are highly competitive for audience analysis systems and, moreover, have very low computational complexity making it well suited for real-time applications. Additionally, novel gender classification algorithm was suggested based on non-linear SVM classifier. It includes several steps, among which are the scale-invariant feature transform, histogram of oriented gradients, Gabor filters, and pre-selection of blocks with the corresponding positive experimental results.

Chapter 8 contains a description of real time eye blink detection method for Android device controlling and its implementation using standard libraries like OpenCV conversion procedures. The Human Computer Interaction (HCI) or Mobile Computer Interaction (MCI) may be realized in different ways. The authors chose the issue of the eye blinking detection that has a significant role in the human mobile interaction, computer interaction, driving safety, and health care. The heuristic propositions concluded from the theory and practice of the HCI were used during design. The eye tracking is the fastest non-invasive method of measuring user attention [17]. The best definition of eye tracking can be the estimation of user gaze's direction, which is very difficult problem especially in the real world. The proposed method consists of five main parts including the mobile camera processing, face detection, eye detection, eye tracking, and blink detection. At the initial step, a video captured from the front camera of any Android device is converted to grayscale frame using OpenCV, and then stored in special mobile folder to be used later for face detection step. For face detection, Haar classifiers are used, while the eyes detection is based on the trained AdaBoost and Haar feature algorithms. The corneal reflection and pupil center of the eyes have been used as the most important eye features in the proposed method in order to track the movement of the eyes. During eye blink detection step, each black color pixel in the grayscale frame represented by 1, while each gray or white color pixel will be represented by 0. The median blur filtering is applied in this step. The mobile activity is controlled in case the eye blinks by several ways, such as the sending a text message, turning on the alarm system, opening a web browser or making a phone call. The experiments with different distances between a human face and phone and various lighting conditions (indoor/outdoor) were conducted. The overall and detection accuracy had reached 98%. For each frame, the average execution time was 12.30 ms that provides a real time execution.

Chapter 9 contains the rich experimental study of the medical image processing and morphological analysis in urology and plastic surgery (hernioplasty). At present, it is necessary not only to identify the localization of the calculus but also to determine the density and configuration of the calculus and evaluate the functional state of the urinary tract above and below the obstruction. For this purpose, the methodology based on a novel method for color coding of contour representation obtained by the digital shearlet transform was developed. Medical images contain the noise of different nature. In order to improve medical images that can have a high resolution the optimized in implementation algorithms of the most frequently used filters, such as the mean filter, Gaussian filter, median filter, and 2D Cleaner filter, were developed. The study of properties of filters' functioning permitted to use all three type of parallelism, viz. the data, algorithmic, and functional parallelism. A comparison of the optimized and ordinary implementations of noise reduction filters shows great speed improvement of the optimized implementations (around 3–20 times). The highest increase in the processing speed was achieved for the median filter. Thus, for the small kernel 3×3 the acceleration was about 8 times and for the large kernel 11×11 is around 70 times. For the parallel implementation, the OpenMP standard was used. The chapter includes the pseudo-code of some program procedures. For contour representation, the simple conventional methods, such as Roberts, Prewitt, Sobel, and more complex ones (LoG and Canny methods) were tested. However, the main attention was paid to use digital shearlet transform in order to obtain the best results in contour extraction [18]. This approach was enforced by the novel color coding algorithm. It is based on the color selection and density distribution of the isolines in an image corresponds to the known technique of building elastic maps using the spatial data. As a result, the image accuracy of estimates in urology and plastic surgery (hernioplasty) was increased up 10–25% in averaged. In urology, the proposed color coding method increased accuracy, especially in complex cases of multiple stones in the kidney. In hernioplasty, the color coding allows to conduct more efficient analysis of the tissue regeneration by controlling the variability of texture with improved accuracy.

Chapter 10 investigates a study of medical image processing and analysis in clinical decision support systems regarding the cervix oncological changes diagnostics [19, 20]. The novelty of the proposed approach is based on the combined two known approaches that analyze the images obtained in white light illumination and fluorescent images separately. Chapter provides the detailed description of that field and rationality of three classes to define the diagnosis, such as Norm, Chronic Nonspecific Inflammation (CNI), and Cervical Intraepithelial Neoplasia in various types of oncological changes (CIN I, CIN II, CIN III) as a single class. First, the ordinary procedures to improve a quality of the images (noise reduction, contrast enhancement, etc.) are executed. Second, the special medical imaging procedures (the matching medical images taken under different lighting conditions, automatic segmentation of regions of interest, and removal of highlights in the images) are developed. It is worth noting that many algorithms were developed, for example shift compensation algorithm for the multispectral images and their analysis in a combination with the images obtained in white light based on a phase correlation in Fourier domain, segmentation of region of interests using Gauss filtering and morphological operations of dilatation and erosion. Practical recommendations for different image processing are included in this chapter. For classification task, some famous strategies were tested in real images provided by South Korea clinics. These investigations show that the best specificity, sensitivity, and accuracy are obtained using the random decision forest strategy. This result is connected with the high degree of image variability obtained from different patients with differences not only due to pathology but also due to differences in the age, menopause, and other features of woman physical condition. The final results in the sensitivity and specificity exceed the results of inexperienced physicians that demonstrates a possibility of practical application of the pathology maps for colposcopist examination.

Chapter 11 involves some results in Foot Progression Angle (FPA) detection, which is an important measurement in clinical gait analysis [21]. The proposed Visual Feature Matching (VFM) model is a solution for long-term real time FPA monitoring in home or community like environments for the patients with movement disorders or abnormal gait. The FPA is calculated as the angle between the foot vector, line joining the heel point center and the second metatarsal head [22]. Thus, the efforts of the authors were directed to the accurate estimation of foot orientation. Image calibration and rectification are the tools used to eliminate two types of distortion, such as the lens distortion and perspective distortion caused by the optical lens and position of the camera relative to the subject. The classic approach for foot feature extraction and matching includes the algorithm for feature extraction, for example, using the SURF descriptor and matching algorithm, in this case, the statistically robust M-estimator sample consensus algorithm. However, the authors proposed an alternative approach, when a pair of paper strips, instead of shoes, is used in the investigations because of different color and textile of participant's shoes. Some estimators for the FPA measurements were obtained.

The authors proposed the original equipment, when the camera is mounted downward on the torso that has a few advantages.

1.3 Conclusions

In this book, reader will find many original and innovative algorithms from many fields of computer vision. All chapters involve great experimental material with the corresponding explanation of results that makes the proposed methods more valuable. Culture, robotics, community interactions, and medicine are not the closed list of real life application, where computer vision helps to improve a quantity of a human life. A variety of the presented innovative algorithms is wide: from the graphological analysis of handwritten texts, perceptually tuned image watermarking, mobile robots surveillance, and ambient audiovisual monitoring of human meetings to the medical image processing considering the current achievements in computer vision.

References

1. Volkov, D.M., Mironovsky, L.A., Reshetnikova, N.N.: Automated system of graphological research of manuscripts. In: XIII International Scientific and Engineering Workshop “Modern Technologies in Problems of Control, Automation and Information processing”, pp. 482–483 (in Russian) (2004)
2. Artemov, I.V., Volkov, D.M., Kondakova, I.A., Nikitina, A.A., Reshetnikova, N.N., Soloviev, N.V.: Information Retrieval System of graphological analysis and identification of handwritten texts. The certificate of registration of the industry development. FSSI “State Coordinating Center for Information Technologies, an industry fund of algorithms and programs.” Moscow, pp. 1–10 (in Russian) (2006)
3. Favorskaya, M.N., Savchina, E.I.: Content preserving watermarking for medical images using shearlet transform and SVD. Int Arch Photogramm Remote Sens Spatial Inf Sci, XLII-2/W4, 101–108 (2017)
4. Favorskaya, M., Oreshkina, E.: Digital gray-scale watermarking based on biometrics. In: Damiani E, Howlett RJ, Jain LC, Gallo L, De Pietro G (eds.) Intelligent Interactive Multimedia Systems and Services, SIST, vol. 40, pp. 203–214 Springer International Publishing, Switzerland (2015)
5. Prozorov, A., Priorov, A.: Three-dimensional reconstruction of a scene with the use of monocular vision. Meas. Tech. **57**(10), 1137–1143 (2015)
6. Pleshkova, Sn., Bekiarski, Al.: Audio visual attention models in the mobile robots navigation. In: Kountchev, R., Nakamatsu, K. (eds.) New Approaches in Intelligent Image Analysis, ISRL, vol. 108 pp. 253–294 Springer International Publishing, Switzerland (2016)
7. Dehkharghani, S.S., Bekiarski, A., Pleshkova, S.: Application of probabilistic methods in mobile robots audio visual motion control combined with laser range finder distance measurements. In: 11th WSEAS International Conference on Circuits, Systems, Electronics, Control & Signal Processing (CSECS’2012), pp. 91–98 (2012)

8. Al, Bekiarski: Visual mobile robots perception for motion control. In: Kountchev, R., Nakamatsu, K. (eds.) Advances in Reasoning-Based Image Processing Intelligent Systems, ISRL, vol. 29, pp. 173–209. Springer, Berlin (2012)
9. Venkov, P., Bekiarski, Al., Dehkharghani, S.S., Pleshkova, Sn.: Search and tracking of targets with mobile robot by using audio-visual information. In: International Conference on Automation and Informatics (CAI'2010), Sofia, pp. 463–469 (2010)
10. Ronzhin Al.L., Budkov V.Yu.: Determination and recording of active speaker in meeting room. In: 14th 17th International Conference Speech and Computer (SPECOM'2011), Moscow, Kazan, Russia, pp. 361–366 (2011)
11. Yusupov, R.M., AnL, Ronzhin, Prischepa, M., AlL, Ronzhin: Models and hardware software solutions for automatic control of intelligent hall. Automat. Remote Control **72**(7), 1389–1397 (2011)
12. Ronzhin AlL.: Audiovisual recording system for e-learning applications. In: International Conference on Computer Graphics Theory and Applications (GRAPP'2012), pp. 515–518 (2012)
13. Ronzhin, A., Budkov, V., Karpov, A.: Multichannel system of audio-visual support of remote mobile participant at e-meeting. In: Balandin, S., Dunaytsev, R., Koucheryavy, Y. (eds.) Smart Spaces and Next Generation Wired/Wireless Networking, LNCS, vol. 6294, pp. 62–71 (2010)
14. Khryashchev, V., Ganin, A., Golubev, M., Shmaglit, L.: Audience analysis system on the basis of face detection, tracking and classification techniques. In: International MultiConference of Engineers and Computer Scientists Hong Kong, LNECS, pp. 446–450 (2013)
15. Khryashchev, V., Priorov, A., Shmaglit, L., Golubev, M.: Gender Recognition via Face Area Analysis. World Congress on Engineering and Computer Science, Berkeley, USA, pp. 645–649 (2012)
16. Nenakhov, I., Khryashchev, V., Priorov, A.: No-reference image quality assessment based on local binary patterns. In: 14th IEEE East-West Design & Test Symposium, pp. 529–532 (2016)
17. Anwar, S., Milanova, M., Bigazzi, A., Bocchi, L., Guazzini, A.: Real time intention recognition. In: 42nd Annual Conference of the IEEE Industrial Electronics Society (IECON'2016). (2016). doi:10.1109/IECON.2016.7794016
18. Cadena, L., Espinosa, N., Cadena, F., Kirillova, S., Barkova, D., Zotin, A.: Processing medical images by new several mathematics shearlet transform. In: International MultiConference of Engineers and Computer Scientists (IMECS'2016), vol. I, pp. 369–371 (2016)
19. Muhuri, S., Bhattacharjee, M. (2014). Automated identification and analysis of cervical cancer. In: 3rd World Conference on Applied Science, Engineering and Technology, pp. 516–520
20. Liang, M., Zheng, G., Huang, X., Milledge, G., Tokuta, A.: Identification of abnormal cervical regions from colposcopy image sequences. In: 21st International Conference on Computer Graphics, Visualization and Computer Vision (WSCG'2013), pp. 130–136 (2013)
21. Hinman, R.S., Hunt, M.A., Simic, M., Bennell, K.L.: Exercise, gait retraining, footwear and insoles for knee osteoarthritis. Curr. Phys. Med. Rehabil. Rep. **1**, 21–28 (2013)
22. Simic, M., Wrigley, T., Hinman, R.S., Hunt, M., Bennell, K.: Altering foot progression angle in people with medial knee osteoarthritis: the effects of varying toe-in and toe-out angles are mediated by pain and malalignment. Osteoarth. Cartil. **21**(9), 1272–1280 (2013)

Chapter 2
Graphological Analysis and Identification of Handwritten Texts

Leonid A. Mironovsky, Alexander V. Nikitin, Nina N. Reshetnikova and Nikolay V. Soloviev

Abstract The problem of recognition of handwriting text is still far from its final solution. The existing systems of recognition of handwritten texts are usually developed for some special applications. The difficulties are caused by recognition of the conjoint writing because a variability of handwritings is the highest and often it is necessary to solve the problem of delimitation of the separate letters. In this chapter, along with to the known methods of the handwritten fragments' analysis, it is offered to use the developed methods of vectorization of raster images and vector dynamic parameterization. Also, a description of the automated information storage and retrieval system for the graphological analysis and identification of unintelligible fragments of handwritten texts is given. The system contains a database of handwriting samples with variants of the author's calligraphy from the Manuscript Department of the Institute of Russian Literature (Pushkin's House) of the Russian Academy of Sciences.

Keywords Graphological analysis · Handwritten text · Text segmentation
Comparison of words · Symbols and ligatures · Dynamic parameterization
Drafts autographs · Automated information retrieval system

L.A. Mironovsky (✉) · A.V. Nikitin · N.N. Reshetnikova · N.V. Soloviev
St. Petersburg State University of Aerospace Instrumentation,
67 Bol. Morskaya St, Saint Petersburg 190000, Russian Federation
e-mail: miron@aanet.ru

A.V. Nikitin
e-mail: nike51@mail.ru

N.N. Reshetnikova
e-mail: reni_07@list.ru

N.V. Soloviev
e-mail: famsol@yandex.ru

M.N. Favorskaya and L.C. Jain (eds.), *Computer Vision in Control Systems-4*,
Intelligent Systems Reference Library 136,
https://doi.org/10.1007/978-3-319-67994-5_2

2.1 Introduction

The first attempts to solve the problem of handwritten or printed text recognition in bitmap images were made more than fifty years ago, almost immediately after appearance of the devices that could upload images into computer memory [1, 2]. In this context, the text recognition means an automated process for obtaining of ASCII codes of symbols (letters, numbers, and punctuation marks). Selection and partial recognition of image segments in the photos, diagrams, plots, mathematical formulas, and tables are also possible.

Today there are a lot of software products that successfully recognize the printed characters [3, 4]. For bitmap images with high resolution (300–600 dpi), contrast, and sharpness, the number of recognition mistakes does not exceed 0.5% [5]. Also, there are a lot of software products for processing of images to be recognized in order to increase their contrast and sharpness with removal possible noise and defects [6]. In this case, the successful recognition can be implemented easily thanks to the fact that a bitmap image of printed text could be easily segmented into images of separate characters and the images of similar characters are nearly identical.

The process of recognition of printed texts is commonly referred to as Optical Character Recognition (OCR) [7] even though the OCR means recognition of any text, printed or handwritten [8]. It is notable that apart from the OCR software, which was designed for image processing using the desktop image scanners, some devices were designed specifically for recognition of the printed texts [9]. The problem of recognition of the mixed printed and handwritten texts is also solved quite successfully [4]. The main difference of such text from printed ones lies in the fact that images of similar characters can significantly differ from each other. To solve this problem, some text recognition software tools include the learning algorithms. In this case, a software tool attempts to learn the specific handwriting style from an image, which significantly increases the chance for successful recognition [10].

The most difficult task is the recognition of joined-up handwriting. This process is known as Hand Writing Recognition (HWR) [11]. The main differences of handwritten texts from the printed ones are described below. First, the lines of handwritten texts are often not parallel, which is especially common for texts written on an unlined sheet. There also can be partial overlapping of the adjacent lines and words/letters in these lines. Second, even though the words in handwritten texts are separated with spaces, some words are not written jointly, which means that some letters or even letter fragments can be separated with spaces. Finally, even one person's handwriting depends on many factors. For example, a readability of a word in the official document can differ a lot from the same word written in a personal note book. Moreover, a way of the letter writing greatly depends on the writing of adjacent letters. The first two of the mentioned above reasons complicate the segmentation of images with handwritten texts into separate words, while the third one complicates the recognition of letters.

The chapter is organized as follows. The related works are reviewed in Sect. 2.2. Section 2.3 provides the analysis and identification of handwritten texts including a problem formulation. A formation of alphabets based on the handwriting samples is considered in Sect. 2.4. The methods of bitmap binary image based on the graphological analysis and identification are represented in Sect. 2.5, while the methods of the graphological analysis and identification based on vectorization of bitmap images are developed in Sect. 2.6. Section 2.7 includes a description of architecture of the designed software tool. Section 2.8 concluded the chapter.

2.2 Related Works

The interest to the problem of recognition of the joined-up handwritten texts is unabated, which is evident from the great amount of publications on this topic [12–33]. There are two main approaches that support two different ways to obtain the handwritten texts called as "on-line" recognition (when the characters are written with a pen in a special screen, for example a tablet PC screen) and "off-line" recognition (when a document with a handwritten text already exists). Generally, the on-line recognition is easier task because the problem of text segmentation into words does not appear due to the on-line mode of capturing a lot of clues that can be successfully used to learn the users' handwritings. The existing on-line hardware and software recognition products are effectively demonstrate this proposition [34].

The off-line text recognition is a more complicated task. As it is listed in many reviews [13, 35–37], the classic methods for selection of the handwriting features, creation of a database with samples, and features identification do not work successfully in some cases [38]. It is known [39] that the artificial neural networks are able to identify the generalized features of recognizable images. In this regard, the application of artificial neural networks to the off-line recognition of handwritten texts seems to be a good idea [40–52]. The reasons for mistakes lie in the facts that the same characters can be written in different ways by different people and it is very hard to identify the separate characters in the handwritten texts.

While recognizing a handwritten text, people consider its context, which means that they take into account the information they obtained from the parts of the same text, which they read before. It is known [53] that a human being can easily read a word if it is a part of a discourse (normal text) but the same person will have problems while indentifying the same word in a text, consisting of random words. It is even more difficult for a human being to identify a separate letter even though that same letter in a text can be read easily. Presumably, this is the recognition systems' ability to understand the meaning of handwritten texts that will allow for a breakthrough in the science but the existing intelligent systems are still far from solving this task.

Sometimes a person, who reads a handwritten text, faces a problem of recognizing a separate word that can be one of several possible similar-looking words, each of which does not violate the rules of grammar and retains the meaning of the text. For example, this problem can be faced by a literary scholar, who works with a handwritten archive of a writer [54]. Such person visually compares an unidentified word fragment with similar-looking fragments of words that are already recognized in order to find the right one. The system described below in this chapter was designed to speed up this process and increase the correctness of recognition results.

It is obvious that the problem of handwritten text recognition is far from being completely solved. The existing handwritten text recognition systems are usually developed for some special cases or applications based on the specific features of the texts to be recognized. These specific features are the quality of handwritten texts and size of vocabulary. The main quality groups of handwritten texts are texts written with the block letters and those written with the cursive handwriting. The cursive handwriting, in its turn, falls into the separate letters handwriting and joined-up handwriting.

The most difficult problem is the recognition of joined-up handwriting because in this case a variability of handwriting samples for different letters is very high and the task of identifying the borders of each letter becomes very complicated. The volume of vocabulary, i.e. a number of different words in the text, is also very important. Vocabularies of authors' handwriting styles can be of great value, when literary texts are recognized. Such vocabularies are not comprised only of a large number of words used in a literary text but can also include sets of the author's calligraphy variants (ways of writing of different letters and links between them—ligatures) in order to define the correctness of this or that word. The use of computers to recognize the illegible handwritten texts can significantly speed up the process of finding the letters, ligatures, and words and increases an objectivity of comparison of the reference fragments of handwritten texts (taking into account the variability of handwriting samples) with ones under recognition.

2.3 Analysis and Identification of Handwritten Texts: Problem Formulation

Reading the handwritten drafts is often a very difficult task. Let us take draft manuscripts written by A.S. Pushkin and A.S. Griboedov as an example. These manuscripts are kept at the Manuscript Department of the Institute of Russian Literature (Pushkin's House) of the Russian Academy of Sciences. Just looking at the reproduction of Pushkin's draft manuscripts in the Big Academic Collection of his works (16 volumes printed in 1937–1949), one can see a huge number of notes and <illegible text> signs, that bespeak of the uncertainty of editors' interpretation of the author's handwriting [54].

There are longstanding academic disputes about some authors' handwritten drafts. Moreover, the samples of writing of some letters can greatly vary depending on the speed of writing and the writing tool used (goose quill, pen point, or pencil). A textual analyst has to create a vocabulary of various samples of the author's calligraphy (ways of writing separate letters and links between them—ligatures) in order to read correctly this or that word. Figure 2.1 shows examples of A.S. Pushkin's handwritten drafts of the "Poltava" poem (courtesy provided by the Manuscript Department of the Institute of Russian Literature of the Russian Academy of Sciences).

Usually, while the analyzed fragment of handwritten text cannot be interpreted unambiguously, a textual analyst performs the following steps:

Step 1. Selection a possible combination of letters in accordance with the grammar rules, semantic content of the text, and known individual specifics of author's writing.

Step 2. Searching the author's fair copies to pick text fragments that matches the selected combination of letters.

Step 3. Finding a fragment that matches the selected example from the analyzed handwritten text.

Step 4. Visually comparison the sample with the recognized/analyzed fragment of handwritten text.

Step 5. If the fragments do not match, repeats Steps 2–4 or maybe even Steps 1–4.

The use of advanced computer means and technologies can significantly speed up the process of finding samples (Steps 1 and 2), facilitate in objective integral assessment of a degree of closeness of compared fragments, and reduce the subjective judgment due to a human factor.

The automatic analysis and identification of illegible fragments of handwritten text requires to solve a number of problems [55–60]:

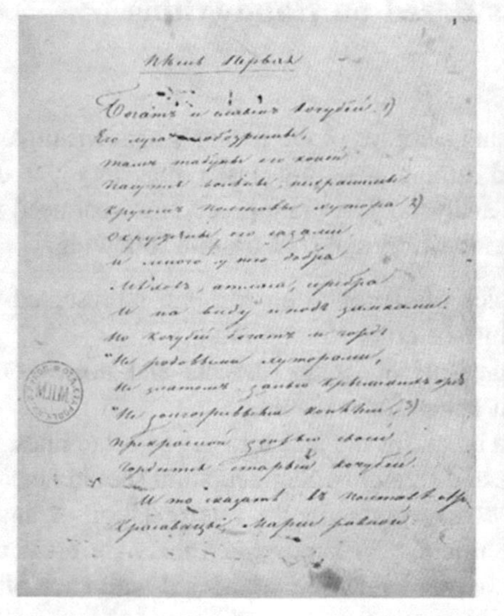

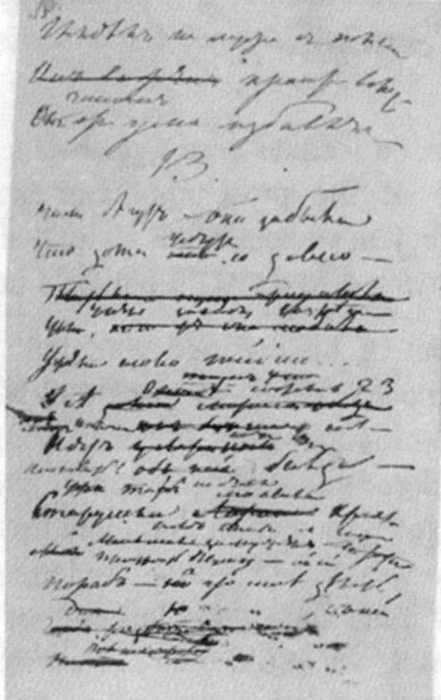

Fig. 2.1 Examples of A.S. Pushkin's handwriting

- Identification of the type of representation of the script fragment using the bitmap binary image with the suppressed brightness distortions and vector representation of an image.
- Analysis and selection of acceptable geometric transformations of an image in order to achieve the best possible matching.
- Development of methods for assessment of a degree of closeness of images during identification.
- Development of architecture for a computer system to store and retrieve information needed for the handwriting analysis and identification of illegible fragments of handwritten text.
- Creation of a database of alphabets based on samples of the author's handwriting taking into account the variability of author's calligraphy and difference of samples of the same letters found in the different author's texts.
- Creation of a bank of efficient methods for the recognition, handwriting analysis, and identification of handwritten texts.
- Design of a system of queries to work with the bank of methods and database of alphabets based on the author's handwriting samples and ensures access to them using prospective technologies of augmented reality through the intuitive interface.

Consider some of the mentioned above tasks in details. We will illustrate studies and experiments with examples of fair copies and drafts of handwritten scripts written by A.S. Pushkin, the classic of Russian literature, whose works always have a big number of non-text fragments and noises.

2.4 Formation of Alphabets Based on Handwriting Samples

Consider a formation of alphabets using samples of the author's handwritings. This is one of the most important and time-consuming tasks that requires direct involvement of specialists in textual analysis. The main procedures that need to be implemented to form an author's handwriting alphabet are the following:

- Input of handwritten text. Samples of handwritten text are represented on a computer screen as black-and-white or color bitmap images.
- Pre-processing of images. Elimination of image defects and noises on the background, and selection of text zones for analysis.
- Text segmentation that includes a breaking the text into the separate lines, lines into words, and words into letters and ligatures. Segmentation should be carried out more than once in cases of controversies in the interpretation of analysis results, since the intervals between letters in handwritten texts are often wider than those between words, whereas words can be connected with each other.

- Saving the letters, ligatures, and words as the author's handwriting samples in a database.

Selection of handwritten text fragments can be made using either widely known computer software (bitmap graphics editors like Adobe Photoshop or GIMP as an open source software) or specialized applications with a range of functions for textual analysis (horizontal line adjustment, selection of different combinations of segments, etc.) with the intuitive interface.

To address the task of alphabet formation, a technique for pre-processing of images of handwritten text fragments with elimination of image defects and background noises was developed and tested. The most commonly used tools for input of static visual information are the digital cameras and scanners (hand and flatbed).

Consider a still image on a flat media—rectangle sheet of paper (Figs. 2.1, 2.2, and 2.5). Our input device is a flatbed scanner that allows to obtain both color and grayscale images of various resolutions. The comparative analysis of the color, grayscale, and binary images of handwritten text showed that the color images do not have noticeable advantages over grayscale ones in terms of use for text recognition [57, 58]. Based on the conducted experiments, it was found that the most suitable resolution for analysis of images of handwritten texts is 200–300 pixels per inch (dpi). Further increase of resolution does not lead to noticeable increase in an image quality, while its decrease impairs image representation on computer screens and sometimes leads to the loss of small text elements.

Preliminary image processing, as a rule, includes the suppression of noises and image binarization. The latter operation is needed for segmentation, i.e. selection of

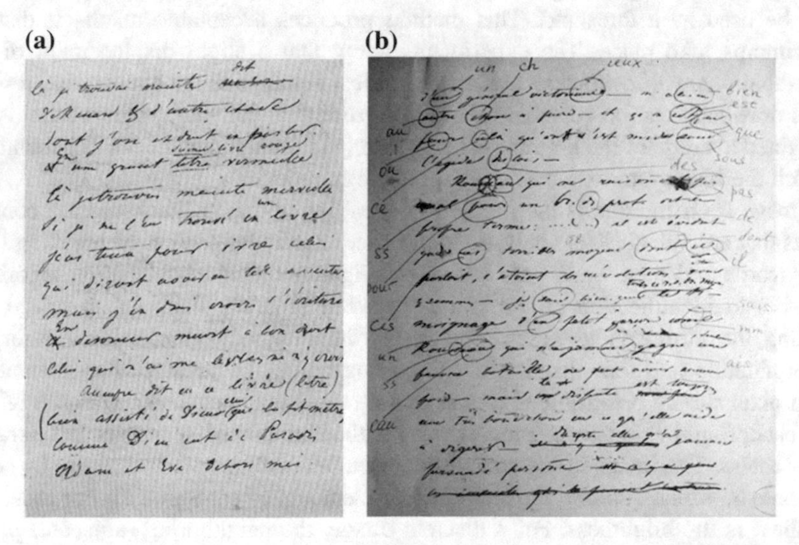

Fig. 2.2 Parts of "Roman de Renard", a French literary classic of the 12–13th centuries: **a** original image, **b** marked image

the separate fragments consisting of adjacent pixels. Noises in the scanned bitmap images can be divided in two groups:

- Technical caused by errors in the operation of image acquisition, transfer, and storage equipment.
- Real ones in the form of sheet of paper defects (spots, fold lines) and author's notes (corrections, strikethroughs, and images).

The analysis of images obtained from a scanner showed virtually complete absence of noises of the first group. On the other hand, the problem of automated identification and elimination of noises in the second group is an extremely challenging task. In the first place, it is due to the need for formalization of attributes that can be used to identify image areas, which should be considered noises. At the stage of testing of a textual image pre-processing technology, the elimination of such noises should be carried out by a specialist in textual analysis in the interactive mode. Any graphics software that allows to change the brightness of pixels in bitmap images can be used for this operation.

Image binarization methods can be considered as non-linear pixel-by-pixel transformations with simple image transformation algorithms, where the main problem is the selection of binarization thresholds. Existing thresholding methods try to adapt to different brightness levels of separate image fragments. Most of these methods are based on the analysis of brightness histograms [61].

Brightness histogram of an image depicting several objects on a uniform background has two maximums, one of which corresponds to the brightness of object pixels and the another reflects the brightness of background pixels. There are many methods of thresholding using image brightness histograms [61]. For example, the histogram's global minimum found between two highest maximums can be used as a threshold. This method produces acceptable results if distinct maximums take place. The experiments show that a slight displacement of the threshold in any direction can lead to a significant change in the binarization results. This drawback can be compensated by a method that calculates the threshold as a weighted average of the brightness histogram [61]. This method does not require to search the histogram extremes that significantly reduces a processing time.

Image segmentation is the process of breaking an image into separate components that are valuable for analysis. In our case, this is the process of breaking lines into words and then words into letters and ligatures. The segmentation algorithm consists in identification of similarities between separate pixels of an image and finding the homogenous areas. The main difficulty in segmentation lies in the determination and formalization of the homogeneity. As a result of segmentation, each pixel should be attributed to a segment (number of segment), which it relates to (background is a zero segment), where the number of segments is unknown beforehand. The image is segmented into a number of homogenous areas by some attribute S, which qualifies the similarity of elements of each area. In our case, this attribute is the brightness. For a discrete image, the neighboring (adjacent) pixels are grouped in the homogenous areas with similar brightness.

The above said is true, when the image fragments corresponding to the separate characters are separated with background color and do not linked with each other. For handwritten texts, only words are separated with background color and even then not in every case. On the other hand, even a way of writing of some letters in a script can sometimes look like separate letters. For example, parts of Cyrillic letters “й” or “ф”, which in A.S. Pushkin’s scripts looks like letters “c” and “p” written one after another, are separated with background, and, therefore, can be recognized as separate symbols if the segmentation is executed in completely automatic mode.

Taking into account the interactive work of textual analysts (in the process of alphabet formation), we propose the following sequence of actions on the example of a small portion of French text (Fig. 2.2). After capturing an image of handwritten text with a flatbed scanner in black and white mode and resolution of 300 dpi, a text analyst uses the graphics editing software in order to select the separate line or word, removes noises (underlines, corrections, etc.), and puts dividers of letters and ligatures in the image changing the brightness of pixels that relate to the text (Fig. 2.2).

The edited text fragment is loaded into the segmentation software that automatically converts the image into binary format using the weighted average threshold of the brightness histogram or random threshold set by the operator, determines the relevance of pixels to separate segments using a two-pass algorithm, and calculates the number of segments. The textual analyst uses the segmentation software to look through the selected segments or groups of segments and saves them as separate images into a database specifying all necessary attributes (uppercase or lowercase letter, number of the page and line in the text, etc.).

The proposed segmentation algorithm includes two passes of each image and minimized linking table. The algorithm is used in the information storage and retrieval system that allows to form and view the script segments, merge segments into groups, and save the letters, combinations of letters (ligatures), and fonts of scripts in the database. The structure of this information storage and retrieval system is described in Sect. 2.6.

2.5 Methods of Bitmap Binary Image Based Graphological Analysis and Identification

Two approaches, representing in Sects. 2.5.1–2.5.2, were applied for analysis and identification of the bitmap binary images. Consider them in details.

2.5.1 Method of Skeleton Transformations of Letters, Ligatures, and Words

Two main approaches are used to construct an image skeleton—the skeletal transformation itself and the so-called "thinning" algorithm. Both of these approaches are applied if the relative position of strokes is the most important thing in an image. Usually, the skeleton transformation is a sequential removal of the maximum number of image pixels that allows preserving the general outline of that image. The process of highlighting the skeleton points is often associated with burning. Let us suppose that an object's outline was set on fire and the flames spread along the normal towards its border. Then the points, where the flame fronts meet, will be considered as parts of the skeleton. This approach is called "thinning".

Another possible definition of the skeleton is the locus of the centers of circles covering the object. The task of processing a handwritten text imposes certain specificity on the skeleton transformation algorithms applied. The experiments conducted to produce skeletons through different methods showed that the "thinning" algorithm is the most accurate way of solving the problem.

The image edge detection algorithms are also of certain interest for textual analysts. There are many of those algorithms but their basic idea is about the same. An original image is considered as a function of the $f(x, y)$ coordinates. If we cut an image in an arbitrary direction, then the edge of the image fragment will match the brightness jump with respect to the function $f(x, y)$ defined on a straight line. With respect to the gradient and Laplacian, this will correspond to maximum and zero values. Most of the existing edge detection methods are reduced to the discrete approximation of gradient or Laplacian. The Prewitt, Sobel, and Canny filters and the direct Laplace approximation were used to obtain the object contour images. The studies have shown that these algorithms produce the similar results. Therefore, the algorithms that use the scale selection, morphological or wavelet transformations do not seem appropriate.

2.5.2 Method Based on Calculation of the Hamming Distance Between Two Binary Images

The software algorithm was implemented to study the method based on the calculation of the Hamming distance. This algorithm allows to compare any two segments, which in the given software implementation should be presented in black and white colors [57, 60]. One black and white image of a handwritten fragment is added into the software application from a file, and another image is taken from the database. When an image is uploaded, the parameters that will later be used to for analysis are calculated. Those are the segment area and shape center. The segment shape center is marked by the intersection of two red lines; its coordinates and area are displayed in a popup window, when a popup window is hovered over the image.

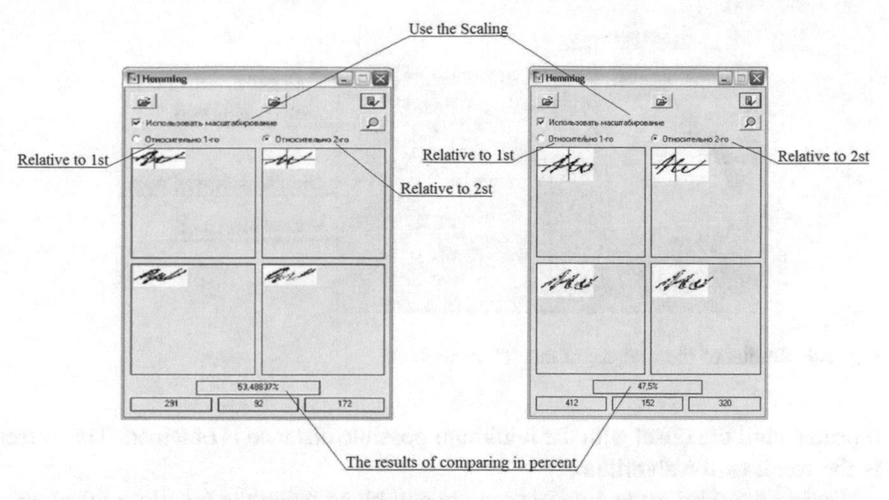

Fig. 2.3 Results of the comparisons of individual symbols and symbol combinations

The main application window, which allows to compare two manuscript fragment, is shown in Fig. 2.3. These fragments can be seen in the autograph by A.S. Pushkin illustrated in Fig. 2.1.

The comparison allows to create two combined images and show their matching degree. The first combined image shows the initial position of the segments. The initial position corresponds to the position with two shape centers combined. The second image (on the right part in Fig. 2.3) shows the result generated by the application. The segment mutual offset providing the maximum overlap area is obtained. The text box below these images displays their matching degree. Of course, it only makes sense to compare the segments that have the similar shapes, otherwise, the result of such comparison can be very poor and a feasibility of such comparison becomes highly questionable.

The algorithm can be divided into three stages:

- Calculation of the image parameters (area and shape center).
- Identification of a mutual image offset value, wherein the Hamming distance between the images is minimal.
- Calculation of the matching degree of the two images.

Two moments of the zero-order and first-order are calculated during the first stage. The zero-order moment corresponds to the area. The shape center is calculated using the OX and OY axes. An offset is calculated during the second stage as a combination of two shape centers. The Hamming distance is calculated based on this offset. Single shifts in different directions on both axes are performed in a cycle afterwards. The Hamming distance is defined for each shift. All distances are compared and the offset corresponding to the shortest distance is chosen. The distance should be less than the one obtained in the previous stage. This operation is

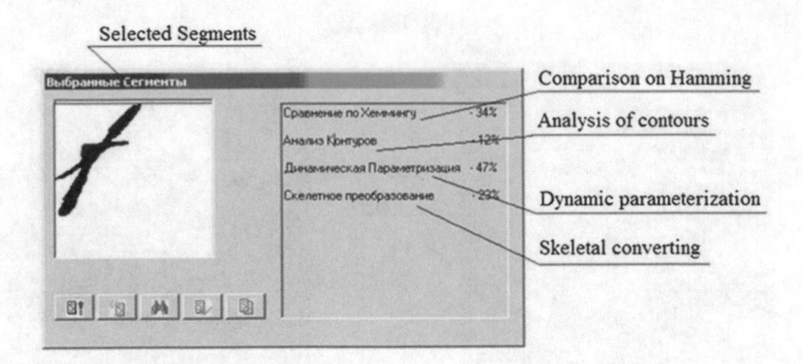

Fig. 2.4 Results of the analysis of the “f” symbol

repeated until the offset with the minimum possible distance is obtained. This offset is the result of the algorithm.

During the third stage, two images are combined using the resulting offset, and their matching degree is determined. The testing showed that the algorithm is fast enough and can be applied to compare image arrays. Further modification of the algorithm will make it capable of comparing not only black and white but any type of images. It should be noted that the segments ought to have a number of matching parameters (scale, color, orientation, etc.) for the algorithm to run properly. Figure 2.4 shows the result of the analysis made in the application specifically designed for the study of the analysis and identification methods. A textual analyst can check the given parameter combination using a special-purpose menu.

Consider the methods of the graphological analysis and identification representing in Sect. 2.6.

2.6 Methods of Graphological Analysis and Identification Based on Vectorization of Bitmap Images

Along with the methods and algorithms of bitmap transformation described above, we propose the following methods of analysis and identification of handwritten fragments [58, 62, 63]:

- Vectorization of bitmap images of handwritten texts.
- Vector dynamic parameterization.

Three groups of methods can be distinguished at the level of symbol identification. Conventionally, they can be called the reference, structural, and feature methods. The first group is formed by the reference methods. They are based on the comparison of a recognized character with a set of prototypes—reference symbol images. The second group of methods uses the information about the structural interrelations of graphemes or elementary parts of a symbol: vertical and horizontal strokes, loops, and their parts. In the third case, the recognition process is based on the analysis of a set of invariant features that characterize the symbol, such as its size, number of loops, position relating to the line, etc.

Note that the feature detection phase is crucial for the effective operation of identification algorithms. First, it includes the significant data compression (matrix, describing the symbol image, is converted into a much smaller set of features) and, second, identification of relations between elements of object images that are characteristic to each class (intraclass invariant search).

Some of the features used in existing detection and identification systems are shown below. Simple geometric feature is the ratio of vertical and horizontal size of an object. Other geometric signs describe the integral properties of top, bottom, left, and right symbol profiles. Statistical features are associated with the average number of intersections of the symbol with vertical and horizontal lines. The symbol image is divided into equal horizontal areas. In each zone, the feature value is calculated as the average number of intersections of the symbol with all horizontal lines in this area. Topological features reflect the presence of jumps (sudden changes) of symbol profiles, as well as the number of internal topological areas (holes) on the object image. In classical handwriting, this number equals 0 for Cyrillic letters “к”, “л”, “м”, “н”, 1 for “а”, “б”, “о”, “я”, and 2 for “в” and “ф”. In practice, these relations are not as strict; however, the statistical dependencies are maintained. Many other features can be offered besides the ones mentioned above, for example, the center of gravity of letters, presence of the vertical, horizontal or another symmetry, direction of the letter bypass, etc. The number of potential features increases if the bitmap letter representation is replaced with vector images and the static representation is substituted by dynamic representation. The proposed methods and algorithms are discussed in Sects. 2.6.1–2.6.2.

2.6.1 Method of Vectorization of Bitmap Handwritten Text Images

Identification of handwritten symbols requires the ability to use the symbol contour bitmap image to create its vector representation, i.e. the polygonal chain or spline that accurately approximates the image. In this regard, the orientation field (direction field)-based method of vectorization of bitmap handwritten text images was designed. Orientation fields, that is, vector fields describing the orientation of image contours, are widely used in the recognition of bitmap images mainly formed by the shape contours. In particular, this technique is used for fingerprint recognition.

Description of the implemented algorithm. Bitmap black and white handwritten text images are the input data for the algorithm. It is assumed that the image resolution is at least as thick as the line or stroke. Besides, it is deemed that a line is darker than the background; otherwise, an inverse (negative) image can be used. A fragment of the manuscript by A.S. Pushkin shown in Fig. 2.5 can be considered example of such image. This is the beginning of the entry to a poem “The Fountain

(a)

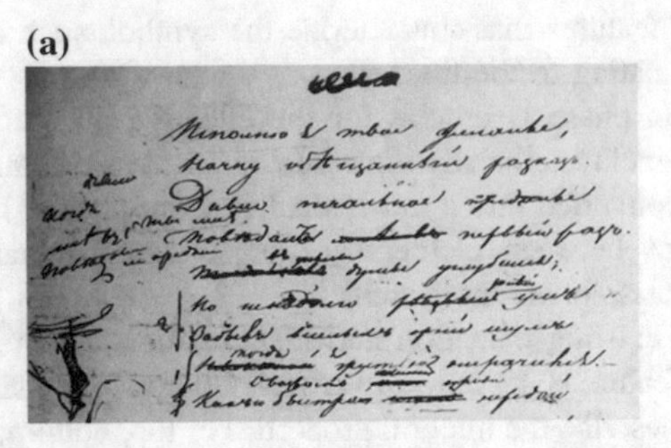

(b)

«Исполню я твое желанье,
Начну обещанный рассказ.
Давно, когда мне в первый раз
Поведали сие преданье,
Мне стало грустно; резвый ум
Был омрачен невольной думой,
Но скоро пылких оргий шум
Развеселил мой сон угрюмый»,
....

Fig. 2.5 Beginning of the entry to a poem "The Fountain of Bakhchisarai": **a** fragment of the manuscript by A.S. Pushkin, **b** printing analogue

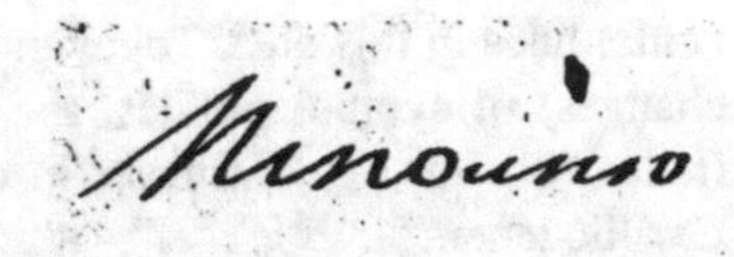

Fig. 2.6 Bitmap image of the word from the manuscript

of Bakhchisarai", which was not included in the final text. Figure 2.6 shows a bitmap image of the first words of this piece "cut out" from the text for further processing and analysis.

The implemented processing algorithm includes the following stages:

- Pre-filtering of the bitmap image.
- Construction of the orientation field for this bitmap image.
- Filtering and extrapolation of the direction field.
- Searching for and tracking of image contours using the created orientation field.

The core idea of the algorithm is to trace the contours in the original (or pre-filtered) image with regard to the orientation field that provides information about the direction of a line or a stroke near a given point.

Image pre-filtering. A standard initial step for all image tracing algorithms includes image normalization and application of blur filter with small blur radius (of about one pixel).

Creation and representation of the orientation field. The filtering and extrapolation require the orientation field in any given point to be a smooth function of the image brightness values near this point.

Mathematically, the orientation field may be described by Eq. 2.1, where $\{A(x, y)\}$ are the tangent angles to level lines, $I(x, y)$ is a brightness function, *angle* is an angle between the vector and the OX axis, *mod* is a calculation of the remainder of division.

$$A(x,y) = \left(\text{angle}(\text{grad}(I(x,y))) + \frac{\pi}{2}\right) \bmod \pi \tag{2.1}$$

Obviously, the direct representation of the field in this form is not continuous at each point. Therefore, the following complex (vector) function $D(x, y)$ is proposed to represent the orientation field:

$$D(x,y) = \left(-\frac{\partial}{\partial x}I(x,y) + i\frac{\partial}{\partial y}I(x,y)\right)^2 \tag{2.2}$$

Here, the squared expression is a perpendicular to the gradient written in the form of a complex number. Squaring is performed in order to eliminate the differences between mutually opposite directions of the gradient, which correspond to the complex gradient values opposite in their signs. The direction field values must be the same.

The level curves angle $A(x, y)$ is associated with the $D(x, y)$ function through Eq. 2.3.

$$A(x,y) = \frac{\arg(D(x,y))}{2} \tag{2.3}$$

When working with bitmap images the partial derivatives of brightness functions are replaced by finite brightness differences of the horizontally and vertically adjacent pixels.

Orientation field extrapolation and filtering. If the direction field is created for handwritten text images similar to monochromatic ones, the vector function $D(x, y)$ is noticeably different from zero only in stroke borders. Therefore, it is necessary to extrapolate the non-zero field values in adjacent areas in order to use the orientation field to track image lines and strokes. Image defects, such as ragged edges, add a noise to the direction field. The linear filtering algorithm with subsequent normalization is used to eliminate such noise.

Contour tracking. The following algorithm is proposed to track the contours:

Step 1. A point lying on the contour is located.

Step 2. A polygonal chain lying inside the contour (vectorized line), moving along the direction field in predetermined increments (about one pixel), is drawn from this point. The chain is completed, when it goes outside the contour.

Step 3. Repeat Steps 1 and 2 until all points of the contour of a given bitmap image are used.

To improve the tracking accuracy along the field line, an additional offset proportional to the negative field gradient is added:

Fig. 2.7 Image of the orientation field and the result of image vectorization

$$r_{n+1} = r_n - \beta \operatorname{grad}(I(r_n)) \tag{2.4}$$

where β is a small positive coefficient, r_n is radius vector of vectorized contour points.

This offset provides a slow drift of the point towards the lowest brightness level, i.e. to the middle of the contour line (assuming that the contour is darker than the background, that is, the brightness function value inside the contour is smaller than it is outside). An example of the results of the algorithm for a scanned fragment with $\beta = 0.01$ is shown in Fig. 2.7.

Thin lines here show the direction of the orientation field, bold lines are based on the image tracing results. The figure shows how the vectorized contours move along the lines of an extrapolated direction field. The direction field away from the contours looks chaotic, as the figure does not display the orientation field vector amplitude. If the Gaussian smoothing is applied to the direction field, the amplitude of randomly oriented areas decreases due to the mutual suppression. The described algorithm was implemented as a Java application that processes the image files of various graphic formats.

Based on the results of the computer experiment, the following properties of the proposed method of image vectorization using the orientation field can be noted:

- The method is insensitive to the additive white noise and small contour irregularities, since such defects correspond to chaotic orientation fields smoothed during the filtering process.
- The method allows to ignore the short contour breaks smaller than the extrapolation radius.
- The method allows to recognize successfully the handwritten lines with the "acute angle" elements, which are often found in handwritten text images. Unlike the skeleton methods generating the Y-shaped contours (Sect. 2.4) for acute angles, the orientation field-based method restores a V-shaped contour that is closer to the original path of a writing tool.

These properties allow to expect a more effective solution to future recognition problems.

2.6.2 *Method of Vector Dynamic Parameterization*

If an image is stored in a bitmap format, such important information as the order of writing of letters, pen speed, and its direction, is lost. The image vectorization method based on the orientation fields shown above provides some possibilities in this direction but it solves the problem only partially. The following is an alternative approach, which may be called the method of vector dynamic parameterization. This method allows to obtain more information about letter images using their vector dynamic representations. It creates additional possibilities for the formation of new diagnostic features, as well as the ways to use the thoroughly developed means of 1D analysis and statistics for processing and analysis of 2D letter images.

The purpose of this method is to restore the dynamic information about the movement of a pen, when writing a letter, including the indication of the start and end points, movement direction, and number of cyclic letter outlines. This method includes two stages [62, 63]. First, the image vectorization is done, during which an operator or a textual analyst uses a light pen or a mouse to outline the letter image as they had been written originally. This approach allows the substantial use of important but poorly formalized information about the ways the letters were written based on the given manuscript and personal and on the personal experience of an expert. Note that a vector image obtained through the orientation field method can be considered as the initial image. Such vectorization creates 2D array of Cartesian coordinates of image points (x, y) arranged in the order they were written. For convenience of further processing, it can be converted to a file, containing ASCII codes of coordinates. In a more complete version, such array can contain a third coordinate representing time. Second, the resulting 2D array can be visualized not only as the letter image but also as two separate curves $x(t)$ and $y(t)$. This allows to create the diagnostic features, such as the number of minima and maxima on the curves, inflection points, and spectral ratios. The possibility of calculating the correlation ratios between the curves $x(t)$ and $y(t)$ should be highlighted. In this case, thoroughly developed means of 1D analysis and statistics can be used to process and analyze 2D letter images. The procedure of performing letter scaling, inclination correction, and affine and projective transformations is greatly simplified.

The MATLAB software was used for practical implementation of the method. 2D numeric arrays created through the dynamic factorization and containing the letter coordinates are processed as standard package files. At the stage of letters' pre-processing, these arrays were standardized, i.e. the differences in the size and inclination of such letters were removed. Letters were rotated and compressed through affine transformations.

Before converting a letter, it is reasonable to calculate the numerical features, such as its width, height, the width to height ratio, as well as the center of gravity and the number of extrema of functions $x(t)$ and $y(t)$. Additionally, it is useful to count the number of internal topological areas ("holes") in the letter. For this purpose, five horizontal lines equidistant from one another were allocated in functions $x(t)$ and $y(t)$ and the number of intersections between the $x(t)$ and $y(t)$ curves and these lines were counted. For convenience of further comparison of letters, they must have the same number of points. The experiments showed that a number of points required to adequately describe a letter ranges from 40 to 240. Accordingly, the standard length for $x(t)$ and $y(t)$ arrays was selected as 150 points. Furthermore, the array points should be evenly distributed throughout the period of writing a letter. The interpolation procedure providing an array of points equidistant from each other was used for this purpose.

The informative diagnostic features of the letter recognition include the correlation ratio between the $x(t)$ and $y(t)$ curves of compared letters. The application that compares a given letter with each letter in the matrix of reference letters was designed to calculate this ratio and identify the best matches. Afterwards, the correlation ratios between the $x(t)$ and $y(t)$ functions of compared letters are identified and added to a special-purpose matrix. This data, among with other diagnostic features, will later be used for statistical analysis.

One of the advantages of the vector dynamic representation is the ability to visualize each letter in the form of a 3D image with x, y, and t coordinates. The corresponding letter images look like a spatial spiral, where the first point corresponds to the beginning of a letter and last point corresponds to its end. An important geometric property of this representation is the absence of self-intersection points even if they were present in the original image. As an example, Fig. 2.8 shows the spatial representation of Cyrillic letter "*a*".

The spatial representation makes it possible to compare two letters by the average distance between the spirals. The distance between the analyzed letter and the current reference letter is then calculated. When identifying the letters, the distances between the analyzed letter and each of the reference letters are calculated and recorded in a separate array. The position of the minimum element of this array is taken as the number of a most similar reference letter. The calculation of the

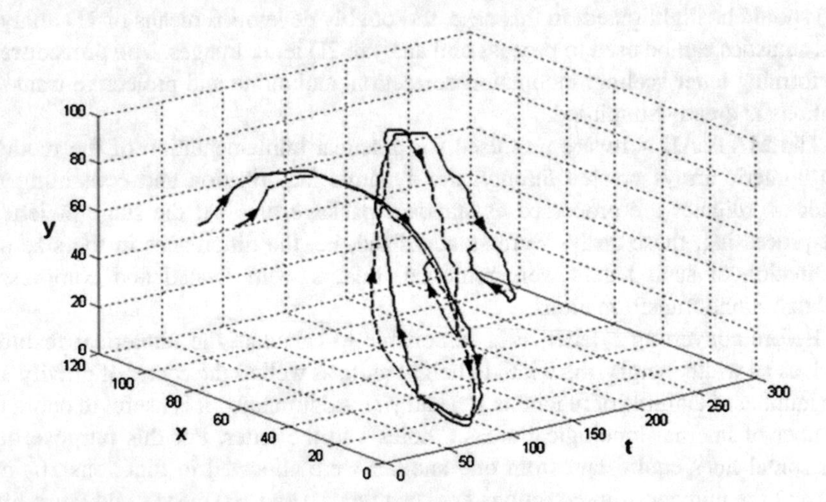

Fig. 2.8 3D image of two examples of Cyrillic letter "*a*"

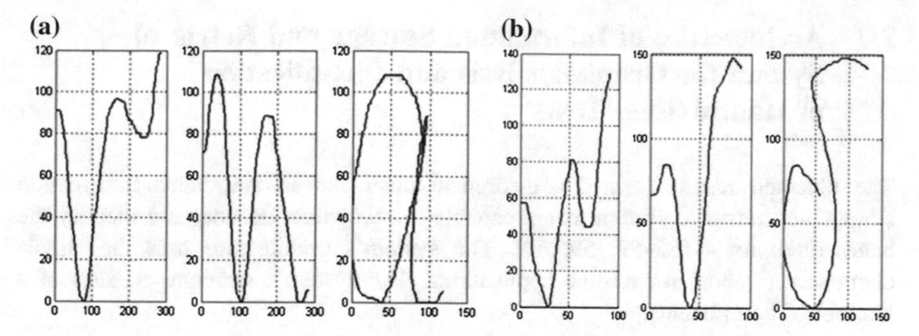

Fig. 2.9 Reference curves: **a** for Cyrillic letter "*a*", **b** for Cyrillic letter "*б*"

correlation ratios and the geometric characteristics of 3D letter images essentially extends the standard set of diagnostic features used to analyze and identify the letters. The calculated diagnostic features are used as a basis for the table, into which the results of the comparison of the given letter with the reference letters are added. The feature table is a matrix, each row of which contains the results of the calculation of different characteristics for each letter. During the computing experiments, this table included the following features: the distance between letter images in the three dimensions, ratio of letter's height to its length, correlation ratios, the number of extrema of $x(t)$ and $y(t)$ functions, and number of intersections between $x(t)$ and $y(t)$ functions and five fixed horizontal levels.

The features of an analyzed letter are identified and stored as separate variables in order to analyze the data contained in the table. They are later compared with similar features of the reference letters. The comparative results are recorded in the feature table consisting of 16 columns and 33 rows (the number of letters in the Cyrillic alphabet). The Euclidean norm is calculated to determine the quality of matching. As a result, the algorithm finds the number of a letter from the reference array that appears as the most similar to the analyzed one.

The reference letter images were created for the purposes of computer experiments (3–4 files for each letter). Figure 2.9 shows the reference curves $x(t)$, $y(t)$, and $y(x)$ for Cyrillic letters "*a*" and "*б*".

The reference array contains references of all letters and different dynamic versions of their writing. The recognition quality during the computer experiments amounted to 85%. It can be increased by the way of more careful selection of diagnostic features and improvement of letter pre-processing, in particular, using the matrixes of affinity, projective, or other types of transformations.

2.7 Architecture of Information Storage and Retrieval System for Graphoanalysis and Identification of Handwritten Texts

The obtained results formed a theoretical basis for an automated information storage and retrieval system that is capable to recognize, decode, and identify the handwritten texts [55–57, 59, 60]. The system's architecture uses the classic client-server model for on-line applications. The system's structure consists of a number of components:

- Database.
- Server side.
- Client side.
- A set of plug-ins.

The database contains the author's handwritings, such as the alphabets, ligatures, lines of handwritten texts, and manuscript codes. The system of queries to database is divided into three groups of stored procedures providing both data access and storage security:

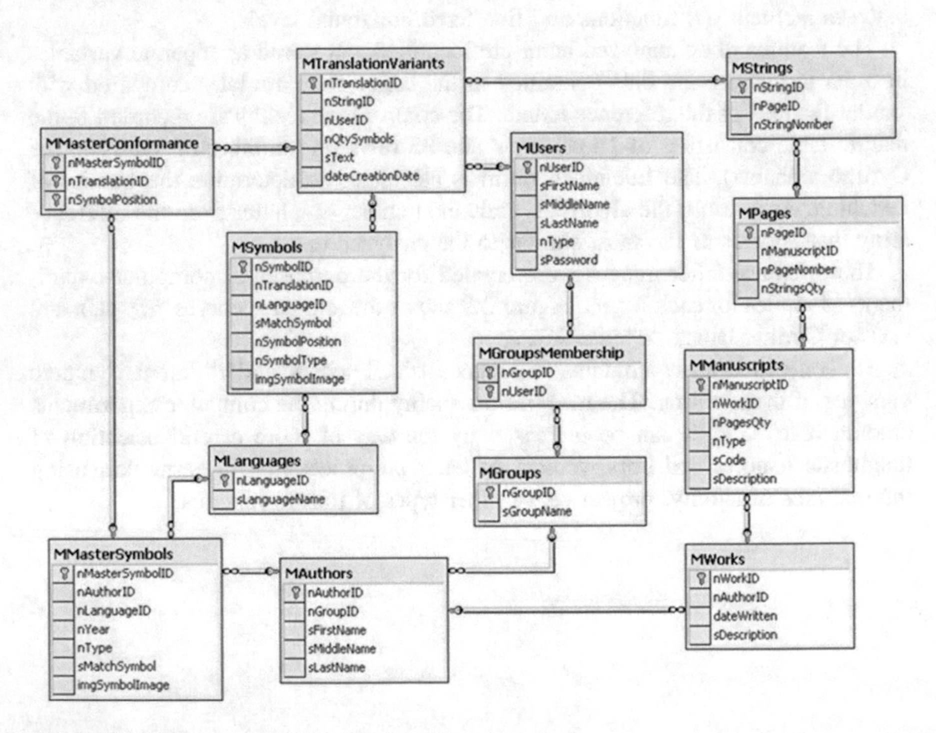

Fig. 2.10 Database structure

- The security group contains the user authentication procedures and interaction dialogs.
- The manuscript codes' group stores the procedures related to processing of manuscript data.
- The handwriting alphabets' group contains the procedures that work with samples of the letters, ligatures, words, and lines of manuscripts.

The handwriting database is worked under Microsoft SQL Server MDBS and configured to store images of the separate letters, ligatures, words, and lines of manuscripts and related information (author, year, name of work, line number, etc.). The database structure is shown in Fig. 2.10. Interaction with the database is carried out through the stored procedures.

Server part is the main component of the system. On the one hand, it works with the database directly through the stored procedures; while on the other hand, it interacts with users. Also, the server part allows users to run various analysis and identification procedures in the form of plug-ins. The user interface allows the registration of remote users and separate storage of the results of their work, as well as a simultaneous access to the general database to assist in reading of illegible manuscript fragments.

The development of a friendly and intuitive interface for specialists in textual analysis is a very important task. Two approaches for creation of the client side of the information system for graphoanalysis and identification of handwritten texts were considered:

- Standard Windows application [57].
- Windows application with the use of the augmented reality technology.

The software interface for textual analyst realized as standard Windows application [57] includes the algorithms for segmentation of handwritten texts into the

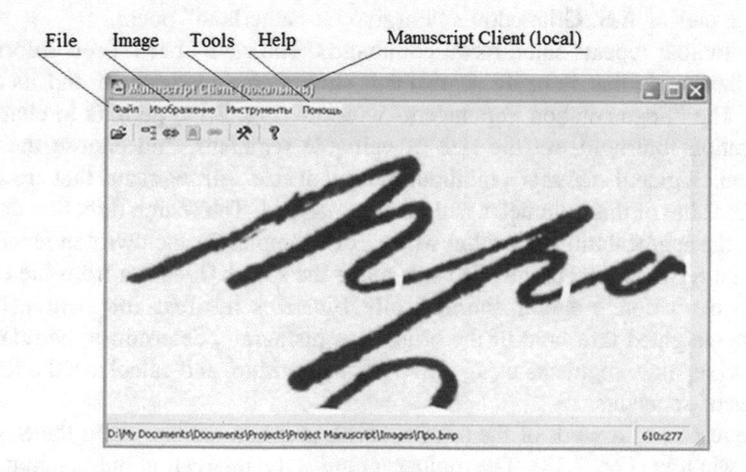

Fig. 2.11 Main window of the system's Windows application

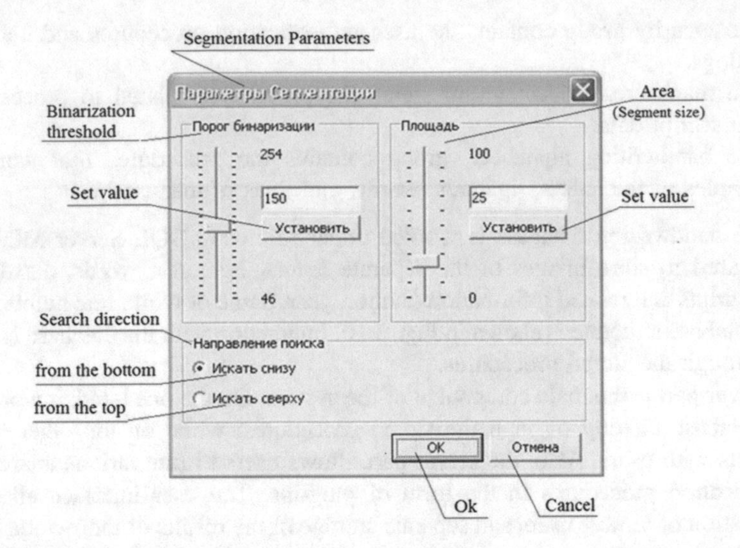

Fig. 2.12 Segmentation parameters

lines, words, ligatures, and letters and also involves an algorithm for preliminary processing that helps to remove the noises and heterogeneity of background of the scanned handwritten texts.

Consider the algorithm of interactive work of a textual analyst, while forming an alphabet based on a real example. The textual analyst uses the segmentation application to look through the selected text segments or groups of segments one by one and saves them as separate images into the database indicating all necessary attributes (uppercase or lowercase letter, line number, page number, etc.). The application's main window for segmentation includes the menu, toolbar, status bar, and area that shows the image that is being segmented (Fig. 2.11). Figure 2.11 shows a part of A.S. Griboedov's "Forgive us Fatherland" poem.

The toolbar repeats some menu commands. The status bar reflects information about the image that is being segmented, such as the image path and its size in pixels. The "Segmentation Parameters" window (Fig. 2.12) permits to change the binarization threshold, set the size of valuable segments, and choose the search direction. Segment size sets a minimum segment size. All segments that are smaller than the value of this parameter will not be analyzed. The search direction defines a way of the segmentation algorithm work. For example, for handwritten texts with a set to the right, it is recommended to choose the search direction from the bottom. The segmentation program automatically binarizes the text fragment using the average weighted threshold of the brightness histogram, determines the relation of pixels to separate segments using a two-pass algorithm, and calculates the resulting number of segments.

The results of a work of the segmentation program are shown in the Segments dialog window (Fig. 2.13). The toolbar includes the navigation buttons that switch from one segment to another, remove segments, and save segments in the database

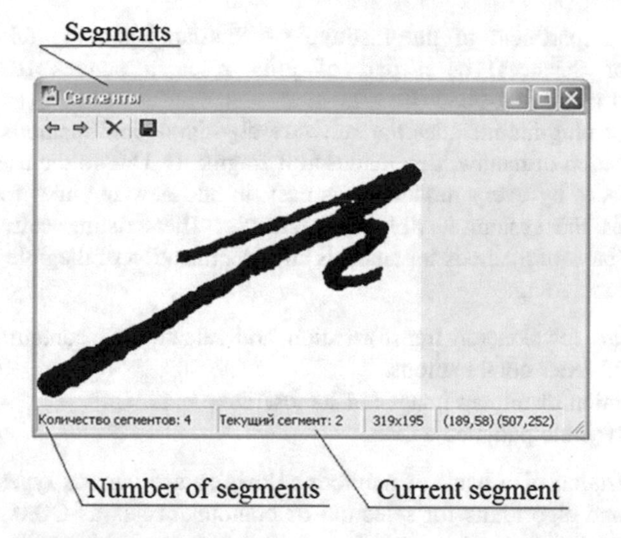

Fig. 2.13 Segments of a text fragment

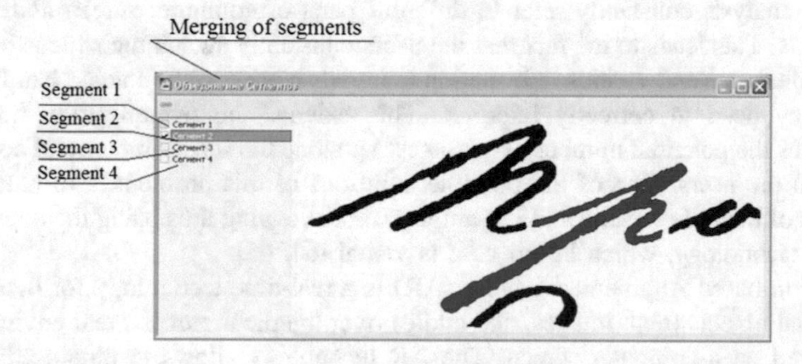

Fig. 2.14 Merging of segments

or file in a hard drive. The Status Bar reflects the information about the number of segments, number of the currently viewed segment, its size and position in the source image.

To merge segments, just select the segments to merge and press the Merge button (Fig. 2.14). The selected segments and the currently viewed segment are highlighted.

As a result, the samples of handwritings of the classics of Russian literature of the nineteenth century A.S. Pushkin and A.S. Griboedov formed using the fair and draft copies of their manuscripts were created and saved in the database. The manuscripts that were used for this research were A.S. Pushkin's "Eugene Onegin" novel, "The Gypsies" poem, early drafts of the "The Fountain of Bakhchisaray" poem, as well as letters and poems by A.S. Griboedov (courtesy provided by the

Manuscript Department of the Institute of Russian Literature of the Russian Academy of Sciences) as a part of joint research studies [RFBR, grants 04-01-00464 and 04-07-90354].

The set of plug-ins includes the software algorithms and methods for analysis and identification of handwritten manuscript fragments. Due to the use of a unified set of interfaces by every module, it is easy to add new modules to the system, which makes the system flexible and scalable. The existing software product comprises a bank of methods for analysis and identification of illegible fragments of manuscripts including:

- Algorithms for skeleton transformation and selection of contours of separate letters and letter combinations.
- Vectorization of bitmap images of handwritten texts.
- Vector dynamic parametrization.

The realization of a bank of methods allows to use various types of skeleton algorithms and algorithms for selection of contour images as COM objects. The augmented reality technology can be applied as a prospective approach to the organization of textual analyst's interface.

While conducing the automated study of handwriting, even highly experienced textual analysts constantly refer to different parts of computer screen and printed materials. This leads to the fact that the users constantly switch their focus between the studied area and various information materials related to the studied handwriting that they need to correctly interpret. This extends the overall study time and increases the potential number of mistakes, let alone the sensorimotor and cognitive load on the users. One of the possible solutions of this problem is to place both images of illegible writing and its samples in the viewing area using the augmented reality technology, which in our case is visual [64, 65].

Vision-based Augmented Reality (AR) is a real-time technology for overlaying of digital objects (text, images, and audio) over the picture of the real environment displayed on a computer screen. The AR technology allows to obtain additional information contextually related to the objects in the environment. The AR interaction is carried out from the point of view of each separate user and in accordance with each user's individual settings. Examples of use of the AR in text recognition are given in [65–67].

The potential advantages of application of the augmented reality technology to handwriting recognition are mentioned below:

- Reduced time consumption and error rate.
- Reduced sensorimotor and cognitive load on the users (movements of head, eyes and hands, data interpretation).
- Possibility of real time cooperation of researchers in the events, where the expert advices are required (multi-user mode).
- Individual and cooperative practical training.
- Possibility to lower the level of skill requirements for experts.

Thus, the problem of handwritten text recognition is reduced to a visual comparison of background images representing a handwritten fragment (letters, ligatures, words, and lines) with overlaid sample images downloaded from the generated database.

The structure of the implemented desktop AR system includes:

- The computer, monitor and web-camera located perpendicularly to a table at a height of 50 cm. The operating zone of the camera is fitted with markers.
- Software includes MS Windows 10, Unity3D [68] for development of scenes to place and manipulate images, and Vuforia [69] to create augmented reality.
- A set of markers is used to study images of fragments of handwritten texts (letters, ligatures, words, and lines) both digitized and processed using analysis and identification methods.
- A set of samples (letters, ligatures, words, and lines) on a transparent background selected by a textual analyst from the database for comparison with the studied fragments of the manuscript.

Two modes called as the preparatory and operating modes are used during a working with the system. The preparatory mode includes:

- Digitizing of handwritten fragments (letters, ligatures, words, and lines) and processing of their images applying the developed methods, which are described in details in Sects. 2.4–2.6.
- Selection of images of handwritten fragments for markers.
- Recognition of markers using the Vuforia platform [69].
- Binding sample images from the database to the markers.

The operating mode involves:

- Placement of the marker under the web-cam.
- After the marker is recognized, it is displayed on the monitor under the ribbon of the sample images from the database.
- Selection of an image from the ribbon that (visually) overlays the analyzed sample.
- Moving, rotating or zooming of the sample images in order to visually fit the marker.
- If we find a matching, the symbol or fragment is considered as recognized; otherwise, another sample is selected from the database.

The experiments that included the analysis of draft autographs by A.S. Pushkin showed that the developed automated information search and retrieval system can be successfully used for the detection, decoding, and identification of handwritten texts given a highly variable author's calligraphy. In particular, it can be used by textual analysts or can serve as an evidence base for the correct recognition of draft autographs. During design the prototype handwriting recognition system based on the AR, the experience of the authors in development of applications in healthcare and cultural heritage was considered [70].

2.8 Conclusions

This chapter provides an overview of known handwriting recognition methods. The tasks of the computer analysis of handwriting and identification of illegible fragments of manuscripts are identified on the base of textual analysts' work with draft autographs of famous writers. The procedures of formation of alphabets composed of handwriting samples of the writers are considered using the developed methods of vectorization bitmap image and vector dynamic parameterization. The prepared alphabets are saved in the database taking into account a variability of calligraphy of the author's. The results of the identification methods for bitmap image and their program implementation on the examples of the literary manuscripts are given.

Also the description of the automated information retrieval system graphological analysis and identification illegible fragments of manuscripts is given. The system interacts with the database, which is pre-filled with a collection of different versions of the author's calligraphy (images of individual letters and their ligaments/ligatures). In the database, the fragments of text from a variety of literary manuscripts are saved. The database size is about 400 Mb. The system allows to use the methods described in the chapter for a quantitative assessment of the degree of coincidence recognizable fragment of manuscript with fragments of handwriting from the database. As a result a textual analyst receives the objective data that help him/her to read illegible handwritten fragment or conclude about affiliation of fragment of a specific author.

Researches were carried out on the example of manuscripts and draft autographs analysis of A.S. Pushkin and A.S. Griboyedov with the participation of the textual analysts from the Institute of Russian Literature (The Pushkin House) of the Russian Academy of Sciences. The use of advanced computer means and technology can significantly speed up the process of finding the handwritten fragments, facilitate the objective integral assessment of a closeness degree of the compared fragments, and reduce the subjective judgment due to a human factor.

Acknowledgements The reported study was funded by the Russian Fund for Basic Researches according to the research projects No. 04-01-00464 and No. 04-07-90354.

References

1. Avrukh, M.L., Birman, N.Y.: Automatic reading of typewritten texts. Proceeding/Automatic Reading of the Text. VINITI (1967) (in Russian)
2. Weinstein, V.S., Grigoryan, A.A.: The input device of graphics information into a computer. Automatic Reading of the Text. VINITI (1967). (in Russian)
3. Text Recognition. Available from: http://mrtranslate.ru/download/textrecognition.html. Accessed 20 June 2016
4. ABBYY Fine Reader Engine 8.1 for Windows. Available from: http://www.interface.ru/home.asp?artId=6911. Accessed 1 July 2017 (in Russian)

5. Lisin, S.: OCR-systems. Available from: http://old.ci.ru/inform16_02/p_22text.htm. Accessed 1 July 2017 (in Russian)
6. Edit of the Scanned Document in Photoshop. Available from: http://photolessons.org/edit-scan. Accessed 1 July 2017 (in Russian)
7. Andrianov, A.I.: Comparison of OCR-systems based on the accuracy of image analysis. Bus. Inform. **4**:44–50 (2009) (in Russian)
8. Bashmakov, A.I., Bashmakov, I.A.: Intelligent Information Technologies: Textbook. MSTU N.E. Bauman Publ. (2005) (in Russian)
9. Pint, E.M., Romanenko, I.I., Elichev, K.A., Semov, I.N.: Designing device for recognition of printed information. Young Sci. **4**, 235–238 (2015)
10. Misyurev, A.V.: Using artificial neural networks for characters recognition handwritten. In: Proceedings/ISA RAS intelligent technology of input and information processing, pp. 122–127 (1998) (in Russian)
11. Edwards, J.A.: Easily adaptable handwriting recognition in historical manuscripts. Technical Report No. UCB/EECS-2007-76, University of California at Berkeley (2007)
12. Dedgaonkar, S.G., Chandavale, A.A., Sapkal, A.M.: Survey of methods for character recognition. Int. J. Eng. Innov. Technol. **1**(5), 180–189 (2012)
13. Gaidukov, N.P , Savkova, E.O.: Review of handwriting recognition methods. In: International Scientific-technical Conference of Students, Graduate Students and Young Scientists (Information Control Systems And Computer Monitoring'2012) (2012) (in Russian)
14. Aggarwal, A., Rani, R., Dhir, R.: Handwritten devanagari character recognition using gradient features. Int. J. Adv. Res. Comput. Sci. Software Eng. **2**(5), 85–90 (2012)
15. Deshpande, P.S., Malik, L., Arora, S.: Handwritten Devanagari Character Recognition Using Connected Segments and Minimum Edit Distance. IEEE TENCON'2007, pp. 1–4 (2007)
16. Dholakia, K.: A survey on handwritten character recognition techniques for various indian languages. Int. J. Comput. Appl. **115**(1), 17–21 (2015)
17. Impedovo, D., Pirlo, G.: Zoning methods for handwritten character recognition: a survey. Pattern Recogn. **47**(3), 969–981 (2014)
18. Jieun, K., Yoon, H.: Graph matching method for character recognition in natural scene images: a study of character recognition in natural scene image considering visual impairments. In: 15th IEEE International Conference on Intelligent Engineering Systems (INES'2011), pp. 347–350 (2011)
19. John, J., Pramod, K.V., Balakrishnan, K.: Unconstrained handwritten malayalam character recognition using wavelet transform and support vector machine classifier. Proc. Eng. **30**, 598–605 (2012)
20. Kato, N., Suzuki, M., Omachi, S., Aso, H., Nemoto, Y.: A handwritten character recognition system using directional element feature and asymmetric Mahalanobis distance. IEEE Trans. Pattern Anal. Mach. Intell. **21**(3), 258–262 (1999)
21. Khobragade, R.N., Nitin, A.K., Mahendra, S.M.: A survey on recognition of devnagari script. Int. J. Comput. Appl. Inf. Technol. **2**(1), 22–16 (2013)
22. Kim, H.C., Kim, D., Bang, S.Y.: A numeral character recognition using the PCA mixture model. Pattern Recogn. Lett. **23**(1–3), 103–111 (2002)
23. Kimura, F., Wakabayashi, T., Tsuruoka, S., Miyake, Y.: Improvement of handwritten Japanese character recognition using weighted direction code histogram. Pattern Recogn. **30**(8), 1329–1337 (1997)
24. Kressel, U., SchÄurmann, J.: Pattern classification techniques based on function approximation. In: Bunke, H., Wang, P.S.P. (eds.) Handbook of Character Recognition and Document Image Analysis. World Scientifc, pp. 49–78 (1997)
25. LeCun, Y., Bottou, L., Bengio, Y., Haffner, P.: Gradient based learning applied to document recognition. IEEE Proc. **86**(11), 2278–2324 (1998)
26. Mohamed, M., Gader, P.: Handwritten word recognition using segmentation-free hidden markov modeling and segmentation-based dynamic programming techniques. IEEE Trans. Pattern Anal. Mach. Intell. **18**(5), 548–554 (1996)

27. Niranjan, S.K., Kumar, V., Kumar, H.G., Aradhya, M.V.N.: FLD based unconstrained handwritten Kannada character recognition. Int. J. Database Theory Appl. **2**(3), 21–26 (2009)
28. Siddharth, K.S., Dhir, R., Rani, R.: Handwritten Gurumukhi character recognition using zoning density and background directional distribution features. Int. J. Comput. Sci. Inf. Technol. **2**(3), 1036–1041 (2011)
29. Starner, T., Makhoul, J., Schwartz, R., Chou, G.: On-line cursive handwriting recognition using speech recognition methods. BBN Systems and Technologies (1995)
30. Suen, C.Y., Kiu, K., Strathy, N.W.: Sorting and recognizing cheques and financial documents. In: Lee, S.W., Nakano, Y. (eds.) Document Analysis Systems: Theory and Practice, vol. 1655, pp. 173–187. Springer, LNCS (1999)
31. Suen, C.Y., Lam, L.: Multiple classifier combination methodologies for different output levels. In: Kittler, J., Roli, F. (eds.) Multiple Classifier Systems, vol. 1857, pp. 52–66. Springer, LNCS (2000)
32. Erosh, I.L., Sergeyev, M.B., Soloviev, N.V.: Methods for rapid character recognition are suitable for hardware implementation. Inform. Control Sys. **4**: 2–6, “Politehnica” Publication, St. Petersburg (2004) (in Russian)
33. Wakabayashi, T., Pal, U., Kimura, F., Miyake, Y.: F-ratio based weighted feature extraction for similar shape character recognition. In: 10th International Conference on Document Analysis and Recognition (ICDAR’2009), pp. 196–200 (2009)
34. Writepad Stylus. Available from: http://androplanet.com/2804-writepad-stylus.html. Accessed 1 July 2017 (in Russian)
35. Mozgovoth, A.A.: Problems of existing methods optical handwriting recognition. Herald Voronezh State Techn. Univ. **8**(7-1), 22–25 (2012) (in Russian)
36. Mori, S., Suen, C.Y., Yamamoto, K.: Historical review of OCR research and development. IEEE Proc. **80**(7), 1029–1058 (1992)
37. Rahman, A.F.R., Fairhurst, M.C.: Multiple classifier decision combination strategies for character recognition: a review. Int. J. Doc. Anal. Recogn. **5**(4), 166–194 (2003)
38. Buchkina, E., Soloviev, S.: Review of research on handwriting recognition problems. Available from: http://www.textolog-rgali.ru/?view=articles&t=article1. Accessed 1 July 2017 (in Russian)
39. Haykin, S.: Neural Networks. A Comprehensive Foundation, 2nd edn. Prentice Hall (1998)
40. Feng, Y., Fan, Y.: Character recognition using parallel BP neural network. In: International Conference on Audio, Language and Image Processing (ICALIP’2008), pp. 1595–1599 (2008)
41. Sahu, V.L., Kubde, B.: Offline handwritten character recognition techniques using neural network: a review. Int. J. Sci. Res. **2**(1), 87–94 (2013)
42. George, A., Gafoor, F.: Contourlet transform based feature extraction for handwritten malayalam character recognition using neural network. Int. J. Ind. Electron. Electr. Eng. **2**(4), 19–22 (2014)
43. Desai, A.A.: Gujarati handwritten numeral optical character reorganization through neural network. Pattern Recogn. **43**(7), 2582–2589 (2010)
44. Liu, C.L., Nakagawa, M.: Evaluation of prototype learning algorithms for nearest neighbor classifier in application to handwritten character recognition. Pattern Recogn. **34**(3), 601–615 (2001)
45. Liu, C.L., Nakashima, K., Sako, H., Fujisawa, H.: Handwritten digit recognition: benchmarking of state-of-the-art techniques. Pattern Recogn. **36**(10), 2271–2285 (2003)
46. Liu, C.L., Sako, H., Fujisawa, H.: Performance evaluation of pattern classifiers for handwritten character recognition. Int. J. Doc. Anal. Recogn. **4**(3), 191–204 (2002)
47. Liu, C.L., Sako, H., Fujisawa, H.: Discriminative learning quadratic discriminant function for handwriting recognition. IEEE Trans. Neural Networks **15**(2), 430–444 (2004)

48. Liu, C.L., Fujisawa, H.: Classification and learning for character recognition: comparison of methods and remaining problems. Available from: http://www.dsi.unifi.it/NNLDAR/Papers/01-NNLDAR05-Liu.pdf (2006). Accessed 1 July 2017
49. Marinai, S., Gori, M., Soda, G.: Artificial neural networks for document analysis and recognition. IEEE Trans. Pattern Anal. Mach. Intell. **27**(1), 23–35 (2005)
50. Oh, I.S., Suen, C.Y.: A class-modular feedforward neural network for handwriting recognition. Pattern Recogn. **35**(1), 229–244 (2002)
51. Pradeep, J., Srinivasan, E., Himavathi, S.: Diagonal based feature extraction for handwritten alphabets recognition system using neural network. Int. J. Comput. Sci. Inf. Technol. **3**(1), 27–38 (2011)
52. Arnold, R., Póth Miklós, P.: Character recognition using neural networks. In: IEEE 11th International Symposium Computational Intelligence and Informatics (CINTI'2010), pp. 311–314. (2010)
53. Potapov, A.S. Artificial intelligence and universal thinking. St. Petersburg. « Politehnica » Publ. (2012) (in Russian)
54. Fomichev, S.A.: "The eyes above the letters glide" (To the problem of Pushkin's textology). In: VII International Conference on Pushkin's "Pushkin and World Culture": Proceedings/ Institute of Russian Literature—NSU named after Yaroslav the Wise, pp. 280–287. St. Petersburg–Velikiy Novgorod (2005). (in Russian)
55. Artemov, I.V., Reshetnikova, N.N., Soloviev, N.V.: The system architecture for the identification of literary handwritten texts. SUAI Scientific session: Proceedings/ Saint-Petersburg State University of Aerospace Instrumentation (SUAI). vol 2, pp. 414–415, St. Petersburg (2005) (in Russian)
56. Artemov, I.V., Reshetnikova, N.N., Soloviev, N.V.: On the tasks of the automated system of graphological research handwritten texts. In: VII International Conference Pushkin's "Pushkin and World Culture": Proceedings/Institute of Russian Literature—NSU named after Yaroslav the Wise. St. Petersburg-Velikiy Novgorod, pp. 287–294 (2005) (in Russian)
57. Artemov, I.V., Volkov, D.M., Kondakova, I.A., Nikitina, A.A., Reshetnikova, N.N., Soloviev, N.V.: Information retrieval system of graphological analysis and identification of handwritten texts. The certificate of registration of the industry development. FSSI "State Coordinating Center for Information Technologies, an industry fund of algorithms and programs", pp. 1–10. Moscow (2006) (in Russian)
58. Mironovsky, L.A., Petrova, K.Y., Reshetnikova, N.N.: Prototyping of system analysis the handwritten documents by means MATLAB. In: XIV International Scientific and Technical Workshop "Modern Technologies in Problems of Control, Automation and Information Processing", p. 299 (2005) (in Russian)
59. Volkov DM, Mironovsky LA, Reshetnikova, N.N.: Automated system of graphological research of manuscripts. In: XIII International Scientific and Engineering Workshop "Modern technologies in problems of control, automation and information processing", pp. 482–483 (2004) (in Russian)
60. Artemov, I.V., Soloviev, N.V.: Analysis of coincidence the fragments handwriting text on the basis calculation of the Hamming distance. SUAI Scientific Session: Proceedings/ Saint-Petersburg State University of Aerospace Instrumentation (SUAI), vol 2, pp. 22–23 St. Petersburg (2006) (in Russian)
61. Gonzalez, R.C., Woods, R.E.:Digital Image Processing, 3rd edn. Prentice Hall (2008)
62. Mironovsky, L.A.: Redundancy, the invariance and symmetry in technical cybernetics. Cybernetics and Informatics (50th anniversary of the section of cybernetics House of Scientists named A.M. Gorky RAS): Proceeding/Polytechnic University, pp. 193–208 (2006) (in Russian)
63. Mironovsky, L.A., Shintyakov, D.V.: Interrelation Hankel singular values and frequency characteristics of linear systems. In: Abstracts 14th International Scientific and Technical Workshop "Modern Technologies in Problems of Control, Automation and Information Processing", p. 300 (2005) (in Russian)

64. van Krevelen, D.W.F., Poelman, R.: A survey of augmented reality: technologies, applications and limitations. Int. J. Virtual Real. **9**(2), 1–20 (2010)
65. Rose, F., Bhuvaneswari, G.: Word recognition incorporating augmented reality for linguistic e-conversion. In: International Conference on Electrical, Electronics, and Optimization Techniques (ICEEOT'2016) (2016)
66. Google Translate Adds 20 Languages to Augmented Reality Application. Available from: http://www.Popsci.Com/Google-Translate-Adds-Augmented-Reality-Translation-App. Accessed 1 July 2017
67. Fragoso, V., Gauglitz, S., Kleban, J., Zamora, S., Turk, M.: Translat AR: A mobile augmented reality translator on the Nokia N900. In: IEEE Workshop Applications of Computer Vision (WACV'2011), pp. 497–502 (2011)
68. Website Unity. Available from: http://unity3d.com/. Accessed 1 July 2017
69. Website Vuforia. Available from: https://developer.vuforia.com/. Accessed 1 July 2017
70. Website of Laboratory Computer Graphics and Virtual Reality SUAI. Available from: http://guap.ru/labvr/projects/. Accessed 1 July 2017

Chapter 3
Perceptually Tuned Watermarking Using Non-subsampled Shearlet Transform

Margarita N. Favorskaya, Lakhmi C. Jain and Eugenia I. Savchina

Abstract Digital watermarking remains the rapidly developed branch of the computer science due to the huge amount of internet resources, requiring a defense. In the past decades, many promising methods for a watermark embedding in frequency domain were elaborated. At the same time, the excellent concept of human visual system properties ought to be applied in these new transforms. The Non-Subsampled Shearlet Transform (NSST) is one the most perspective techniques, providing a high level of visibility and payload for the host image. The perceptual channel decomposition is used for detection of textural regions or regions with the expressed edges that have high contrast values. The embedding process is executed using the NSST and the Singular Value Decomposition (SVD), when the last significant bits in a sequence of eigenvalues are replaced by the embedded watermark in a binary representation. The algorithm is reinforced by the Arnold's transform of a watermark, the parameters of which are stored in a secret key. The quality of the extracted watermarks under typical attacks, such as the scaling, disproportionate scaling, rotation, translation, and cropping, was estimated by the Bit Error Rate (BER) and the Peak Signal to Noise Ration (PSNR) metrics. The proposed method indicates the highest robustness to rotation and proportional scaling (the BER mean values are 1.2–2.7%) and the medium robustness to translation and cropping (the BER mean values are 10.9–12.4%).

M.N. Favorskaya (✉) · E.I. Savchina
Institute of Informatics and Telecommunications, Reshetnev Siberian State University of Science and Technology, 31, Krasnoyarsky Rabochy ave., 660037 Krasnoyarsk, Russian Federation
e-mail: favorskaya@sibsau.ru

E.I. Savchina
e-mail: savchina.ei@gmail.com

L.C. Jain
Faculty of Education, Science, Technology and Mathematics, University of Canberra, Canberra, ACT 2601, Australia
e-mail: jainlakhmi@gmail.com

L.C. Jain
Bournemouth University, Bournemouth, UK

M.N. Favorskaya and L.C. Jain (eds.), *Computer Vision in Control Systems-4*, Intelligent Systems Reference Library 136,
https://doi.org/10.1007/978-3-319-67994-5_3

Keywords Digital watermarking · Blind technique · Non-subsampled shearlet transform · Internet attacks · Human visual system · Perceptual channel decomposition · Arnold's transform · Robust watermarking

3.1 Introduction

Digital watermarking as a technique of content authorization and copyright protection has been developed since 1990s, when the internet became the reality of a human life. A watermark can be used for such purposes as copyright protection, fingerprinting, copy protection, broadcast monitoring, data authorization, indexing, medical safety, and data hiding. The watermarking methods for various multimedia contents differ due to the properties of the host object like image, video, and audio that define the type of the embedding procedures as well as a payload of the embedded information.

At present time, a family of the watermarking techniques is very wide but not all of them satisfy to the high requirements regarding to their traits and the possible attacks through the internet. Depending on the security purposes, the watermarking techniques are classified as the robust, semi-fragile, or fragile. The robust watermarking methods are resilient to the main types of distortions, such as cropping, shrinking, noise, brightness, and geometric modifications. These methods are used for the copyright protection. In contradiction to this, the fragile watermarking methods are sensitive to any image modifications and protect against the malicious manipulations with low probability. The semi-fragile schemes occupy the middle position; they are tolerant to some image-processing operations like a quantization noise from a lossy compression. The investigations in fragile watermarking techniques are continued [1]. However, the robust watermarking techniques cause the major interest. Other significant criteria regarding to the domain, watermark, scheme, and information types are mentioned below:

- *Domain type*: the space domain (the pixels values of a host image are modified) and the frequency domain (the transform coefficients are modified).
- *Watermark type*: the pseudo random number sequence (has a normal distribution with zero mean and unity variance) and the visual watermark (without encoding by the generator with a secret seed).
- *Scheme type*: the reversible scheme (the exact restoration of the host non-watermarked image is possible) and the irreversible scheme (the distortion in the watermarked image is irreversible).
- *Information type*: the non-blind type (both the host image and the secret key(s) are necessary), the semi-blind type (the watermark and the secret key(s) ought to be known), and the blind type (only the secret key(s) is required).

It is strongly difficult to develop the generalized watermarking algorithm with the high indicators because of growing types of the tampering attacks during a transmission through internet and broadcasting channels. The tampering attacks modify the original digital content in the intentional or unintentional manner [2]. The goal of the intentional tampering is to modify the content maliciously or to remove the copyright sign. The unintentional tampering means various digital processing distortions, such as the format conversion, brightness correction, and resizing. For video content, the tampering attacks are classified as spatial and temporal, while for image content, only spatial tampering attacks are possible. The spatial attacks can be local and global. The local attacks change the fragment of image including the color changes, removal of pixels' blocks, and overlapping a set of pixels. The global attacks modify a full content of an image, i.e. the brightness adjustment, format conversion, resizing, and rotating. The most popular spatial attacks are mentioned below (note that only two first ones are caused by the local attacks):

- Composite content attack consists in imposing of two and more additional images on an image.
- Cropping attack means a removal of some parts from an image.
- Noise addition attack brings any type of noise in an image; the simplest one is a salt-and-pepper noise.
- Flipping attack rotates an image in such manner that a watermark remains untouched. Sometime a row or column flipping occurs. This type of attack is difficult to detect.
- JPEG compression attack executes a re-coding of image with a jpeg-codec. A visibility becomes worse but typical for jpeg format. This type of attack may be missed.
- Brightness attack is one of the most common types of attack, when a brightness is decreased or increased on some level, for example ±25%, ±50%.
- Scaling attack impacts on the image sizes with the following boundaries' adding or discarding in order to save the original image sizes.
- Translation attack shifts an image along OX and/or OY axes.
- Rotating attack is a typical geometrical attack, often meeting with a scaling attack.

Note that the most difficult cases occur, when several types of attacks are implemented simultaneously. The examples of attacks are depicted in Fig. 3.1.

The contradiction between a wish to embed the maximum data volume (an image payload) and a visibility leaded to the appearance of the multiplicative watermarking techniques, when a balance of the signal magnitude (fidelity) and the quality of the watermarked image (robustness) is supported at an acceptable level, employing the properties of the Human Visual System (HVS) [3]. The modelling of the transform coefficients can be achieved by two general ways:

- The multiplicative approach based on such transforms as the discrete cosine transform, discrete Fourier transform, discrete wavelet transform, conjugate symmetric sequency-ordered complex Hadamard transform, curvelet transform,

Fig. 3.1 Examples of the tampering attacks: **a** composite, **b** cropping, **c** noise, **d** flipping, **e** jpeg, **f** brightness, **g** scaling, **h** rotating + scaling

ridgelet transform, contourlet transform, digital shearlet transform, among others.

- The probabilistic approach, when a maximum likelihood is estimated using the popular Gaussian distribution, general Gaussian distribution, Laplacian distributions, Cauchy distributions, Gauss-Hermite expansion, Weibull distribution, etc.

In order to combine the advantages of the both approached, the novel approaches based on the non-subsampled multiplicative transforms were developed [4, 5]. Wang et al. [6] mentioned that the NSST is the effective multi-scale and multi-direction analysis method, which besides the accurate multi-resolution analysis provides nearly an optimal approximation for a piecewise smooth function using a corresponding probability density function. For this goal, the Bessel K form modelling was implemented successfully.

Our contribution deals with a development of efficient and robust watermarking scheme based on the prerequisites of the HVS. This means that the edge detection with a high sensitivity impacts substantially on a watermarking process. Nowadays, the NSST provides it better than the well-known transforms, particularly the Discrete Cosine Transform (DCT) and the Discrete Wavelet Transform (DWT). However, the natural redundancy of the NSST causes some challenges, requiring to be solved. Also, the perceptual channel decomposition for the NSST is proposed. The enhancement of the secrecy and security was achieved by use of the Arnold's transform in the global and multi-regional versions. The SVD scheme was chosen for a watermark embedding.

The chapter is organized as follows. The frequency techniques in a watermarking are briefly reviewed in Sect. 3.2. The shearlet theory is given in Sect. 3.3. A perceptual watermarking is discussed in Sect. 3.4. Section 3.5 presents a digital watermark embedding, while a digital watermarking extraction is discussed in Sect. 3.6. The experimental results, including a behavior of the watermarks under typical attacks, are reported in Sect. 3.7. Finally, the conclusions are drawn at the end of this chapter.

3.2 Overview of Frequency Techniques in Watermarking

Two domain techniques called as the spatial and frequency ones are available to embed a watermark in an image. The spatial methods use a direct changing the pixel values of a host image. It is reasonably considered that the spatial methods are simple and non-robust for the most types of attacks. At present, these methods are applied very rarely in serious applications. The main idea of the frequency methods is to alter the transform coefficients [7]. The transform methods are more robust against different attacks in comparison to the spatial methods. Since 2000s, three families of the transform methods, such as the DCT [8–12], the Discrete Fourier Transform (DFT) [13–16], and the DWT [17–22], are developed. The multiple experiments show that the DWT methods have the best frequency localization properties. Consider this approach in details. It is worth noting that many algorithms exploit the HVS model reinforcing it by various constructions, e.g. the Fuzzy Inference System (FIS) in order to obtain better results [23, 24].

Bhatnagar et al. [25] proposed a robust wavelet based grayscale logo watermarking technique using various frequency sub-bands. A watermark was embedded in the selected blocks made by a ZIG-ZAG sequence. The blocks were selected based on the statistical estimators, particularly their variances using a slicing window with 3×3 or 5×5 pixels. These authors considered a noise as an inner property of an image and did not try to compensate a noise. A grayscale logo image has very small sizes compared to a host image, and thanks to that a watermark logo becomes robust against the intentional and unintentional attacks. During a watermark extraction, the watermark logos (more than 1) can be extracted considering the distortions in the surrounding pixels and the sub-band level of the corresponding block. The total time of such watermarking scheme (embedding + extraction) is prevailed 11 s.

The grayscale logo watermarking technique with the biometrics inspired key generation was suggested by Bhatnagar et al. [26]. This method used the Fractional Wavelet Packet Transform (FrWPT), non-linear chaotic map, and the SVD based on the biometrically generated keys. For this purpose, the minutia features extracted from a fingerprint image, features obtained by the independent component analysis from an iris image, features extracted by the principal or independent component analysis from a facial image, or extended shadow codes from a signature image can be used for a key generation. It should be noted that a uniform procedure for all possible biometrics does not exist. Therefore, a watermarking approach becomes specific to some particular biometrics. First, a grayscale logo is embedded in the FrWPT domain of a host image. Second, the embedded host image is randomized using a non-linear chaotic map by modifying the singular values of the randomized image. Third, an authentication biometrical key is formed. During the extraction process, the watermarked image that is attacked possibly is authenticated by the authentication key. The logo watermark is extracted only after successful authentication step.

The generalized concept of a color watermarking is reduced to the decomposition of RGB image into three layers and representation of each layer as a grayscale analogue with non-negative values in interval [0–255] [27–29]. Another way is to transform RGB color space to YCbCr, YUV, HSV, etc., with the following analysis of Y component. The reasons for watermark embedding in Y (luminance) component rather than in chrominance components are explained by that the HVS is more insensitive to luminance than to the chrominance components and the JPEG and MPEG standards typically use higher density to Y than other two components [30–33].

Recently, the DWT techniques have been strengthened by a paradigm of finding the optimal values of multiple scaling factors during a watermark embedding. Some artificial intelligence techniques especially from a category of evolutionary algorithms, such as the genetic algorithm [34, 35], particle swarm optimization [36, 37], differential evolution [38, 39], and firefly algorithm [40, 41], had been contributed significantly in the field of a digital watermarking.

Ali et al. [42] proposed a robust image watermarking method in the DWT domain based on the SVD and the Artificial Bee Colony (ABC) algorithm. The embedding blocks are selected based on the HVS propositions using the criterion of sum of visual and edge entropies. The optimal parameters are evaluated by the ABC algorithm. However, the authors claimed the high discrepancies in estimators between the proposed algorithm and the SVD, the HVS + SVD, and the discrete fractional Fourier transform implementations.

Mishra et al. [41] focused on optimizing the trade-off between the imperceptibility and robustness of a grayscale image watermarking. They proposed the hybrid transform based on the DWT and the SVD and computed the multiple scaling factors using an evolutionary technique called as firefly algorithm. This algorithm reflects a relation between a variation of the light intensity and attractiveness. It implies the execution of three rules:

- All fireflies are unisexual and will be attracted to the light regardless of their sex.
- The attractiveness is proportional to the brightness of fireflies. A flashing firefly with less brightness will move towards the brighter firefly.
- If the brightness of a firefly is the same as in surrounding, then it will move randomly.

The attractiveness of a firefly was determined by its brightness that is proportional to the objective function as a linear combination of Peak Signal to Noise Ration (PSNR) and Normalized Cross Correlation (NCC).

Another branch of investigation is to build the hybrid frequency based methods using a fuzzy logic. Tamane and Deshmukh [43] used the Mamdani-type Fuzzy Inference System (FIS) available in the Matlab fuzzy logic toolbox. Their FIS involved the membership functions, operators and 15 "If Then" rules. It was used to calculate the weight factors that help to make an automatic decision, where to insert the watermark data in the 3D model (3D due to a multi-resolution analysis of the wavelet coefficients). The FIS consists of the fuzzification of the input parameters, application of fuzzy operators, normalization of the weight values, aggregation of the outputs, and defuzzification of the output parameters.

Mardanpour and Chahooki [44] suggested a watermarking method based on the Digital Shearlet Transform (DST) and bidirectional singular value decomposition factorization. First, the host image is decomposed by the DST with specific parameters. Second, one of the sub-bands is selected and a watermark image is inserted in its bidiagonal singular values. Third, a watermark is embedded directly from the bidiagonal singular values.

Wang et al. [45] studied the influence of the geometric distortions on the embedding and detection of a watermark. They proposed the Fuzzy Support Vector Machine (FSVM) based on the NSST and Polar Harmonic Transforms (PHTs) in order to compensate the possible geometric distortions. First, the optimal NSST provides nearly optimal approximation for 2D image function during the embedding procedure. Second, the PHTs and the FSVM are exploited to estimate the geometric distortions parameters during a watermark extraction. The robustness for the rotation, scaling, cropping, and translation was proved by the experiments. At the same time, this watermarking algorithm is fragile relative to the local geometrical distortions, such as the random bending, column or line removal.

3.3 Shearlet Theory

In the past decades, many new tools for multi-scale decomposition analysis of the images, such as wavelet [46], ridgelet [47, 48], curvelet [49], brushlet [50], wedgelet [51], beamlet [52], contourlet [53], bandelet [54], directionlet [55], and shearlet [56] methods, had been developed. Not all of them are suitable for a watermarking but applied in image denoising, image improvement, among others. In wavelet analysis [57], an orthonormal set of functions as well as a non-orthogonal but linearly independent set of function are utilized. Wavelets are approximated by the discontinuous functions with a fewer number of functions than in the Fourier techniques. Due to the benefits, such as an easy detection of the local properties and the simultaneous signal analysis in the both time and frequency domains, a wavelet transform became very popular instrument in many fields of computer vision including a watermarking.

In a multi-resolution analysis, a projection of a continuous signal $f \in L^2(R)$, having a finite energy on the basis $\{\phi_{j,k}, \{\psi_{j,k}\}_{j \leq 1}\}_{k \in Z}$, is studied. The basis function $\phi_{j,k}(x) = 2^{-j/2}\phi(2^j x - k)$ is obtained from a translation and a dilation of a scaling function $\phi(x)$, $\int \phi(x)dx = 1$. The family $\{\phi_{j,k}\}_{k \in Z}$ covers a subspace $V_j \subset L^2(R)$ and the projection of function $f(x)$ on a subspace V_j provides an approximation $\{a_{j,k} = \langle f(x), \phi_{j,k}\ n \rangle\} k \in Z$ of function $f(x)$ at the scale 2^j. The function $\psi_{j,k}(x) = 2^{-j/2}\psi(2^j x - k)$ is obtained from a translation and a dilation of a mother wavelet function $\psi(x)$, $\int \psi(x)dx = 0$. The family $\{\psi_{j,k}\}_{k \in Z}$ covers a subspace $W_j \subset L^2(R)$ and the projection of function $f(x)$ on a subspace W_j provides an approximation $\{w_{j,k} = \langle f(x), \psi_{j,k}\ n\rangle\} k \in Z$ of function $f(x)$, representing the details between two successive approximations. A multi-resolution l-leveled analysis in a subspace V_j yields the decomposition provided by Eq. 3.1, where the dual functions $\bar{\phi}(x)$ and $\bar{\psi}(x)$ are defined for the best reconstruction.

$$f(x) = \sum_k a_{j,k}\bar{\phi}_{j,k}(x) + \sum_{j \le l}\sum_k w_{j,k}\bar{\psi}_{j,k}(x) \tag{3.1}$$

The 2D wavelet transform is a separable transform, thus 2D transform can be obtained by applying 1D transform along the both OX and OY directions. As a result, four components, involving the approximation coefficients that are represented the base image and horizontal, the vertical, and diagonal coefficients, can be computed. The levels of decomposition of an image I are provided by Eq. 3.2, where I_{LLn} represents the base image at level n and I_{LHn}, I_{HLn}, and I_{HHn} are high frequency along the vertical, horizontal, and diagonal directions of the image at the level n, respectively.

$$I_{LL_{n-1}} = I_{LL_n} + I_{LH_n} + I_{HL_n} + I_{HH_n} \quad n = 1, 2, 3\ldots \tag{3.2}$$

On the nth level of decomposition, a sequence of $(3n + 1)$ sub-images will be received. In practice, the wavelet low-pass and high-pass filters are employed in order to obtain the decomposition with various resolutions.

Wavelet transform is an optimal tool for processing of 1D piecewise smooth signals but not for processing of high-dimensional signals. For example, 2D wavelet transform can capture the limited directional information since it decomposes an image into three directional sub-bands. The estimation of the derivative vectors describing the blobs, corners, junctions, and peaks in an image using the traditional wavelet-based approach decreases the estimation accuracy. The rigelet, curvelet, contourlet, shearlet and other transforms from this group preserve the contours and texture in an image better than the wavelet transform could do [58]. In this case, the anisotropic estimation is more efficiently than the wavelet transform. The contourlet transform combines the sub-band decomposition using the Laplacian pyramid filters [59] that provide a downsampled low-pass of the original image, as well as the difference between the original and the predicted images, and the directional transform using the directional filter banks [60] that generate a decomposition of the original image in different directions and scales. The sub-bands of the wavelet and contourlet transforms are depicted in Fig. 3.2.

The use of the DST provides a variety of sub-bands for inserting the secret data due to the shearlets' multi-resolution property. These sub-bands are correlated in the DWT but not correlated in the DST that makes a watermarking more secure procedure [61]. More, the NSST is a fully shift-invariant, multi-scale, and multi-directional expansion of the shearlet transform [6].

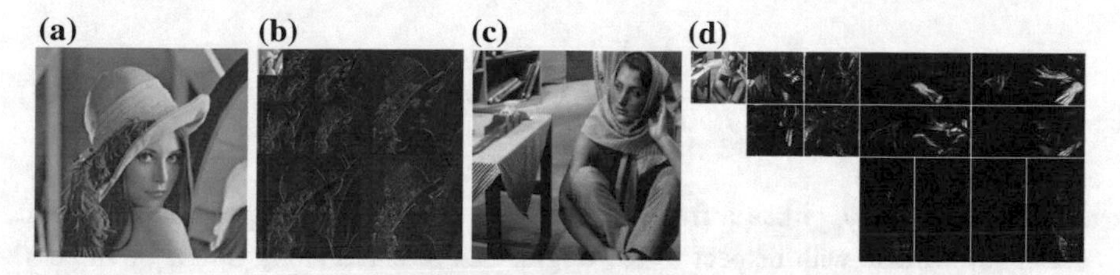

Fig. 3.2 3-level wavelet and contourlet decomposition: **a** the original Lena image, **b** a wavelet transform of the Lena image, **c** the original Barbara image, **d** a contourlet transform of the Barbara image

In 2D space, the continuous shearlets are identical to the affine systems with composite dilations and defined by Eq. 3.3, where $\mathbf{A}_s$ is an anisotropic dilation, $\mathbf{B}_s$ is a shearing matrix, ψ is a generating function.

$$\psi_{a,s,t}(x) = \left|\det \mathbf{M}_{a,s}\right|^{-1/2} \psi\left(\mathbf{M}_{a,s}^{-1} x - t\right) : a, s \in Z, \quad t \in Z^2, \tag{3.3}$$

where

$$\mathbf{M}_{a,s} = \begin{pmatrix} a & \sqrt{as} \\ 0 & \sqrt{a} \end{pmatrix} = \mathbf{B}_s \mathbf{A}_s \quad \mathbf{A}_s = \begin{pmatrix} a & 0 \\ 0 & \sqrt{a} \end{pmatrix} \quad \mathbf{B}_s = \begin{pmatrix} 1 & s \\ 0 & 1 \end{pmatrix}. \tag{3.4}$$

The elements of Eq. 3.3 are called the composite wavelets if Eq. 3.3 forms a Parseval frame (tight frame)

$$\sum_{a,s,t} \left|\langle f, \psi_{a,s,t} \rangle\right|^2 = \|f\|^2 \tag{3.5}$$

for all $f \in R^2$. Note that the Parseval frames allow to construct such elements that range not only at various scales and locations (like wavelets) but also at various orientations [62].

The shearlet transform of function f is determined by Eq. 3.6.

$$SH_{\psi} f(a, s, t) = \langle f, \psi_{a,s,t} \rangle \tag{3.6}$$

If the generative function ψ satisfies the following conditions in frequency domain, such as $\hat{\psi}(\xi_1, \xi_2) = \hat{\psi}_1(\xi_1)\, \hat{\psi}_2(\xi_2/\xi_1)$, where $\hat{\psi}_1$ and $\hat{\psi}_2$ are smooth functions inside the sets [–2, –1/2] ∪ [1/2, 2] for $\hat{\psi}_1$ and [–1, 1] for $\hat{\psi}_2$, then each function $f \in R^2$ can be reconstructed by Eq. 3.7.

$$f(x) = \int_{R^2} \int_{-2}^{2} \int_{0}^{1} \langle f, \psi_{a,s,t} \rangle \, \psi_{a,s,t}(x) \, \frac{da}{a^3} \, ds \, dt \tag{3.7}$$

Each element $\psi_{a,s,t}$ has a frequency support on a pair of trapezoids at several scales, symmetric with respect to the origin, and oriented along line a of slope s. Therefore, the shearlets $\psi_{a,s,t}$ create a collection of localized waveforms at various scales a, orientations s, and locations t as it is depicted in Fig. 3.3a.

The continuous shearlet transform can be discretized by its main parameters, such as the scale, shear and translation parameters. Usually the following parameters are chosen [63]: $a = 2^{-j}$, $s = -d$, where $j, d \in Z$, and $t \in Z^2$ is replaced by a point $k \in Z^2$. In this case, the discrete form of the shearlets is provided by Eq. 3.8:

$$\psi_{j,d,k} = |\det \mathbf{A}_0|^{\frac{j}{2}} \psi\left(\mathbf{B}_0^d \mathbf{A}_0^j x - k\right), \tag{3.8}$$

where

$$\mathbf{A}_0 = \begin{pmatrix} 4 & 0 \\ 0 & 2 \end{pmatrix} \quad \mathbf{B}_0 = \begin{pmatrix} 1 & 1 \\ 0 & 0 \end{pmatrix}. \tag{3.9}$$

The discrete shearlets (Eqs. 3.8–3.9) provide a non-uniform angular covering of the frequency plane under the restricted finite discrete parameters. It is reasonable to reconstruct the shearlets in order to obtain the horizontal cone

$$C_0 = \{(\xi_1, \xi_2) \in R^2 : |\xi_1| \geq 1, |\xi_2/\xi_1| \geq 1\}, \tag{3.10}$$

vertical cone

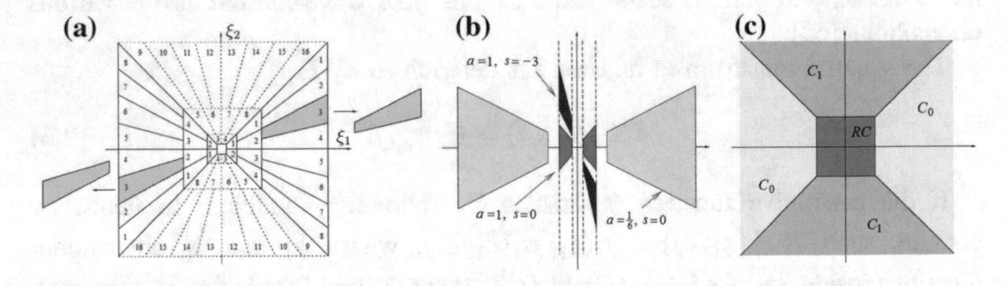

Fig. 3.3 Support of shearlets $\psi_{a,s,t}$ in the frequency domain that is portioned into trapezoidal tiles: **a** the continuous variant—the tilling of the frequency plane R^2, where the tilling of the horizontal cone $C_0 = \left\{(\xi_1, \xi_2) \in \hat{R}^2 : |\xi_1| \geq 1/8, |\xi_2/\xi_1| < 1\right\}$ is illustrated in a solid lines and the tilling of the vertical cone $C_1 = \left\{(\xi_1, \xi_2) \in \hat{R}^2 : |\xi_2| \geq 1/8, |\xi_1/\xi_2| < 1\right\}$ is illustrated in a dashed lines, **b** the continuous variant—with different values of a and s, **c** the discrete variant—the cones C_0 and C_1 and the centered rectangle R in frequency domain for a Parseval frame (Eq. 3.5)

$$C_1 = \{ (\xi_1, \xi_2) \in \mathrm{R}^2 : |\xi_2| \geq 1, |\xi_1/\xi_2| \geq 1 \}, \tag{3.11}$$

and centered rectangle

$$RC = \{ (\xi_1, \xi_2) \in \mathrm{R}^2 : \|(\xi_1, \xi_2)\|_\infty < 1 \}, \tag{3.12}$$

as it is shown in Fig. 3.3b.

The NSST is a shift-invariant version of the shearlet transform [4]. The NSST differs from the shearlet transform by eliminating of the down-samplers and up-samplers. The NSST is a fully shift-invariant, multi-scale and multi-directional transform that combines the non-subsampled Laplacian pyramid transform with different combinations of the shearing filters [62]. The analysis of the non-subsampled Laplacian pyramid can be done through the iterative procedure provided by Eq. 3.13, where f is an image, $NSLP_{j+1}$ is the detail coefficients at scale $j+1$, Ah_j^1 and Ah_k^0 are the low-pass and high-pass filters at scales j and k, respectively.

$$NSLP_{j+1} = A_j f = \left(Ah_j^1 \prod_{k=1}^{j-1} Ah_k^0 \right) f \tag{3.13}$$

The details of the NSST coefficients computation are discussed in [6].

The mathematical properties of shearlets were summarized by Easley et al. [62] as mentioned below:

- The shearlets are well localized. They are compactly supported in the frequency domain and have fast decay in the spatial domain.
- The shearlets satisfy the parabolic scaling. Each element $\hat{\psi}_{a,s,t}$ is supported on a pair of trapezoids, which are approximated by a box with sizes $2^j \times 2^{2j}$.
- The shearlets demonstrate highly directional sensitivity. The elements $\hat{\psi}_{a,s,t}$ are oriented along lines with slope given by $-l2^{-j}$, while the elements $\psi_{a,s,t}$ are oriented along lines with slope given by $l2^{-j}$.
- The shearlets are spatially localized. For any fixed scale and orientation, the shearlets are obtained by translations on the grating Z^2.
- The shearlets are optimally sparse.

It is reasonable to apply the watermarking techniques depending to the content of images with different degree of embedding. Consider the perceptual basics of a watermarking in Sect. 3.4.

3.4 Perceptual Watermarking

The trade-off achievement between three parameters—the payload, fidelity, and robustness, is a cornerstone of any watermarking algorithm. The payload defines a volume of embedded bits. The fidelity shows a degree of signal degradation. The robustness is an ability of the watermark to remain readable under various vulnerabilities. These parameters are restricted not only by technical limitations of the input images and transmission devices but also the HVS requirements.

The cortex theory leaded to the understanding that the retina of the eye splits the visual stimulus composing an image in several components. These components circulate by different tuned channels from the eye to the cortex and each channel is tuned to a corresponding component. A component has the following features: the localization in the image, spatial frequency (the amplitude in polar coordinates), and orientation (the phase in polar coordinates). A perceptive channel can be stimulated only by such component, whose features coincide to its own features. The components, having different characteristics, are independent. Olzak and Thomas [64] show that similar components use the same channels from the eye to the cortex. This causes the non-linear effects if the signals interact and their magnitudes overlaps the minimum level, below which a signal cannot be seen. Masking is one of those effects.

The watermark embedding ought to be implemented with knowledge of the HVS paradigms. Such factors as the luminance, texture style, proximity to the edges, and frequency bands have a direct impact on the distortion sensitivity. For example, the intensity level of a watermark has to be below that the intensity level of a host image due to the invisibility constraint. The HVS is less sensitive to the changes in regions of high and low brightness. Nearby the edges and in textured regions, the HVS is less sensitive to distortions than in smooth areas. Watson et al. [65] mentioned that a human eye is less sensitive to noise in high frequency sub-bands and the bands that have $\pm 45^{\circ}$ orientation. Kundur and Hatzinakos [66] developed a conventional HVS model to produce a visual mask for multi-resolution based image watermarking for grayscale images. Also the human eye is the least sensitive to the blue channel. Thereafter, Vahedi et al. [32] suggested the similar HVS model for color images. The mentioned above propositions may be used during the preliminary choice of regions for watermark embedding with following creating of a spatial mask. Such spatial mask can be applied directly with the gain factor to the host image and indirectly to the partitioned host image in spatial and frequency domains, respectively.

The HVS models can be applied in the both spatial or frequency domains (in the last case for the decomposition coefficients). In general, a watermark embedding is represented by Eq. 3.14, where *WI* is a watermarked image, *HI* is a host image, *W* is a watermark image, α is a coefficient, which controls the embedding strength.

$$WI = HI(1 + \alpha W) \tag{3.14}$$

The goal of the HVS model and the perceptual mask is to optimize the embedding strength α in order to obtain a trade-off between the robustness and invisibility of the watermark.

Thereafter, some modifications of Eq. 3.14 were suggested, Thus, Autrusseau et al. in [67] reinforced this expression by a power function with index k, $k = \{1, 2, 0.5\}$.

$$WI = HI\left(1 + \alpha^k W\right) \tag{3.15}$$

If $k = 1$, then a linear embedding scenario is implemented. If $k = 2$, then the high amplitude wavelet coefficients are strengthened. If $k = 0.5$, then the low amplitude wavelet coefficients are strengthened.

Delaigle et al. [68] developed an additive watermarking technique for the DFT. This perceptual model is derived from Michelson's contrast C that is defined by Eq. 3.16, where $L_{\max}$ and $L_{\min}$ are the maximal and minimal luminance values of grating, respectively.

$$C = \frac{2(L_{\max} - L_{\min})}{L_{\max} + L_{\min}} \tag{3.16}$$

Also, these authors derived a formula for detection threshold contrast C_{th} depending from the contrast C_{ms}, frequency f_{ms}, and orientation θ_{ms} of a masking signal in a view of Eq. 3.17, where f and θ are a frequency and orientation of a testing signal, respectively, C_0 is the visibility threshold without a masking effect with parameters f_0 and θ_0.

$$C_{th}(C_{ms}, f, \theta) = C_0 + k_{(f_0,\theta_0)}(f, \theta)\left(C_{th_{(f_0,\theta_0)}}(C_{ms}) - C_0\right)c \tag{3.17}$$

The coefficient $k(\cdot)$ is defined by Eq. 3.18, where $F(f_0)$ and $\Theta(\theta_0)$ are the parameters spreading the Gaussian function.

$$k_{(f_0,\theta_0)}(f, \theta) = \exp\left(-\left(\frac{\log^2(f/f_0)}{F^2(f_0)}\right) + \frac{(\theta - \theta_0)^2}{\Theta^2(f_0)}\right) \tag{3.18}$$

The threshold contrast has a view of Eq. 3.19, where ε is an index that depends on (f_0, θ_0), $0.6 \leq \varepsilon \leq 1.1$.

$$C_{th_{(f_0,\theta_0)}}(C_{ms}) = \max\left(C_0, C_0\left(\frac{C_{ms}}{C_0}\right)^{\varepsilon}\right) \tag{3.19}$$

A contrast threshold for the DCT can be estimated by an empirical expression for contrast sensitivity as a function of spatial frequency—the Contrast Sensitivity

Function (CSF) [69] in a view of Eq. 3.20, where f is the spatial frequency in cycles/degree of visual angle.

$$CSF(f) = 2.6(0.0192 + 0.114f)\, e^{-(0.114f)^{1.1}} \tag{3.20}$$

A watermarking embedding exploits the weakness of the HVS to make a watermark imperceptible with maximal strength. Watson [70] was the first, who modelled the DCT coefficients employing the weakness of the HVS, such as the frequency sensitivity function, luminance sensitivity function, and contrast sensitivity function.

A watermarking algorithm with perceptually tuned parameters uses the conventional decomposition into three levels of the DWT/DST but involves a possibility to choose the perceptually significant coefficients or just noticeable differences by thresholding in each sub-band. By contrast, Voloshynovskiy et al. [71] proposed and verified a general perceptual mask referred to as the Noise Visibility Function (NVF). Note that factually a watermark is a noise regarding the transmitted information. This mask is based on the maximum aposteriori probability estimation and Markov random fields. The properties of an image denoising are defined by a multiplicative term b, Eq. 3.21, where σ_{im}^2 is a local variance of a host image, σ_n^2 is a variance of a noise, w is a weighting function, a view of which depends on the underlying assumptions about the statistics of a watermarking image.

$$b = \frac{\sigma_{im}^2}{w\,\sigma_n^2 + \sigma_{im}^2} \tag{3.21}$$

The local variance is a good indicator of the local image activity. The large values of variance indicate a presence of edges or highly textured areas, while the small values of variance mean a flat regions. Thus, Eq. 3.21 determines a level of image smoothing. For flat regions, $\sigma_{im}^2 \to 0$ and $b \to 0$, while for edges and textural regions, $\sigma_{im}^2 \gg \sigma_n^2$ and $b \to 1$, remaining a host image without any changes practically. Such adaptive image filtering is well matched with a textural masking property of the HVS, when a noise is more visible in flat areas and less visible in regions with edges and textures.

Other authors continued these investigations. Kim et al. [72] proposed a method based on the computation of the local NVFs. According to this concept, a watermark is inserted into the textural and edge region stronger than the flat region. In the case of stationary generalized Gaussian model, the NVF is provided by Eq. 3.22, where $\sigma^2(s, t)$ is a variance of a host image, w(s, t) is a weight value, (s, t) are the image coordinates.

$$NVF(s,t) = \frac{w(s,t)}{w(s,t) + \sigma^2(s,t)} \tag{3.22}$$

The weight value is calculated using Γ-function. This model is represented by Eq. 3.23, where $I(s, t)$ is an intensity function, $I^{*}(s, t)$ is a local average of intensity, γ is a shape factor, $\gamma \in [0.3\text{–}1.0]$.

$$w(s,t) = \Gamma\left[\left(\sqrt{\Gamma\left(\frac{3}{\gamma}\right)\Big/\Gamma\left(\frac{1}{\gamma}\right)}\right)^{\gamma}\right]\frac{1}{\left\|\frac{I(s,t)-I^{*}(s,t)}{\sigma(s,t)}\right\|^{2-\gamma}} \quad (3.23)$$

One of the challenging issues in a watermarking process is to tune the strength coefficient α (Eq. 3.14) in order to achieve a reasonable trade-off between invisibility and robustness. Autrusseau and Callet [73] resumed their experiments on sine-wave gratings leaded to the CSF concept that provides the just noticeable contrast threshold at a given spatial frequency. In first CSF models [74], it was assumed that the HVS behavior is considered as a single channel and its modulation transfer function is the CSF. However, this proposition cannot be confirmed even for grayscale images. Several physiological studies show that the most number of cells in the HVS are tuned to color, orientation or frequency information [75, 76]. This indicates that the HVS has a multi-channel structure, which can be modelled by a filter bank separating each perceptual channel (called as a perceptual sub-band). Such multi-channel visual decomposition is perfectly suitable to model the masking effects.

The multi-channel visual decomposition is usually modelled as a polar representation of separable channels by the radial frequency bands and orientation selectivity. Autrusseau and Callet [73] assigned the HVS orientation selectivity as 30° and 45° depending on the radial band. This means that 30° and 45° oriented signals do not interact with 160° and 135° oriented signals, respectively, since they are in different channels. However, in the DCT and the DWT, the energies of such oriented signals are represented in the same component. Thus, the horizontal and vertical information are mixed and it is impossible to estimate the energies in the real visual bands. These authors designed so call Perceptual Channel Decomposition (PCD) filters that are similar to the cortex filters developed by Watson [77]. The cortex filters are defined as the product between dom (difference-of-mesa) filters and fan filters providing the radial selectivity and the angular selectivity, respectively. Watson claimed that the median orientation bandwidth of visual cortex cells is about 45° with a rather broad distribution about this value. Note that Goresnic and Rotman [78] recommended the average orientation bandwidth of the cells is about 40° with a large spread from cell to cell according to the cortex transform. Therefore, it is reasonable to follow the shearlet theory with 45° orientation bandwidth.

The PCD (defined in cycle/degree) of Autrusseau and Callet in the frequency DWT domain is depicted in Fig. 3.4a, while our proposed PCD for three-leveled DST is given in Fig. 3.4b.

The detection of the PCD permits to find the most suitable regions for a watermark embedding.

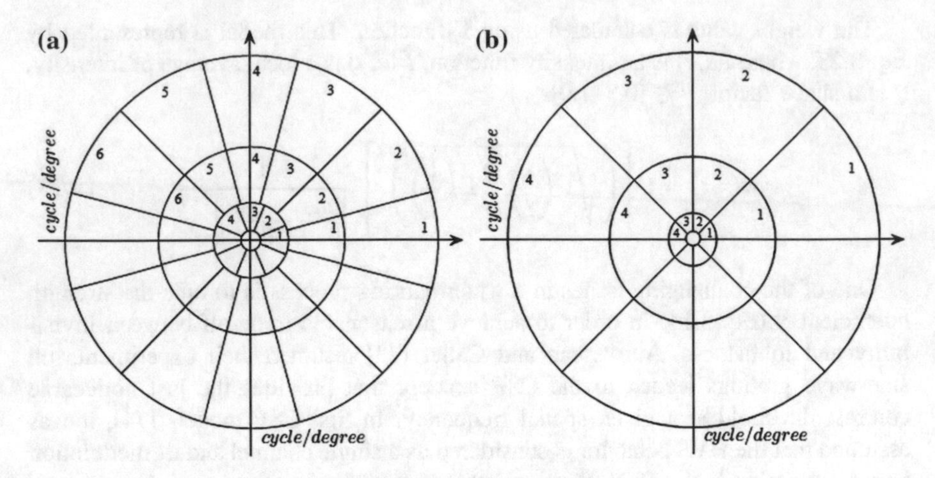

Fig. 3.4 The PCD available for: **a** the DWT, **b** the DST

3.5 Digital Watermark Embedding

Let a host grayscale image $\mathbf{HI} = \{I_{s,\,t}\}_{L \times N}$ with sizes $L \times N$, where $I \in [0, 1, \ldots, 255]$ is an intensity, $(s, t) \in [1, 2, \ldots, L] \times [1, 2, \ldots, N]$, be divided by a grid into patches with sizes $L_p \times N_p$. Thus, a host image HI is segmented into $l \times n$ patches, where $l = [L/L_p]$ and $n = [N/N_p]$. All patches do not overlap each other. Introduce a local coordinate system into each patch in such manner that, for example, $p_{14}(r, c)$ denotes an intensity value of a pixel at the position (r, c), where $(r, c) \in [1, 2, \ldots, L_p] \times [1, 2, \ldots, N_p]$ in the patch p_{14}. The values of L_p and N_p are chosen equal 8. An image patch $p_{z,q}$, $(z, q) \in [1, 2, \ldots, l] \times [1, 2, \ldots, n]$ can be decomposed into the corresponding shearlet sub-band $\psi_{j,d,k}$, where j, d, k mean the scale, orientation, and location, respectively, through the DST. In current research, the third level of decomposition was chosen.

In the case of color image, a host image has a decomposition $\mathbf{HI} = \{\mathbf{I}_{s,t}\}_{L \times N}$, where $\mathbf{I}_{s,t}$ is 3×1 column vector, $\mathbf{I}_{s,t} = \{R_{s,t}, G_{s,t}, B_{s,t}\}$, $(R_{s,t}, G_{s,t}, B_{s,t}) \in \{0, 1, \ldots, 255\}$ are red, green, and blue values in pixel (s,t), respectively. However, it is reasonable to transform the RGB color space to the YUV color space and analyze a luminance component Y.

Further, a watermarking scrambling via Arnold's transform as a way to enhance the secrecy and security will be discussed in Sect. 3.5.1. The basics of the SVD are represented in Sect. 3.5.2. The proposed algorithm of a watermark embedding is described in Sect. 3.5.3.

3.5.1 Watermark Scrambling Via Arnold's Transform

The Arnold's transform refers to the chaotic transforms [79] and is successfully applied as a technique to scramble the watermark in order to increase the robustness of a model against the cropping attacks. This transform disorders a watermark matrix and makes a watermark obscured. The 2D Arnold's transform has a view of Eq. 3.24, where Γ is a transformation, (x, y) and (x', y') are pixel's coordinates of the original and scrambled watermark, $\{a_1, \ldots, a_4\}$ are the coefficients of transform, $a_1 \cdot a_4 - a_2 \cdot a_3 = \pm 1$, $N_w \times N_w$ are the sizes of a watermark (values of the width and length ought to be equaled).

$$\Gamma\begin{pmatrix} x' \\ y' \end{pmatrix} \rightarrow \begin{pmatrix} a_1 & a_2 \\ a_3 & a_4 \end{pmatrix}\begin{pmatrix} x \\ y \end{pmatrix} (\text{mod}\, N_w) \tag{3.24}$$

In practice, the Arnold's Cat Map is often used provided by Eq. 3.25.

$$\Gamma\begin{pmatrix} x' \\ y' \end{pmatrix} \rightarrow \begin{pmatrix} 1 & 1 \\ 1 & 2 \end{pmatrix}\begin{pmatrix} x \\ y \end{pmatrix} (\text{mod}\, N_w) \tag{3.25}$$

The mechanism of the Γ transformation includes three steps during the one iteration:

- Shear in the OX direction by a factor of 1.

$$\begin{pmatrix} x' \\ y' \end{pmatrix} \rightarrow \begin{pmatrix} x+y \\ y \end{pmatrix}$$

- Shear in the OY direction by a factor of 1.

$$\begin{pmatrix} x' \\ y' \end{pmatrix} \rightarrow \begin{pmatrix} x \\ x+y \end{pmatrix}$$

- Evaluate the modulo.

$$\begin{pmatrix} x' \\ y' \end{pmatrix} \rightarrow \begin{pmatrix} x \\ y \end{pmatrix} \text{mod}\, N_w$$

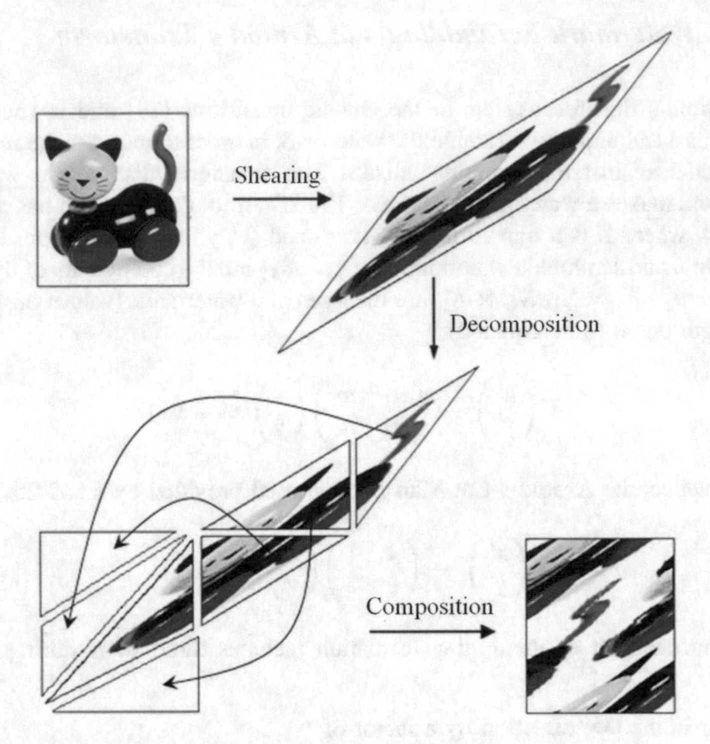

Fig. 3.5 The Arnild's Cat Map decomposition during the single iteration

The illustration of the Arnold's Cat Map decomposition is depicted in Fig. 3.5.

The Arnold's transform is the periodic and invertible mapping. The number of iterations taken is known as the Arnold's period that provides a secret key. During a decryption, if a scrambled image is iterated enough times (the Arnold's period minus a value of secret key), the original image reappears. For color watermarks, the YCbCr color space (Y is a luminance, Cb and Cr are chrominance components) is recommended [80]. In this case, only Y-channel may be processed by the Arnold's transform.

This procedure can be complicated by the multi-regional scrambled (with different parameters) parts of a watermark.

3.5.2 *Basics of Singular Value Decomposition*

The Singular Value Decomposition (SVD) is one of effective numerical analysis tools used to analyze the matrices based on a theory of linear algebra [81]. This technique has a wide application in many fields including the watermarking schemas [28, 82–85]. The SVD transforms the correlated variables into a set of uncorrelated ones. A rectangular matrix B of order $L_p \times N_p$ can be partitioned into the product of three matrices: an orthogonal matrix U, a diagonal matrix S, and the transpose of an orthogonal matrix V, such that

$$B = \begin{bmatrix} b_{1,1} & b_{1,2} & \cdots & b_{1,N_p} \\ b_{2,1} & b_{2,2} & \cdots & b_{2,N_p} \\ \vdots & \vdots & \ddots & \vdots \\ b_{L_p,1} & b_{L_p,2} & & b_{L_p,N_p} \end{bmatrix} = USV^{\mathrm{T}} = \begin{bmatrix} u_{1,1} & u_{1,2} & \cdots & u_{1,L_p} \\ u_{2,1} & u_{2,2} & \cdots & u_{2,L_p} \\ \vdots & \vdots & \ddots & \vdots \\ u_{L_p,1} & u_{L_p,2} & & u_{L_p,L_p} \end{bmatrix}$$
$$\times \begin{bmatrix} \lambda_{1,1} & 0 & \cdots & 0 \\ 0 & \lambda_{2,2} & \cdots & 0 \\ \vdots & \vdots & \ddots & \vdots \\ 0 & 0 & & \lambda_{L_p,N_p} \end{bmatrix} \times \begin{bmatrix} v_{1,1} & v_{1,2} & \cdots & v_{1,N_p} \\ v_{2,1} & v_{2,2} & \cdots & v_{2,N_p} \\ \vdots & \vdots & \ddots & \vdots \\ v_{N_p,1} & v_{N_p,2} & & v_{N_p,N_p} \end{bmatrix}, \tag{3.26}$$

with the matrices U and V are $L_p \times L_p$ and $N_p \times N_p$ real unitary matrices with small singular values, respectively, $UU^{\mathrm{T}} = I_{L_p}$ and $VV^{\mathrm{T}} = I_{N_p}$, where I_{L_p} and I_{N_p} are the identify matrices of order L_p and N_p, respectively, while a matrix S is a $L_p \times N_p$ diagonal matrix with larger singular values, which entries satisfy Eq. 3.27, where r is rank of the matrix.

$$\lambda_{1,1} \geq \lambda_{2,2} \geq \cdots \geq \lambda_{r,r} > \lambda_{r+1,r+1} = \lambda_{r+2,r+2} = \cdots = \lambda_{L_p,N_p} = 0 \tag{3.27}$$

The SVD has several advantages. First, the size of the matrices is not fixed and they can be the square or rectangle matrices. Second, the singular values are less affected during image processing. Third, the singular values contain the intrinsic algebraic properties.

3.5.3 Algorithm of Watermark Embedding

The input information for a watermark embedding is the host grayscale image and grayscale watermark. A functional flow chart of this process is illustrated in Fig. 3.6. In this research, the blind scheme of a watermarking is implemented. The algorithm for embedding of a grayscale watermark in a grayscale host image is formulated as follows.

Step 1. Build the perceptual channel decomposition of the host image providing the appropriate regions for embedding.

Step 2. Apply the Arnold's transform to a watermark, obtain the scrambled watermark, and design a binary representation of a watermark. The parameters of the Arnold's transform are the part of secret key information.

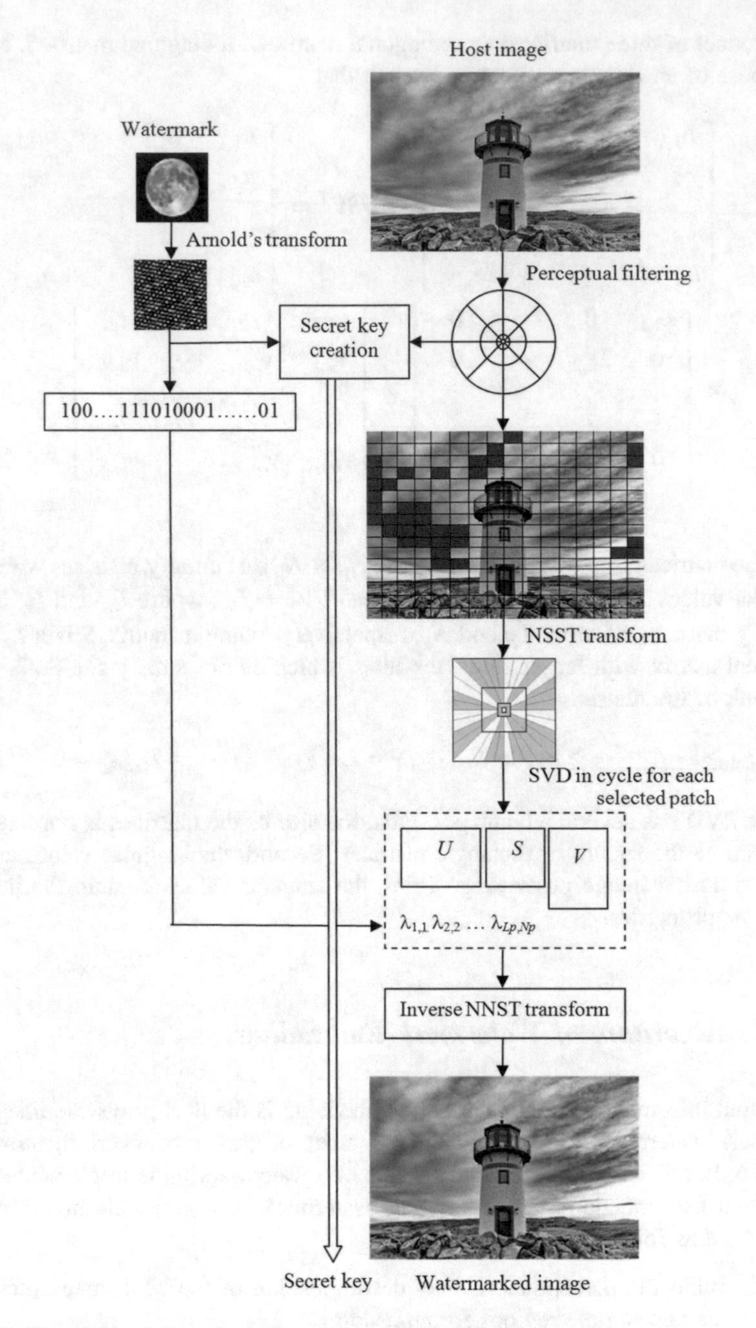

Fig. 3.6 The proposed flow chart of a watermark embedding

Step 3. Apply the non-subsampled shearlet transform by the third sub-band to the selected regions, which are divided into patches with $L_p \times N_p$ elements. For experiments, $L_p = N_p = 8$. Define the order of the processing of the selected regions. This information is a part of secret key information.

Step 4. Loop//Processing of a current patch//.

Step 4.1.
Apply the single value decomposition to the coefficients of the NNST in a patch.

Step 4.2.
Replace the last significant bits in a sequence of eigenvalues $\lambda_{1,2}, \lambda_{2,2}, \ldots, \lambda_{L_p,N_p}$ by a binary sub-sequence of a watermark.

Step 5. Apply the inverse NNST to the watermarked image.

The output information is the watermarked image and secret key. The proposed procedure can be expanded for the color host images and color watermarks.

3.6 Digital Watermark Extraction

The input information for a watermark extraction is the watermarked image and secret key. A functional flow chart of this process is illustrated in Fig. 3.7. The algorithm for extraction of a watermark from a watermarked image is formulated as follows.

Step 1. Select the corresponding regions in the watermarked image using a secret key.

Step 2. Apply the non-subsampled shearlet transform by the third sub-band to the selected regions, which are divided into patches with $L_p \times N_p$ elements. For experiments, the values $L_p = N_p = 8$ were chosen.

Step 3. Extract the order of processing of the selected regions from a secret key.

Step 4. Loop//Processing of a current patch//.

Step 4.1.
Apply the single value decomposition to the coefficients of the NNST in a patch.

Step 4.2.
Extract the last significant bits in a sequence of eigenvalues $\lambda_{1,2}, \lambda_{2,2}, \ldots, \lambda_{L_p,N_p}$ and put them in a binary sub-sequence of a watermark, forming the scrambled image.

Step 5. Apply the inverse NNST to the extracted host image.

Step 6. Apply the Arnold's transform to the scrambled watermark using a secret key.

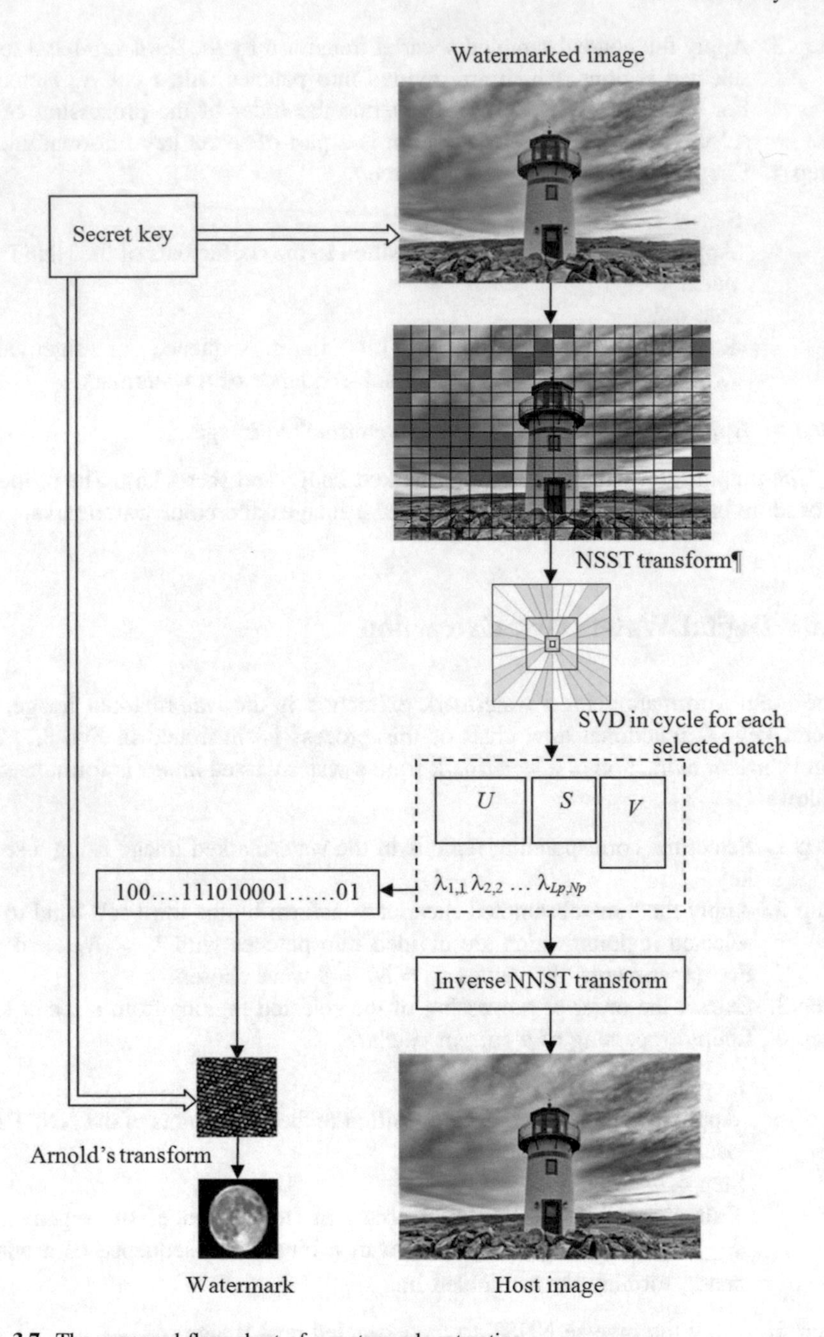

Fig. 3.7 The proposed flow chart of a watermark extraction

The output information is the extracted watermark and extracted host image. Section 3.7 presents the experimental results of the extracted watermark evaluation as well as the influence of typical internet attacks.

3.7 Experimental Results

For experiments, the database Miscellaneous of University of Southern California was used [86]. It involves 44 color and monochrome images including such famous pictures as Lena, Peppers, Barbara, among others.

The DWT method was implemented as a three-level decomposition using the Daubechies filters. A watermark is embedded in the coefficients, which magnitudes do not exceed a threshold value [22]. The NNST method was realized according to its description in Sect. 3.3. The both methods employed the selected regions for embedding (Sect. 3.4). The comparative payload of the DWT and the NNST methods were investigated under a proposition that a watermark has a square shape. Some results are mentioned in Table 3.1.

For images with sizes 512 × 512 pixels, the NNST provides an embedding of a watermark with standard sizes 128 × 128 pixels, while the DWT cannot be used for this goal with the same quantity. The NNST permits to embed the information on 45–50% more than the DWT allows.

The experiments on a robustness of a watermark included the following procedures: the watermark embedding in a host image, simulation of attack, extraction of the transformed watermark, and its comparison with the original watermark. The attacks, which were chosen for experiments, are mentioned below:

- Scaling.
- Disproportionate scaling.
- Rotation + scaling.
- Translation.
- Cropping.

Since a watermark has a bit-sequence representation, the BER estimator can be successfully applied. This is very simple estimator but it allows to calculate a

Table 3.1 The payload of the watermarking methods

Image	Description	Sizes, pixels	Payload of container, pixels			
			By use of DWT	Watermark sizes	By use of NNST	Watermark sizes
4.2.05.tiff	Airplane	512 × 512	66,703	91 × 91	138,190	131 × 131
4.2.06.tiff	Sailboat on lake	512 × 512	72,932	95 × 95	162,034	142 × 142
5.1.12.tiff	Clock	256 × 256	23,413	53 × 53	39,511	70 × 70
5.3.01.tiff	Man	512 × 512	63,008	88 × 88	148,261	136 × 136

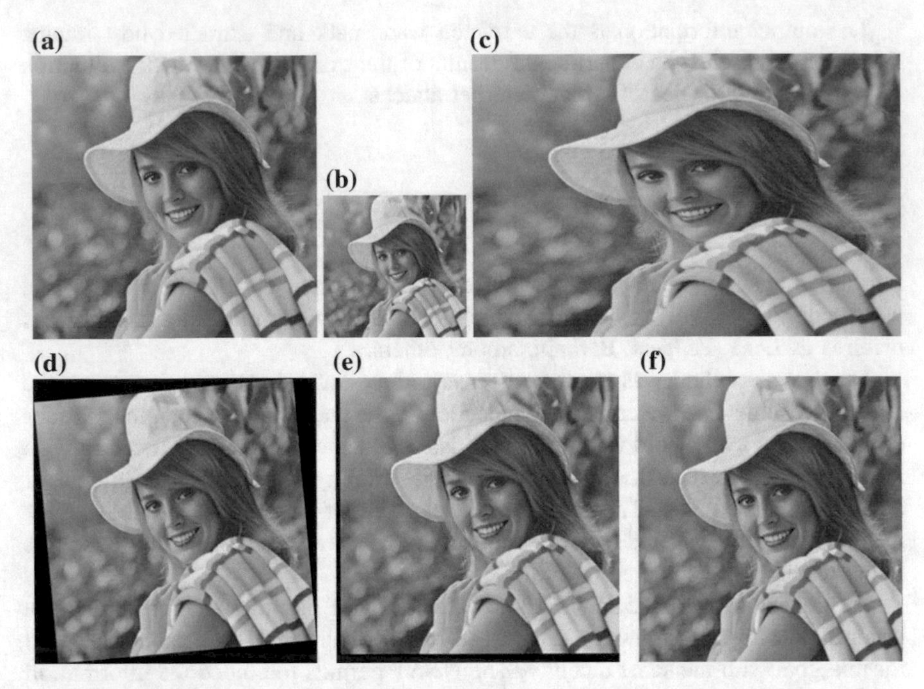

Fig. 3.8 Watermarked image under attacks: **a** original image, **b** proportionate scaling, **c** disproportionate scaling, **d** rotation transform, **e** translation, **f** cropping (the vertical boundaries of image are cropped)

Table 3.2 The description of the applied attacks

Type of attack	Description and feature
Scaling	Resizing to 256 × 256 pixels
Disproportionate scaling	Resizing to 450 × 300 pixels
Rotation	Rotation of 5° and scaling to the original sizes 512 × 512 pixels
Translation	Shift an image by 8 pixels right and by 16 pixels up
Cropping	Crop an image by 16 pixels on each vertical side

number of the error bits very accurately. The test image Elaine.tiff and its watermarking versions under the corresponding attacks are depicted in Fig. 3.8, while a description of the applied attacks are listed in Table 3.2. A view of the extracted watermark under some attacks (with following size normalization) is depicted in Fig. 3.9.

The results of a watermarking testing for robustness to the main types of attacks for the host image Elaine.tiff are specified in Table 3.3. The BER provides a ratio of error bits to the common bits of a watermark. A visibility of the extracted watermark was estimated using the PSNR metric. The watermark sizes were 64 × 64 pixels.

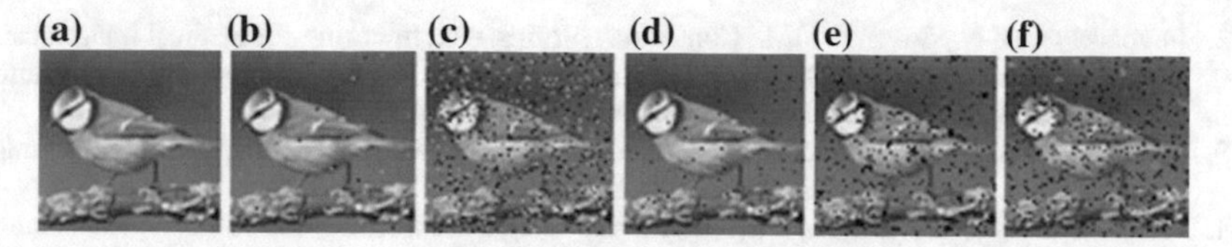

Fig. 3.9 Watermark under attacks: **a** original watermark, **b** proportionate scaling, **c** disproportionate scaling, **d** rotation transform, **e** translation, **f** cropping

Table 3.3 The impact of the attacks on watermark

Attack	BER (%)	PSNR (dB)
Scaling	1.2	48.22
Disproportionate scaling	22.6	23.07
Rotation	2.7	45.90
Translation	10.9	39.13
Cropping	12.4	31.85

The obtained results demonstrate a high robustness of a watermark to the scaling, rotation, and shift that is determined by the NSST properties. The influence of cropping attacks is directly determined by a cropped area.

3.8 Conclusions

In this chapter, a watermarking method based on the promising NSST was developed. A payload of the NSST exceeds a payload of the DWT on 45–50% in average for the test set of images. The experiments show that a perceptual tuning of the embedding process provides better results in a comparison to the conventional scheme. The proposed method was tested on 44 color and monochrome images. The experiments show the highest robustness to the rotation and proportional scaling (the BER mean values are 1.2–2.7%) and the medium robustness to the translation and cropping (the BER mean values are 10.9–12.4%). The disproportionate scaling demonstrated greater impact on the digital watermark. At the same time, the PSNR metric could not provide the explicit results respect to the attacks' influence. In future, the proposed algorithm will be studied in details on the subject of robustness to internet attacks.

References

1. Caragata, D., Mucarquer, J.A., Koscina, M., Assad, S.E.I.: Cryptanalysis of an improved fragile watermarking scheme. Int. J. Electron. Commu. (AEÜ) **70**(6), 777–785 (2016)

2. Favorskaya, M.N., Savchina, E.I.: Content preserving watermarking for medical images using shearlet transform and SVD. The International Archives of the Photogrammetry, Remote Sensing and Spatial Information Sciences, vol. XLII-2/W4, pp. 101–108 (2017)
3. Qi, H.Y., Zheng, D., Zhao, J.Y.: Human visual system based adaptive digital image watermarking. Signal Process. **88**(1), 174–188 (2008)
4. Hou, B., Zhang, X., Bu, X., Feng, H.: SAR image despeckling based on nonsubsampled shearlet transform. IEEE J. Sel. Top Appl. Earth Obs. Remote Sens. **5**(3), 809–823 (2012)
5. Sadreazami, H., Ahmad, M.O., Swamy, M.N.S.: Multiplicative watermark decoder in contourlet domain using the normal inverse Gaussian distribution. IEEE Trans. Multimedia **18** (2), 196–207 (2016)
6. Wang, X.Y., Liu, Y.N., Xu, H., Wang, A.L., Yang, H.Y.: Blind optimum detector for robust image watermarking in nonsubsampled shearlet domain. Inf. Sci. **372**, 634–654 (2016)
7. Langelaar, G.C., Setyawan, I., Lagendijk, R.L.: Watermarking digital image and video data: a state-of-the-art overview. IEEE Signal Process. Mag. **17**(5), 20–46 (2000)
8. Barni, M., Bartolini, F., Cappellini, V., Piva, A.: A DCT domain system for robust image watermarking. Signal Process. **66**(3), 357–372 (1998)
9. Huang, J., Shi, Y.Q., Shi, Y.: Embedding image watermarks in DC components. IEEE Trans. Circ. Syst. Video Technol. **10**(6), 974–979 (2000)
10. Sahail, M.A., Obaidat, M.S.: Digital watermarking-based DCT and JPEG model. IEEE Trans. Instrum. Meas. **52**(5), 1640–1647 (2003)
11. Briassouli, A., Strintzis, M.G.: Locally optimum nonlinearities for DCT watermark detection. IEEE Trans. Image Process. **13**(12), 1604–1617 (2004)
12. Briassouli, A., Tsakalides, P., Stouraitis, A.: Hidden messages in heavy-tails: DCT-domain watermark detection using alpha-stable models. IEEE Trans. Multimedia **7**(4), 700–715 (2005)
13. Solachidis, V., Pitas, L.: Circularly symmetric watermark embedding in 2-D DFT domain. IEEE Trans. Image Process. **10**(11), 1741–1753 (2001)
14. Ganic, E., Dexter, S.D., Eskicioglu, A.M.: Embedding multiple watermarks in the DFT domain using low- and high-frequency bands. In: Proceedings of the SPIE: Security, Steganography, and Watermarking of Multimedia Contents, vol. 5681, pp. 175–184 (2005)
15. Cedillo-Hernandez, M., Garcia-Ugalde, F., Nakano-Miyatake, M., Perez-Meana, H.: Robust watermarking method in DFT domain for effective management of medical imaging. SIViP **9** (5), 1163–1178 (2015)
16. Ganic, E., Eskicioglu, A.M.: A DFT-based semi-blind multiple watermarking scheme for images. 4th New York Metro Area Networking Workshop (NYMAN'2004), pp 1–5 (2004)
17. Ng, T., Garg, H.: Maximum-likelihood detection in DWT domain image watermarking using Laplacian modeling. IEEE Signal Process. Lett. **12**(4), 285–288 (2005)
18. Wang, J., Liu, G., Sun, J., Wang, Z., Lian, S.: Locally optimum detection for Barni's multiplicative watermarking in DWT domain. Signal Process. **88**(1), 117–130 (2008)
19. Rahman, S.M.M., Ahmad, M.O., Swamy, M.S.S.: A new statistical detector for DWT-based additive image watermarking using the Gauss-Hermite expansion. IEEE Trans. Image Process. **18**(8), 1782–1796 (2009)
20. Kwitt, R., Meerwald, P., Uhl, A.: Lightweight detection of additive watermarking in the DWT-domain. IEEE Trans. Image Process. **20**(2), 474–484 (2011)
21. Makbol, N.M., Khoo, B.E.: Robust blind image watermarking scheme based on redundant discrete wavelet transform and singular value decomposition. Int. J. Electron. Commun. (AEU) **67**, 102–112 (2012)
22. Favorskaya, M., Oreshkina, E.: Digital gray-scale watermarking based on biometrics. In: Damiani, E., Howlett, R.J., Jain, L.C., Gallo, L., De Pietro, G. (eds.) Intelligent Interactive Multimedia Systems and Services, SIST, vol. 40, pp. 203–214. Springer International Publishing Switzerland (2015)
23. Motwani, M.C., Harris, F.C.: Fuzzy perceptual watermarking for ownership verification. Int Conf Image Processing, Computer Vision, and Pattern Recogn (IPCV'2009), pp 321–325

24. Dumpa, G., Meenakshi, K.: Fuzzy perceptual watermarking for ownership verification in DCT domain. Int. J. Mag. Eng. Technol. Manage. Res. **3**(2), 360–367 (2015)
25. Bhatnagar, G., Wu, Q.M.J., Raman, B.: Robust gray-scale logo watermarking in wavelet domain. Comput. Electr. Eng. **38**(5), 1164–1176 (2012)
26. Bhatnagar, G., Wu, Q.M.J., Atrey, P.K.: Robust logo watermarking using biometrics inspired key generation. Expert Syst. Appl. **41**(10), 4563–4578 (2014)
27. Su, Q., Niu, Y., Zhao, Y., Pang, S., Liu, X.: A dual color images watermarking scheme based on the optimized compensation of singular value decomposition. Int. J. Electron. Commun. (AEU) **67**(8), 652–664 (2013)
28. Jia, S.I.: A novel blind color images watermarking based on SVD. Optik **125**(12), 2868–2874 (2014)
29. Botta, M., Cavagnino, D., Pomponiu, V.: A modular framework for color image watermarking. Sig. Process. **119**, 102–114 (2016)
30. Wang, M.S., Chen, W.C.: A majority-voting based watermarking scheme for color image tamper detection and recovery. Comput. Stand. Interfaces **29**(5), 561–570 (2007)
31. LiuKC, Kuo-Cheng: Wavelet-based watermarking for color images through visual masking. Int. J. Electron. Commun. (AEU) **64**(2), 112–124 (2010)
32. Vahedi, E., Zoroofi, R.A., Shiva, M.: Toward a new wavelet-based watermarking approach for color images using bio-inspired optimization principles. Digit. Signal Proc. **22**(1), 153–162 (2012)
33. Shao, Z., Shang, Y., Zeng, R., Shu, H., Coatrieux, C., Wu, J.: Robust watermarking scheme for color image based o nquaternion type moment invariants and visual cryptography. Sig. Process. Image Commun. **48**, 12–21 (2016)
34. Kumsawat, P., Attakitmongcol, K., Srikaew, A.: A new approach for optimization in image watermarking by using genetic algorithms. IEEE Trans. Signal Proc. **53**(12), 4707–4719 (2005)
35. Shih, F.Y., Wu, Y.T.: Enhancement of image watermark retrieval based on genetic algorithms. J. Vis. Commun. Image Represent. **16**(2), 115–133 (2005)
36. Wang, Y.R., Lin, W.H., Yang, L.: An intelligent watermarking method based on particle swarm optimization. Expert Syst. Appl. **38**(7), 8024–8029 (2011)
37. Tsai, H.H., Jhuang, Y.J., Lai, Y.S.: An SVD-based image watermarking in wavelet domain using SVR and PSO. Appl. Soft Comput. **12**(8), 2442–2453 (2012)
38. Aslantas, V.: SVD and DWT-SVD domain robust watermarking using differential evolution algorithm. In: Ao, S.I., Gelman, L. (eds.) Advances in Electrical Engineering and Computational Science, LNEE, vol. 39, pp. 147–159. Springer, Heidelberg (2009)
39. Ali, M., Ahn, C.W., Pant, M.: A robust image watermarking technique using SVD and differential evolution in DCT domain. Optik—Int. J. Light Electron. Opt. **125**(1), 428–434 (2014)
40. Yang, X.S.: Nature-inspired metaheuristic algorithms, 2nd edn. Luniver Press, UK (2010)
41. Mishra, A., Agarwal, C., Sharma, A., Bedi, P.: Optimized gray-scale image watermarking using DWT–SVD and firefly algorithm. Expert Syst. Appl. **41**(17), 7858–7867 (2014)
42. Ali, M., Ahn, C.W., Pant, M., Siarry, P.: An image watermarking scheme in wavelet domain with optimized compensation of singular value decomposition via artificial bee colony. Inf. Sci. **301**, 44–60 (2015)
43. Tamane, S.C., Deshmukh, R.R.: Blind 3D model watermarking based on multi-resolution representation and fuzzy logic. Int. J. Comput. Sci. Inf. Technol. **4**(1), 117–125 (2012)
44. Mardanpour, M., Chahooki, M.A.Z.: Robust transparent image watermarking with Shearlet transform and bidiagonal singular value decomposition. Int. J. Electron. Commun. (AEÜ) **70** (6), 790–798 (2016)
45. Wang, X.Y., Liu, Y.N., Li, S., Yang, H.Y., Niu, P.P.: Robust image watermarking approach using polar harmonic transforms based geometric correction. Neurocomputing **174**(Part B), 627–642 (2016)
46. Mallat, S.G.: A theory for multiresolution signal decomposition: the wavelet representation. IEEE Trans. Pattern Anal. Mach. Intell. **11**(7), 674–693 (1989)

47. Candes, E.J.: Ridgelets: theory and applications. Ph.D. thesis, Department of Statistics, Stanford University (1998)
48. Campisi, P., Kundur, D., Neri, A.: Robust digital watermarking in the ridgelet domain. IEEE Signal Process. Lett. **11**(10), 826–830 (2004)
49. Candes. E.J., Donoho, D.L.: Curvelets: a surprisingly effective nonadaptive representation for objects with edges. Tech report, Department of Statistics, Stanford University (2000)
50. Meyer, G.F., Ronald, R.C.: Brushlets: a tool for directional image analysis and image compression. Appl. Comput. Harmon. Anal. **4**(2), 147–187 (1997)
51. Donoho, D.L.: Wedgelets: nearly minimax estimation of edges. Ann. Stat. **27**(3), 859–897 (1999)
52. Donoho, D.L., Huo, X.: Beamlets and multiscale image analysis. Technical report, Stanford University (2001)
53. Do, M.N., Martin, V.: Contourlets. Stud. Comput. Math. **10**, 83–105 (2003)
54. Le Pennec, E., Mallat, S.: Sparse geometric image representations with bandelets. IEEE Trans. Image Process. **14**(4), 423–438 (2005)
55. Velisavljevic, V.: Directionlets: anisotropic multidirectional representation with separable filtering. IEEE Trans. Image Process. **15**(7), 1916–1933 (2006)
56. Labate, D., Lim, W.Q., Kutyniok, G., Weiss, G.: Sparse multidimensional representation using shearlets. In: Papadakis, M., Laine, A.F., Unser, M.A. (eds) Wavelets XI, vol. 5914, pp. 254–262 (2005)
57. Daubechies, I.: Ten lectures notes on wavelet. University of Lowell, Philadelphia, vol. 61. Society for Industrial and Applied Mathematics (SIAM) (1992)
58. Do, M.N., Martin Vetterli, M.: The contourlet transform: an efficient directional multiresolution image representation. IEEE Trans. Image Process. **14**(12), 2091–2106 (2005)
59. Burt, P.J., Adelson, Edward H.: The Laplacian pyramid as a compact image code. IEEE Trans. Commun. **31**(4), 532–540 (1983)
60. Bamberger, R.H., Smith, M.J.T.: A filter bank for the directional decomposition of images: theory and design. IEEE Trans. Signal Process. **40**(4), 882–893 (1992)
61. Kumar, S.: A comparative study of transform based on secure image steganography. Int. J. Comput. Commun. Eng. **4**(2), 107–116 (2015)
62. Easley, G., Labate, D., Lim, W.Q.: Sparse directional image representations using the discrete shearlet transform. Appl. Comput. Harmon. Anal. **25**(1), 25–46 (2008)
63. Kutyniok, G., Labate, D.: Resolution of the wavefront set using continuous shearlets. Trans. Am. Math. Soc. **361**(5), 2719–2754 (2009)
64. Olzak, L.A., Thomas, J.P.: Seeing Spatial Patterns. In: Boff, K.R., Kaufman, L., Thomas, J. P. (eds.) Handbook of Perception and Human Performance, vol 1: Sensory Processes and Perception. Wiley, New York (1986)
65. Watson, A.B., Yang, G.Y., Solomon, J.A., Villasenor, J.: Visibility of wavelet quantization noise. IEEE Trans. Image Process. **6**(8), 1164–1175 (1997)
66. Kundur, D., Hatzinakos, D.: Toward robust logo watermarking using multiresolution image fusion principles. IEEE Trans. Multimed. **6**(1), 185–198 (2004)
67. Autrusseau, F., David, S., Pankajakshan, V.: A subjective study of visibility thresholds for wavelet domain watermarking. In: IEEE International Conference on Image Processing (ICIP'2010), pp. 201–204 (2010)
68. Delaigle, J.F., De Vleeschouwer, C., Macq, B.: Watermarking algorithm based on a human visual model. Sig. Process. **66**(3), 319–335 (1998)
69. Levicky, D., Foris, P.: Human visual system models in digital image watermarking. Radioengineering **13**(4), 38–43 (2004)
70. Watson, A.B.: DCTune: a technique for visual optimization of DCT quantization matrices for individual images. Society for Information Display Digest of Technical Papers XXIV, pp. 946–949 (1993)
71. Voloshynovskiy, S., Herrigel, A., Baumgaertner, N., Pun, T.: A stochastic approach to content adaptive digital image watermarking. In: Pfitzmann, A. (ed.) LNCS, vol. 1768, pp. 212–236. Springer, Heidelberg (2000)

72. Kim, J.H., Kim, H.J., Kwon, K.R.: Multiwavelet image watermarking using perceptually tuned model. Int. J. Comput. Sci. Netw. Secur. **6**(12), 233–238 (2006)
73. Autrusseau, F., Le Callet, P.: A robust image watermarking technique based on quantization noise visibility thresholds. Sig. Process. **87**(6), 1363–1383 (2007)
74. Barten, P.G.J.: Contrast sensitivity of the human eye and its effects on image quality. SPIE Optical Engineering Press, Bellingham (1999)
75. Campbell, F.W., Robson, J.G.: Application of Fourier analysis to the visibility of gratings. J. Physiol. **197**(3), 551–566 (1968)
76. Sachs, M.B., Nachmias, J., Robson, J.G.: Spatial frequency channels in human vision. J. Opt. Soc. Am. **61**(9), 1176–1186 (1971)
77. Watson, A.B.: The cortex transform: Rapid computation of simulated neural images. Comput. Vis. Graph. Image Process. **39**(3), 311–327 (1987)
78. Goresnic, C., Rotman, S.R.: Texture classification using the cortex transform. Graph. Models Image Process. **54**(4), 329–339 (1992)
79. Arnold, V.I., Avez, A.: Ergodic Problems of Classical Mechanics. Math Physics Monograph Series. Benjamin, New York (1968)
80. Elahian, A., Khalili, M., Shokouhi, S.B.: Improved robust DWT-watermarking in YCbCr color space. Global J. Comput. Appl. Technol. **1**(3), 300–304 (2011)
81. Liu, R., Tan, T.: An SVD-based watermarking scheme for protecting rightful ownership. IEEE Trans. Multimed **4**(1), 121–128 (2002)
82. Mohammad, A.A., Alhaj, A., Shaltaf, S.: An improved SVD-based watermarking scheme for protecting rightful ownership. Sig. Process. **88**(9), 2158–2180 (2008)
83. Lai, C.C.: An improved SVD-based watermarking scheme using human visual characteristics. Opt. Commun. **284**(4), 938–944 (2011)
84. Run, R.S., Horng, S.J., Lai, J.L., Kao, T.W., Chen, R.J.: An improved SVD-based watermarking technique for copyright protection. Expert Syst. Appl. **39**(1), 673–689 (2012)
85. Yao, L., Yuan, C., Qiang, J., Feng, S., Nie, S.: A symmetric color image encryption based on singular value decomposition. Opt. Lasers Eng. **89**, 80–87 (2017)
86. The USC-SIPI Image Database “Miscellaneous”. Available from: http://sipi.usc.edu/database. Accessed 2 Jan 2017

Chapter 4
Unscented RGB-D SLAM in Indoor Environment

Alexander Prozorov, Andrew Priorov and Vladimir Khryashchev

Abstract The research considers the implementation of simultaneous localization and mapping algorithm based on the FastSLAM technique and specific problems that are typical for the RGB-D sensor-based solutions. An improvement of the classical FastSLAM algorithm has been obtained by replacing the method of landmarks' observations filtering with unscented Kalman filters. Instead of linearizing, the nonlinear models through the first order Taylor series expansion at the mean of the landmark state were applied. The proposed algorithm computes a more accurate mean and uncertainty of the landmarks, which are moving nonlinearly. Various data preprocessing issues are discussed, such as the method of calibration of Kinect-like cameras, depth map restoration using a modified interpolation technique, and filtering the noise in the RGB images for more accurate detection of key features. Additionally, the chapter presents an improved resampling algorithm for the particle filtering through the adaptive thresholding based on the data of the effective particle number evolution. The proposed algorithm runs in real time and shows good accuracy and robustness in comparison with other modern SLAM systems using all the advantages and disadvantages of the RGB-D sensors.

Keywords Image processing · Computer vision · SLAM · RGB-D
Unscented Kalman filter · Particle filter

A. Prozorov (✉) · A. Priorov · V. Khryashchev
P.G. Demidov Yaroslavl State University, 14 Sovetskaya st.,
Yaroslavl 150000, Russian Federation
e-mail: alexprozoroff@gmail.com

A. Priorov
e-mail: andcat@yandex.ru

V. Khryashchev
e-mail: vhr@yandex.ru

M.N. Favorskaya and L.C. Jain (eds.), *Computer Vision in Control Systems-4*,
Intelligent Systems Reference Library 136,
https://doi.org/10.1007/978-3-319-67994-5_4

4.1 Introduction

At the current stage of computer vision evolution, the Simultaneous Localization And Mapping (SLAM) algorithms became very popular. In most cases, the object of the application of these algorithms is a mobile robotic platform, which is equipped with a set of sensors. Required parameters of the system, such as the desired accuracy of the mapping, lighting conditions, and geometry of space, indicate the types of sensors, which should be used. A variety of sensors includes the laser rangefinders, digital cameras (visible or infrared), sonars, and so on. Accuracy of the calculated map is entirely dependent on the characteristics of utilized sensors and conditions of the surrounding space. However, despite this more often the problem is reduced to finding specific areas of space, the so-called landmarks, which can be stably identified within the data stream coming from the sensor [1–3].

In recent years, solutions based on the use of digital cameras as the main sensors, became especially popular. The main reason is the rapid growth of the computing power and reduced cost of digital signal processing devices. The number of algorithms that can be applied in real time is increasing, while the limitations on the resolution of source images are weakening. Although the use of high resolution images is still associated with certain difficulties, the low cost of digital cameras compared to other sensors that are traditionally used in the SLAM algorithms allows them to find a wide range of applications.

Over the last decade, the area of application of the SLAM algorithms has grown tremendously. The most common argument in favor of these methods is the need for local autonomous navigation of mobile robots in the absence of the possibility of using global positioning systems, as well as in the cases, when the map should be more accurate than satellite solutions can provide.

In general terms, the localization task implies the assessment of the current position of the camera in space depending on the history of observations, measurements and the available control commands. The result of the whole set of algorithms is a map of the surrounding area, as well as full or partial path of movement within the received map. As a rule, the location of the robot is considered in a local coordinate system related to the initial position since a priori information about the initial coordinates is absent. Requirements to the map are dictated by the possibility of its future use for localization. Without these conditions, the construction of the map and calculation of the location are performed independently resulting in continuous growth of the error [3]. Figure 4.1 shows an example of the algorithm work, which depicts two paths: the real camera movement trajectory and trajectory calculated using the SLAM algorithm.

As one can see, the error in this case is accumulated with each subsequent step. In this case, the resulting map is a 2D set of the observed landmarks. Globally, the SLAM problem identifies three paradigms:

- Extended Kalman Filter (EKF) based solutions.
- Particle filter based solutions.
- Graph optimization based solutions.

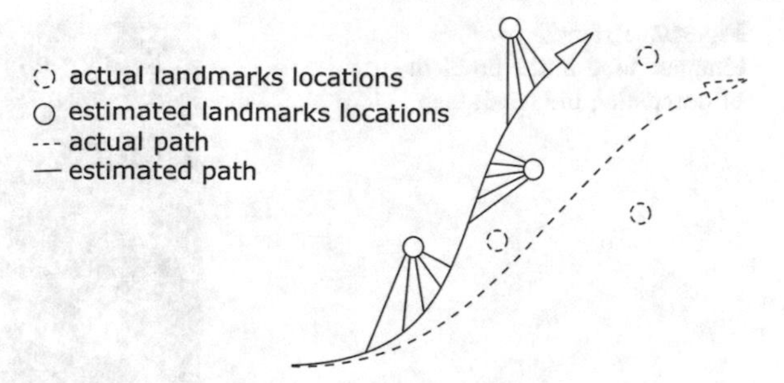

Fig. 4.1 Example of the SLAM algorithm process

As a rule, the choice of a paradigm depends on the technical characteristics of the robot implementation, the external environment, as well as the performance requirements of software. The first two approaches are most commonly used to solve the problem of simultaneous localization and mapping based on the analysis of the video stream, so-called Visual SLAM (VSLAM) algorithm. Solutions based on the graph optimization are more common in systems equipped with laser sensors and operating indoors in a small space [4, 5]. In this case, the landmarks are higher-level objects, rather than the local feature points of the image. For example, a plane (wall) and the angles of the indoor space can be used as such high-level landmarks. The number of landmarks in this case is much less than in the case of natural local image features. Primarily, this is due to the fact that the complexity of the optimization problem for the constructed graph depends on the size of the surrounding space and length of the trajectory of the camera movement. Therefore, the graph based solutions are not considered in this research, as the most typical for systems equipped with laser sensors that cannot work in real time.

The chapter is organized as follows. The concepts of the RGB-D system functioning are discussed in Sect. 4.2. The improvement of FastSLAM algorithm is presented in Sect. 4.3. Section 4.4 includes the probabilistic properties of the SLAM algorithm. The role of the particle filter in the SLAM task is considered in Sect. 4.5. Evaluation of the RGB-D SLAM testing results contains in Sect. 4.6. Section 4.7 provides the feature points detection and description. Unscented Kalman filter for the landmarks tracking is discussed in Sect. 4.8. The depth map preprocessing is described in Sect. 4.9, while the adaptive particles resampling is represented in Sect. 4.10. Conclusions are given in Sect. 4.11.

4.2 Obtaining Depth Maps in RGB-D System

The concept of 3D camera implies a device that allows to obtain information about the remoteness of the observed scene points (depth value) (Fig. 4.2). This component is a system of at least two processing devices—the infrared projector and infrared camera. The source emits a contrast light-shadow pattern with a permanent

Fig. 4.2 Example of template used in the problem of computing the depth map

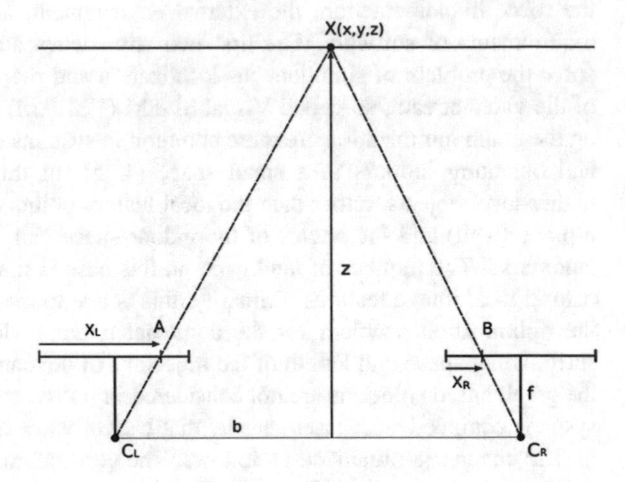

Fig. 4.3 The task of triangulation for the pinhole camera model

structure, consisting of light and dark areas. In current research, the Microsoft Kinect camera was used. The optical characteristics of the device obtained during the calibration are listed in [6].

This pattern is created from a set of diffraction gratings considering the suppressing of zero-order maximum. Reflected pattern captured by the infrared camera corresponds to a reference projection, which is obtained by capturing projection at a certain distance from the sensor. That base image is stored in memory. When an image is projected onto an object, the distance to the sensor is different from the reference plane. Offset of an infrared image will be determined by the size of the stereo base—a straight line connecting the optical center of the projector and camera. The task of extracting information about the depth from a data received via the infrared camera is widely covered in [6].

A method for producing a final depth map of the observed scene is also considered in [6]. In this approach, the infrared projector and infrared camera are considered as a stereo pair. Thus, the measurement of depth process is a problem of triangulation (Fig. 4.3).

For the given case, we can write the equation based on the similarity of triangles C_LC_RX and ABX provided by Eq. 4.1, where $b = |C_LC_R|$ is a stereobasis of the system, f is a focal length.

$$\frac{|C_LC_R|}{z} = \frac{|AB|}{z-f} \tag{4.1}$$

Disparity $d_s = X_L - X_R$ is equal to the distance $b - |AB|$. Therefore, the depth z of the point X is calculated by Eq. 4.2.

$$z = \frac{bf}{d_s} \tag{4.2}$$

An extended version of this relationship is presented in the research [7]. These authors considered the difference in the real disparity and raw value, which was calculated by comparing the images by Eq. 4.3, where c_0 and c_1 are the coefficients of the polynomial transformation from value d to d_s.

$$d_s = c_1 d + c_0 \tag{4.3}$$

Taking into account the additional term that characterizes the error of image offset variation, the final equation becomes in the form of Eq. 4.4.

$$z = \frac{bf}{c_1 d + c_0} + Z_\delta(u, v) \tag{4.4}$$

Similar conclusions are given in [8], wherein the depth model is defined by Eq. 4.5.

$$z = \frac{1}{c_1 d_k + c_0} \tag{4.5}$$

Here, parameters b and f are the part of the polynomial and do not considered explicitly. The value d_k describes the disparities in the point based on the distortion provided by Eq. 4.6.

$$d_k = d + D_\sigma(u, v)\exp(\alpha_0 - \alpha_1 d) \tag{4.6}$$

This ratio is confirmed by experimental data in using the set of the plane wall images. Without distortion, we can perform a reverse conversion applying Eq. 4.7.

$$d_k = \frac{1}{c_1 z_d} - \frac{c_0}{c_1} \tag{4.7}$$

In this case, it is necessary to take into account the effect of distortion for d_k. This expression becomes much more difficult due to the exponential dependence. This problem can be solved using Lambert functions [8] (Eqs. 4.8–4.12).

$$y = \exp(\alpha_0 - \alpha 1 d_k + \alpha_1 D_\delta(u, v) y) \tag{4.8}$$

$$y = \frac{d_k - d}{D_\delta(u, v)} \tag{4.9}$$

$$y = \exp(\alpha_1 D_\delta(u, v) y) \exp(\alpha_0 - \alpha 1 d_k) \tag{4.10}$$

$$\frac{-y}{\alpha_1 D_\delta(u, v)} = \exp(-y) \exp(\alpha_0 - \alpha 1 d_k) \tag{4.11}$$

$$y \cdot \exp(y) = -\alpha_1 D_\delta(u, v) \exp(-y) \exp(\alpha_0 - \alpha 1 d_k) \tag{4.12}$$

After substituting Lambert function $W(.)$ (equation $W(z)\exp(W(z)) = z$), the solution takes the form of Eq. 4.13.

$$d = d_k + \frac{W(-\alpha_1 D_\delta(u, v) \exp(\alpha_0 - \alpha 1 d_k))}{\alpha_1} \tag{4.13}$$

As we know, Lambert function $W(.)$ has no analytical solution, and the problem can be solved approximately using the recurrence relation [9].

4.3 FastSLAM Algorithm

The first and most popular approaches to solve the problem of simultaneous localization and mapping are the methods based on the EKF [10]. Algorithms of this group apply a probabilistic approach to the solution using a form of Bayesian filters. The main idea of this direction is to provide an initial, a priori assessment of the state, which is based on the observation model, accumulated measurements, and a set of methods used to calculate a posteriori refined evaluation after direct measurements.

The main disadvantage of solutions based on the EKF is a serious dependence of the computational complexity of the algorithm on the number of the considered landmarks. This is because of the covariance matrix ε of the filter that has a dimension $m \times n$, where n is a number of landmarks. At each stage of matrix ε update, each element should be recalculated. Therefore, the complexity of the algorithm is $O(n^2)$. To solve this problem, a new approach for solving the simultaneous localization and mapping problem has been developed in 2002 [11]. The FastSLAM method divides the task into multiple equivalent subtasks using the independence of the state of individual elements in the SLAM model.

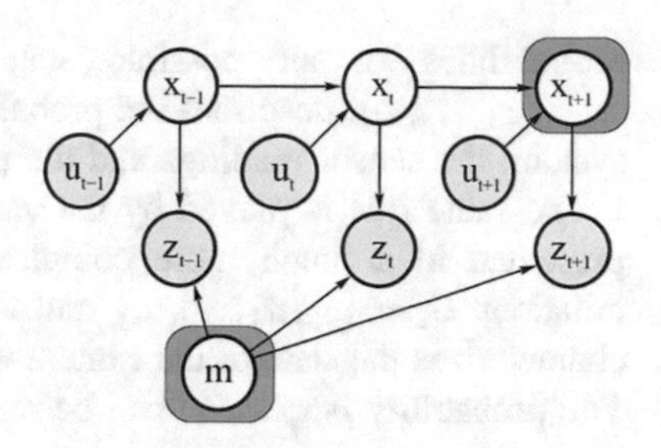

Fig. 4.4 SLAM process. From the position x_{t-1}, a movement occurs under the control actions $u_{t-1}, \ldots, u_{t+1}$. Observations of the landmark m are indicated as z_{t-1}, z_{t+1}

The FastSLAM algorithm is based on the use of Bayesian probabilistic. The diagram of this process is represented schematically in Fig. 4.4. Each measurement $z_1, \ldots, z_n$ is a function of the landmark θ_i coordinates and the robot position at the measuring. As we can see in the diagram, the SLAM problem implies that measurements are independent of each other. Thus, this problem can be considered as n independent evaluation computational problems of estimation landmarks coordinates. This observation is discussed in detail in [12] in order to develop an effective particle filter.

Based on the above conclusion, it is possible to consider a problem of mobile robot simultaneous localization and mapping as a challenge, which consists of two parts: the evaluation of the trajectory of the robot movement and evaluation of the locations of landmarks, which depend on the coordinates of the robot at the time of each measurement. In real conditions, a position of the robot can never be known exactly; this is the problem of simultaneous localization and mapping itself. However, the independence of landmarks movements from each other allows to handle each landmark individually that has been used in the FastSLAM algorithm.

In the original version of the FastSLAM algorithm, the modified particle filter for the posterior estimation of the robot position is used. Each of the particles is characterized by a certain weight determined by the states of n Kalman filters, which are used to estimate the landmarks' locations. This algorithm is the Rao-Blackwellized particle filter [13, 14]. A naive implementation of considered ideas has algorithmic complexity, which is equal to $O(MK)$, where M is a number of particles in the filter and K is a number of landmarks, which is observed during navigation. The use of tree data structures reduces a complexity to a value $O(M \log K)$ that results in a significant gain in performance compared to the SLAM solutions based on the ERF.

4.4 Probabilistic Properties of SLAM Problem

Assume that the camera moves in 1D space and its position is characterized by a single variable x. Then $p(x)$ is a probability distribution x that has a Gaussian form. If a probability distribution x reflects the position of the camera and the relative positions of the landmarks in a multidimensional space, it will determine the probabilities of all possible state variables. Thus, equation $p(x|u_0, u_1, \ldots, u_i, z_0, z_1, \ldots, z_i)$. describes the probabilities of all values of the current state of the system: the sensor readings and the movement of the camera received at the time i. The same role is played by the values P_i and X_i in the EKF but there they are presented in a much more complex form. For convenience, we introduce the notations $U_i = \{u_0, u_1, \ldots, u_i\}$ and $Z_i = z_0, z_1, \ldots, z_i$. The variable x in their turn characterizes the state of the camera v and locations of the landmarks $p_0, p_1, \ldots, p_m$. The probability $p(x|U_i, Z_i)$ can be represented by Eq. 4.14.

$$p(x|U_i, Z_i) = p(v, p_0, p_1, p_m|U_i, Z_i) \tag{4.14}$$

To simplify the SLAM task, we can use the laws of the foundations of the probability theory. Suppose that there are two independent random variables A and B. We can say that $p(A, B) = p(A) * p(B)$. However, this expression is unfair, when A depends on B. In this case, it will have the form $p(A, B) = p(A) * p(B|A)$. As is known, the estimation of landmarks depends on the position of the robot. This means that Eq. 4.14 can be represented by Eq. 4.15.

$$p(v, p_0, p_1, p_m|U_i, Z_i) = p(v|U_i, Z_i) \cdot p(p_0, p_1, p_m|U_i, Z_i) \tag{4.15}$$

Since the observation landmarks are independent from each other, which is observed in real conditions in most cases, the expression $p(p_0, p_1, \ldots, p_m|U_i, Z_i, v)$ can be divided into m independent expressions in the form of Eq. 4.16.

$$\begin{aligned} &p(v|U_i, Z_i) \cdot p(p_0, p_1, p_m|U_i, Z_i) \\ &= p(v|U_i, Z_i) \cdot p(p_0|U_i, Z_i) \cdot p(p_1|U_i, Z_i), \ldots, p(p_m|U_i, Z_i) \end{aligned}. \tag{4.16}$$

Final expression to describe the probability distribution is defined by Eq. 4.17.

$$p(x|U_i, Z_i) = p(v|U_i, Z_i) \cdot \Pi_m p(p_m|U_i, Z_i) \tag{4.17}$$

If we look at this expression, it becomes obvious that the problem of SLAM algorithm is divided into $m + 1$ tasks, and none of the landmark location estimates does not depend on others. This, in its turn, allows to solve the problem of polynomial complexity of the EKF and avoid it in the FastSLAM algorithm. The only price we must pay for this simplification is the risk to reduce a precision associated with ignoring the correlation of landmarks estimation errors. The FastSLAM algorithm simultaneously track multiple possible paths, while the EKF does not keep even one but only works with the position of the robot—the last step of the current path. In its original Form, the FastSLAM algorithm saves routes but in the calculation it uses only the previous step.

4.5 Particle Filter in the SLAM Task

For the considered practical task, a more accurate estimation of the system state requires a move away from the assumption that the noise has a Gaussian distribution [15]. In this case, it is proposed to introduce the concept of a multi-modal distribution of noise as a distribution, which has several local maxima. To simulate such systems, we can use the particle filters. Particle filters are a more common approach for solving visual tracking problem using probabilistic methods [16, 17]. The basic algorithm of particle filter, which is used to build most of similar computer vision systems, is the algorithm of conditional density propagation [18]. Consider in general terms how it works.

Suppose that the system may be in states $X_t = (x_1, x_2, \ldots, x_t)$ and at a time t it is characterized by a certain value of probability density. As well as, when using a Kalman filter, the sequence of observations is $Z_t = (z_1, z_2, \ldots, z_t)$. At the same time, we introduce the assumption that the state of the system depends in its previous state x_{t-1} satisfied to the Markov chain condition. Thus, we obtain a system with an independent set of observations. Particle filtering technique represents the probability distribution as a collection of weighted samples called particles. Generation of such samples is regulated by introducing weights. We define a probability density function x_t for the given number as Eq. 4.18.

$$S_t = \{(S_i^t, \pi_i^t), i = \overline{1, N}, \sum_{i=1}^{N} \pi_i^t = 1\} \tag{4.18}$$

The problem reduces to the construction of method that recovers the set S_t using the set S_{t-1}. Formally, the algorithm can be represented as a series of defined steps [16, 18]:

1. Suppose we have a certain collection of weighted samples at a time $t - 1$.

$$S_{t-1} = \{(S_i^{t-1}, \pi_i^{t-1}), i = \overline{1, N}, \sum_{i=1}^{N} \pi_i^{t-1} = 1\}$$

2. Calculation of integral weights.

$$c_i = c_{i-1} + \pi_i^{t-1} \quad i = \overline{1, N} \quad c_0 = 0$$

3. Getting nth item of the set S_t. To do this, we randomly select the number r from the range [0–1] and calculate $j = argmin_i(x_i > r)$. Now the current estimation of the state S_j^{t-1} is known.

4. The prediction of the next state. Action Kalman filter similar excepts that in this case there are no restrictions related to the linearity of the system and the form of the noise distribution.

$$S_n^t = F_{t-1}S_i^{t-1} + w_{t-1}$$

5. Correction using the current observation Z_t and its distribution. Weight are defined.

$$\pi_n^t = p(z_t | x_t = x_n^t)$$

6. Normalization of the sequence of weights π_i^t for equality.

$$\sum_{i=1}^{N} \pi_i^t = 1$$

7. The calculation of the best estimate for the state x_t as a linear convolution of the resulting sample set.

$$x_t = \sum_{i=1}^{N} \pi_i^t s_i^t$$

The described process can be interpreted graphically using the particles concepts (Fig. 4.5).

At the input, this algorithm has a set of particles (s_i^{t-1}, π_i^{t-1}) (first step of diagram). As a result of the random selection of the particles out of the set S_{t-1} of new instances is obtained (second step of diagram). Prediction step leads to the formation of of the set of estimation states of the particles (third step of diagram). Then for each estimation, a correction is performed on the basis of available observations. As a consequence, a set of particles (s_i^t, π_i^t) are generated at the next step.

4.6 Evaluation of Accuracy

This study applies the RGB-D system that was implemented in C# using several libraries for the digital image processing, numerical methods, and data visualization. The OpenCV library is used as the main tool for working with the images in computer vision algorithms. This is the most powerful tool of its type today. Input data in this system are the video streams captured by the color and infrared cameras,

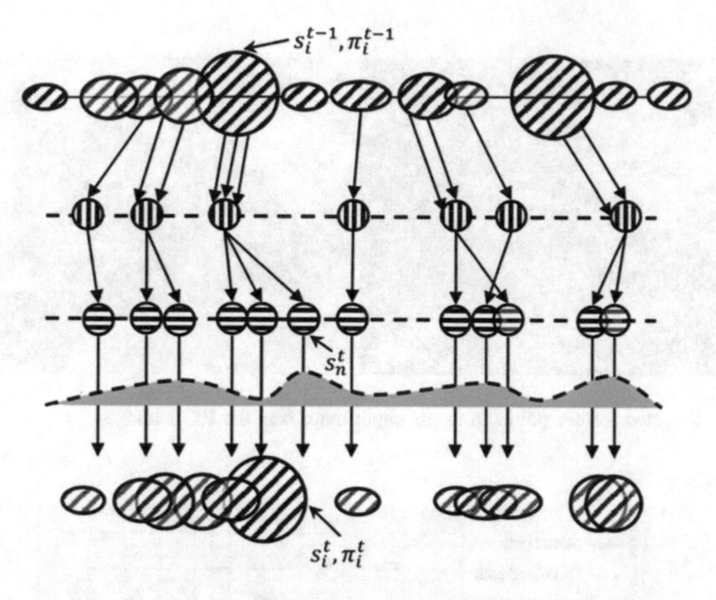

Fig. 4.5 Steps of the algorithm for the conditional density propagation

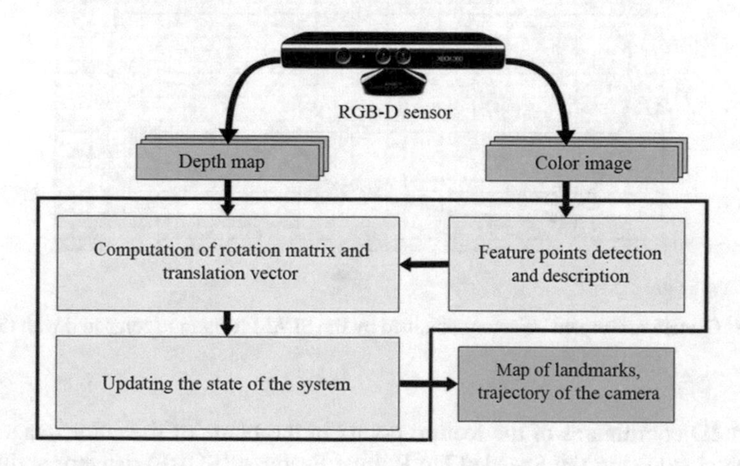

Fig. 4.6 Data flow in the RGB-D system

which are a part of the RGB-D sensor. The diagram in Fig. 4.6 shows the structure of the program modules in the context of data flow [19].

The depth map is a grayscale image encoded with 16 bits/pixel. Colored images from the RGB camera have a color depth encoded with 8 bits/channel. Various test datasets obtained under different lighting conditions and types of the surrounding area were used in the development and testing the algorithm [20–22]. Figure 4.7 shows the depth map and RGB image obtained by the algorithm. The images

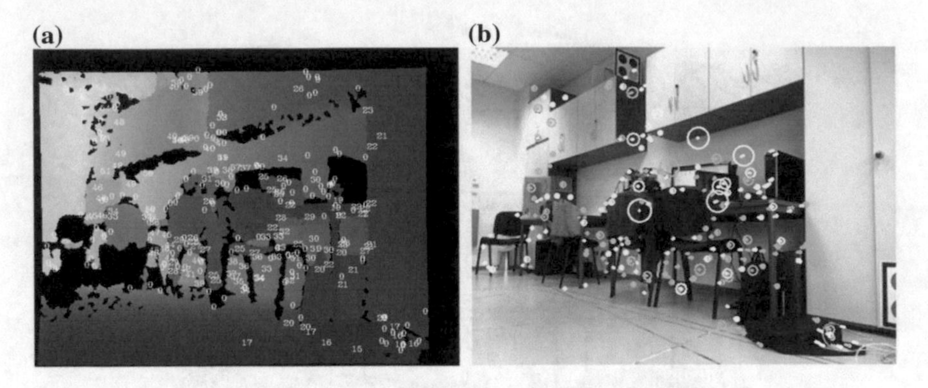

Fig. 4.7 Detected feature points: **a** in the depth map, **b** in the RGB image

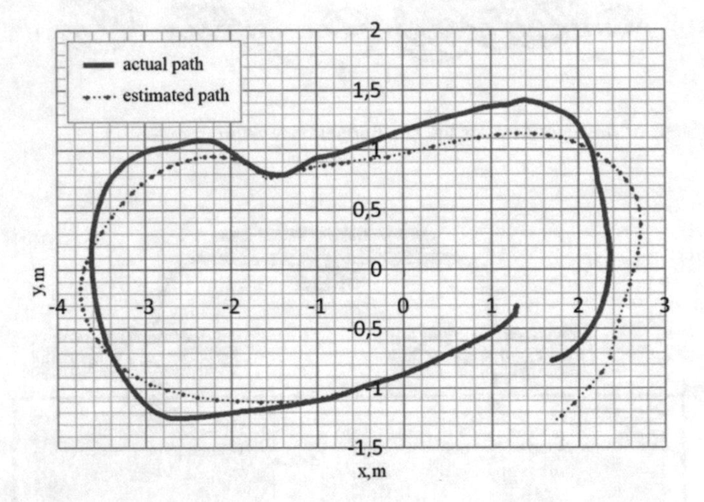

Fig. 4.8 Camera movement trajectory obtained by the SLAM technique compared with the actual path

contain 2D coordinates of the feature points in the plane of the color image. They were obtained using the Speeded Up Robust Features (SURF) detector and natural landmarks. The depth map contains the distances to the landmarks.

Figure 4.8 shows the trajectory of camera movement for the sequence №1 of test data from [21].

A set of correspondences between the feature points and position of the camera in space represent one of the states of the particle filter. The state vector is generated in accordance with various ways to compare the landmarks, which are determined by the method of composing their collections. Thus, the set of detected feature points obtained for the current frame is compared sequentially with the previous sets of feature points. Together with the updated camera position, these data form

Table 4.1 The RGB-D SLAM testing results

Dataset	Map size	RMSE (m)
PUTKK sequence 1 Kin v1	5.98 × 2.66	0.29
PUTKK sequence 2 Kin v1	5.97 × 2.93	0.25
PUTKK sequence 3 Kin v1	6.18 × 2.85	0.22
PUTKK sequence 4 Kin v1	6.77 × 2.28	0.34
fr1/xyz	0.52 × 0.78	0.07
fr1/room	2.53 × 2.17	0.16
fr2/desk	3.95 × 4.40	0.22
fr2/slam	6.96 × 6.52	0.26

one particle, so during each cycle of the algorithm a certain number of the states of the particle filter are updated.

Testing of the implemented algorithm was carried out in single-threaded mode on a PC running Intel Core i5-4210 processor with a frequency of 1.7 GHz. The error was calculated as mean squared deviation of camera coordinates estimation, which was obtained by calculating an affine transformation of landmarks' coordinates for the data stream within the entire movement [23, 24]. The results for a set of test images [20, 21] are shown in Table 4.1.

Thus, the average standard deviation of the camera coordinate related to the actual values for a given datasets is 0.23 ± 0.08 m.

4.7 Feature Points Detection and Description

Comparison of popular methods for detection and description of feature points was carried out. Comparison research includes the detectors, such as Features from Accelerated Segment Test (FAST), Good Features To Track (GFTT), Maximally Stable Extremal Regions (MSER), Oriented fast and Rotated BRIEF (ORB), Scale-Invariant Feature Transform (SIFT), the SURF, Star detectors, and descriptors, such as the ORB, the SURF, the SIFT, and Binary Robust Independent Elementary Features (BRIEF). In this research, we use well known dataset for the feature matching evaluation [25]. The points' detectors investigation shows a significant advantage of the ORB detector over the SURF in terms of speed. In addition, the pair ORB detector-SURF descriptor has a great potential (Fig. 4.9). Comparing the speed results of the algorithm using the SURF and the ORB detectors points show 34% reduction in processing time, when using the ORB detector. Figure 4.10 shows Root-Mean-Square (RMS) errors of feature points' detection for both methods.

Experimental studies show the relative similarity of the two methods in accuracy with a large number of feature points (Table 4.2). The difference in the results about 10% begins to appear in the cases, where the number of processed feature points less than 100. In this case, the SURF detector shows the best results.

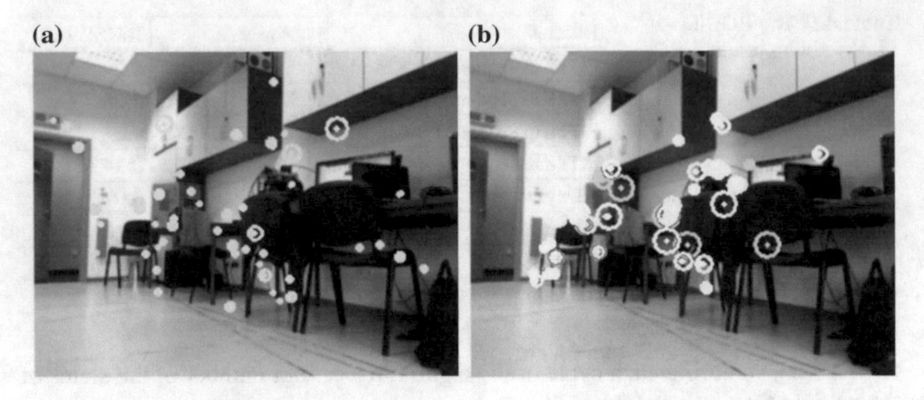

Fig. 4.9 Feature points obtained by: **a** the SURF detector, **b** the ORB detector

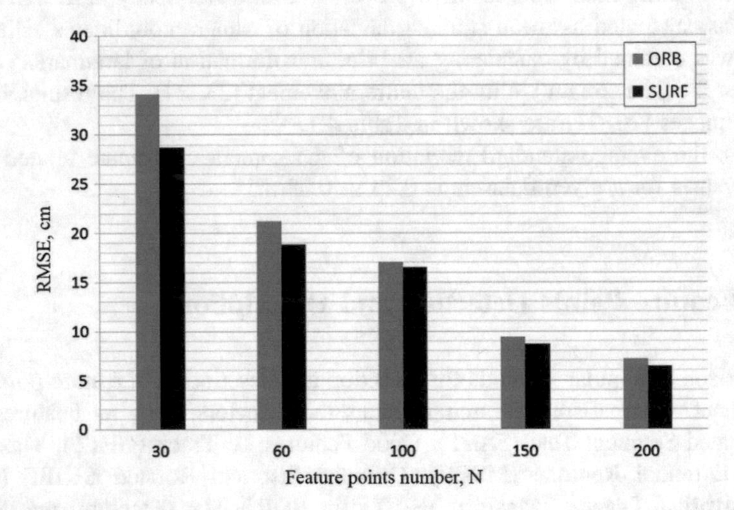

Fig. 4.10 RMS error of the camera motion estimation using the ORB and the SURF detectors in the SLAM algorithm with the landmarks' observation

Table 4.2 RMSE error of the camera motion estimation (cm)

Detector	Feature points number				
	30	60	100	150	200
ORB	34.12	21.21	17.10	9.41	7.15
SURF	28.76	18.82	16.54	8.72	6.47

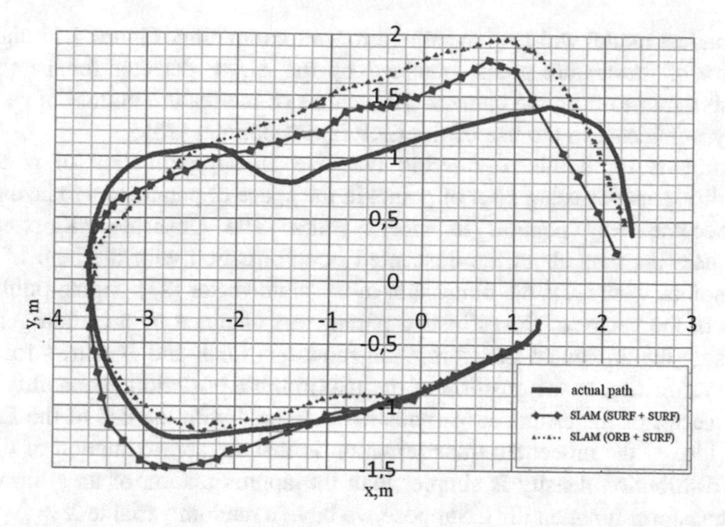

Fig. 4.11 The trajectories of the camera, which were obtained using the ORB and the SURF detectors

Figure 4.11 shows the results of calculating the trajectory of the camera under artificial limitation of the number of detected landmarks for the sequence №1 of test data set [21].

In the first case, the ORB detector is used for detecting feature points, while the second case shows the use of the SURF detector. In both cases, the SURF descriptor is used to describe the found points. Numerical results show an average of 6% worse accuracy of camera motion trajectory for the ORB based algorithm, unlike the case of using the SURF technique. Hence, the ORB detector can be one of the solutions to increase the performance of the SLAM algorithm significantly considering 6% decrease in accuracy.

4.8 Unscented Kalman Filter for Landmarks Tracking

The classic approach for the landmarks tracking in the FastSLAM algorithm is the EKF. This is particularly true for the use of the SLAM algorithm based on the video stream analysis, where the landmarks are the feature points on the images obtained by the camera. However, the estimates of the landmark state vector obtained by the EKF are not optimal, since they are calculated by expanding the non-linear function of the evolution process F and H in the Taylor series, when the members of the higher degrees are dropped.

In other words, the EKF creates a layer of linearization for strictly nonlinear dynamical system. This is the main reason, why this approach may not be effective in the case of strongly nonlinear dynamic model of the system. In this case, the problem is not enough conditioned, i.e. a small error in parameters settings of a mathematical model will lead to large errors in computing. Figure 4.11 shows the trajectory of the feature points obtained by the SURF detector for a number of successive frames from the camera. The nature of the feature points' movement is strongly nonlinear, so the use of the EKF is not optimal [26].

Algorithms for nonlinear filtering that use an approximation of conditional probability densities using a set of points in the space of parameters to be estimated have become very popular in recent years. The Unscented Kalman Filter (UKF) uses the idea about the unscented transformation with the help of $2L + 1$ sigma-points, where L is the dimension of the state vector [27]. Sigma-point means a value of the vector of the estimated parameters obtained by a certain generation law. Using the unscented transformation functions for F and H allows for a more reliable estimation of the position of the maximum a posteriori probability density for the vector of the estimated parameters compared with the use of the EKF.

The idea of the unscented transformation is that the approximation of the probability distribution density is simpler than the approximation of an arbitrary non-linear transform function [28]. Suppose we have a random variable $X = (x_1, x_2, \ldots, x_n)^T$ with a known mean $\bar{X}$ and covariance matrix P_x undergoing a non-linear transformation $y = f(x)$. When using the UKF in the task of visual tracking coordinates of the landmarks, a random variable means 3D coordinates of the landmark in the space in order to estimate the mean $\bar{Y}$ and covariance P_y for the variable y. For this set is formed from $2L + 1$ sigma points so that their mean and covariance are $\bar{X}$ and P_x, respectively (Fig. 4.12).

Then, each point i in this list undergoes the transformation $y_i = f(x_i)$ and it corresponds to a set of scalar weights provided by Eq. 4.19, where α is a coefficient that characterizes the dispersion of the sigma-points around mean, β is a parameter that determines the type of random variable distribution. Coefficient λ is defined as the $\alpha^2(L + k) - L$, where k regulates the spread of sigma points. The value $\beta = 2$ characterizes the normal distribution. A more detailed description of the sigma-point transformation parameters can be found in [29].

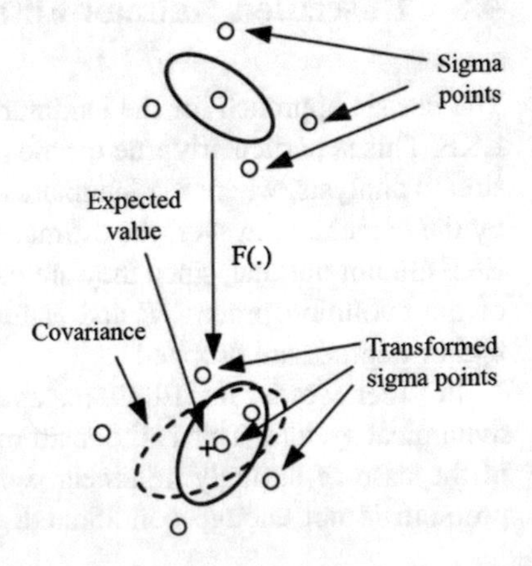

Fig. 4.12 Transformation of sigma-points for the case $L = 2$

$$\begin{aligned} W_0^{(m)} &= \frac{\lambda}{(L+\lambda)} \\ W_0^{(c)} &= \frac{\lambda}{(L+\lambda)} + (1-\alpha^2+\beta) \\ W_i^{(c)} &= W_i^{(m)} = \frac{1}{2(L+\lambda)} \end{aligned} \tag{4.19}$$

Then, the mean is estimated on the basis of the transformed sigma points' values:

$$\bar{y} = \sum_{i=0}^{2L} W_i^{(m)} y_i \tag{4.20}$$

and the covariation

$$P_y = \sum_{i=0}^{2L} W_i^{(c)} [y_i - \bar{y}_i][y_i - \bar{y}_i]^{\mathrm{T}} \tag{4.21}$$

for the random variable y.

Suppose that the state vector x has length n: $x = (x_1, x_2, \ldots, x_n)^{\mathrm{T}}$. In addition, the process noise vector has the form $v = (v_1, v_2, \ldots, v_n)^{\mathrm{T}}$. The measurement and noise vectors z and q have the length m: $z = (z_1, z_2, \ldots, z_m)^{\mathrm{T}}$, $q = (q_1, q_2, \ldots, q_m)^{\mathrm{T}}$. The dynamics of the system is described by Eq. 4.22.

$$\begin{aligned} x_k &= f(x_{k-1}, v_{k-1}) \\ z_k &= h(x_k, q_k) \end{aligned} \tag{4.22}$$

In the most general case, when it is known how the process and measurement noises affect the dynamics of the system, the system state vector can be extended by a respective noise. Then, the complete state vector with the number of components $2n + m$ will have a view of Eq. 4.23.

$$x' = (x_1, x_2, \ldots, x_n, v_1, v_2, \ldots, v_n, q_1, q_2, \ldots, q_m) = (x^T, v^T, q^T)^T \tag{4.23}$$

In this case, the equation for the transition of the state from the step k–1 to step k can be represented as $x'_k = F(x'_{k-1})$. Measurement prediction function for the extended state vector x'_k at step k can be described as $z'_k = H(x'_k)$. The covariance matrix P'_k of the state vector takes the form of Eq. 4.24.

$$P'_k = \begin{bmatrix} \mathrm{cov}(x,x)_k & \mathrm{cov}(x,v)_k & \mathrm{cov}(x,q)_k \\ \mathrm{cov}(v,x)_k & \mathrm{cov}(v,v)_k & \mathrm{cov}(v,q)_k \\ \mathrm{cov}(q,x)_k & \mathrm{cov}(q,v)_k & \mathrm{cov}(q,q)_k \end{bmatrix} \tag{4.24}$$

However, generally, the amount of noise is not correlated, so taking into account the covariance matrix of process $P_{v,k}$ and measurement $P_{q,k}$ the total covariance matrix can be written as Eq. 4.25, where zero elements are the appropriate size zero matrices.

$$P'_k = \begin{bmatrix} P_k & 0 & 0 \\ 0 & P_{v,k} & 0 \\ 0 & 0 & P_{q,k} \end{bmatrix} \tag{4.25}$$

As classical Kalman filter implementation, the UKF comprises the prediction and correction. Before the beginning the basic filter procedures, the main parameters of the algorithm are specified in the initialization step. Consider these steps:

- Initialization. At the step $k = 0$, the initial values for the mean and covariance matrix are defined by Eqs. 4.26 and 4.27.

$$\bar{x}'_0 = M[x'_0] = M \begin{bmatrix} x_0 \\ v_0 \\ q_0 \end{bmatrix} = \begin{pmatrix} M[x_0] \\ M[v_0] \\ M[q_0] \end{pmatrix} = \begin{pmatrix} M[x_0] \\ 0 \\ 0 \end{pmatrix} \tag{4.26}$$

$$P'_k = \begin{bmatrix} P_k & 0 & 0 \\ 0 & P_{v,k} & 0 \\ 0 & 0 & P_{q,k} \end{bmatrix} \tag{4.27}$$

- Prediction. Unscented transformation is performed to estimate the mean of the state vector in order to calculate the list with $2L + 1$ sigma points χ'_k using Eq. 4.28.

$$\begin{aligned} \chi'_{k,0} &= x'_k \\ \chi'_{k,0} &= x'_k + \gamma(\sqrt{P'_k})^i, \quad i = 1, \ldots, L \\ \chi'_{k,0} &= x'_k - \gamma(\sqrt{P'_k})^i, \quad i = L, \ldots, 2L \end{aligned} \tag{4.28}$$

For transformation of sigma points, the transfer function $\chi'^{\mathrm{T}}_{k,i} = F(\chi'_{k,i})$, $i = 0, \ldots, 2L$, estimation the mean value $\bar{x}'_k = \sum_{i=0}^{2L} W_i^{(m)} \chi'^{\mathrm{T}}_{k,i}$, and covariation $P'^{\mathrm{T}}_k = \sum_{i=0}^{2L} W_i^{(c)} [\chi'^{\mathrm{T}}_{k,i} - \bar{x}'_k][\chi'^{\mathrm{T}}_{k,i} - \bar{x}'_k]^{\mathrm{T}}$ are used. Similarly, a list of predicted values is created based on the set of transformed sigma points $z'^{\mathrm{T}}_{k,i} = H(\chi'_{k,i}), i = 0, \ldots, 2L$. Estimation of the mutual covariance of the state and measurements is computed as $P'_{xz,k} = \sum_{i=0}^{2L} W_i^{(c)} [\chi'^{\mathrm{T}}_{k,i} - \bar{x}'_k][\chi'^{\mathrm{T}}_{k,i} - \bar{z}_k]^{\mathrm{T}}$.

- Correction. The gain of the Kalman filter is calculated as $K_k = P'_{xz,k}P'_{z,k}$, the optimal estimate is calculated basing on the new measurement z_k as $\hat{x}'_k = \bar{x}'_k + K_k(z_k - \bar{z}_k)$. Also, the covariance matrix is updated in the form of $P'_k = P'^{\mathrm{T}}_k - K_k P'_{z,k} K^{\mathrm{T}}_k$.

To solve the problem of estimating the coordinates of landmarks, the movement of which is characterized by highly nonlinear nature, the UKF has been implemented on the C# language using Math.NET Numerics library to perform the operations of linear algebra. In the problem of computing the covariance matrix, Cholesky decomposition is used [3]. The vector of estimated parameters of landmarks involves six variables including 3D coordinates of a landmark in the space, and three projections of velocity.

Since the generation of the sigma-points has a constant computational complexity, the total complexity of the FastSLAM algorithm for simultaneous localization and mapping using the UKF is also determined by the number of particles, and it depends linearly. Thus, the computational complexity of the algorithm is the same as in the classical approach of the FastSLAM algorithm. Created UKF shows about 6% worse performance regarding the implementation of the EKF that is present in the OpenCV library. However, the results of such comparisons are not representative enough, since, in fact, the speed of a software implementation depends crucially on a variety of low-level optimizations are not produced in this study. Analysis of a number of publications shows that the EKF and the UKF performances are equal in most cases [26].

Testing was performed using the popular datasets [21]. Tests show that in the case of strictly nonlinear nature of the tracking values, the use of the UKF can significantly (in some cases, several orders of magnitude) reduce the mean square error in comparison with the EKF. Figure 4.13 shows the plots of the average estimation error depending on the number of detected feature points or landmarks. The values are normalized to the maximum value of the error.

Table 4.3 shows the results of evaluation of full trajectory camera applying a particle filter for the cases of the EKF and the UKF using during estimation of the landmarks' coordinates.

Thus, the greatest advantage of the UKF in the SLAM task is in the cases, where a number of feature points in the images is small. This is typical in cases, where an image has not enough high-contrast key features that can occur if the surrounding space is filled with large flat areas. Also, it is typical for the visual SLAM systems, where many landmarks may not contain the depth data if the object is outside the working range of the depth map. Also, if a hardware and software of the system meet the strict performance requirements, it can also result in the need to reduce the number of the tracked landmarks.

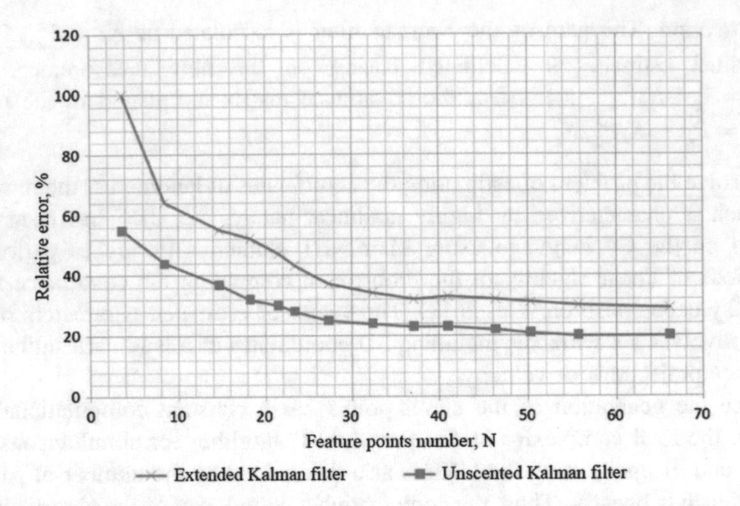

Fig. 4.13 Dependence of the error in estimating the camera motion based on the number of detected points for the EKF and the UKF

Table 4.3 Comparison of the UKF and the EKF accuracy

Feature points number		10	20	50	100
Relative error (%)	FastSLAM + EKF	25.6	20.8	12.7	10.5
	FastSLAM + UKF	14.1	12.1	9.7	8.4

4.9 Depth Map Preprocessing

The actual conditions of the SLAM algorithm involve many aspects that lead to a decrease in the accuracy of the camera localization and construction of the space map. Depth map obtained at the output of the RGB-D camera is exposed to various defects caused by a way of obtaining data about the distance, insufficient light conditions, and materials of the scene objects (Fig. 4.14). Therefore, vision tasks in most cases require the depth map preprocessing to minimize the impact of such defects. Classical solution based on the EKF is typically used in the FastSLAM algorithm that involves a linearization process during modeling landmarks movement. This is the main reason that this approach may not be effective in the case of strongly non-linear motion model.

The depth map obtained in the visual 3D SLAM system using an artificial light texturing typically comprises a plurality of various distortions and defects. Among several different factors that lead to such distortions are the following phenomena [30]:

1. Absorbing and reflecting surfaces give rise to defects in the depth map, which are caused by the inability to determine the pattern at a given point of the scene.
2. Nonzero stereo base of the projection system results in overlapping objects in images, so part of the projected pattern is out of the camera sight.

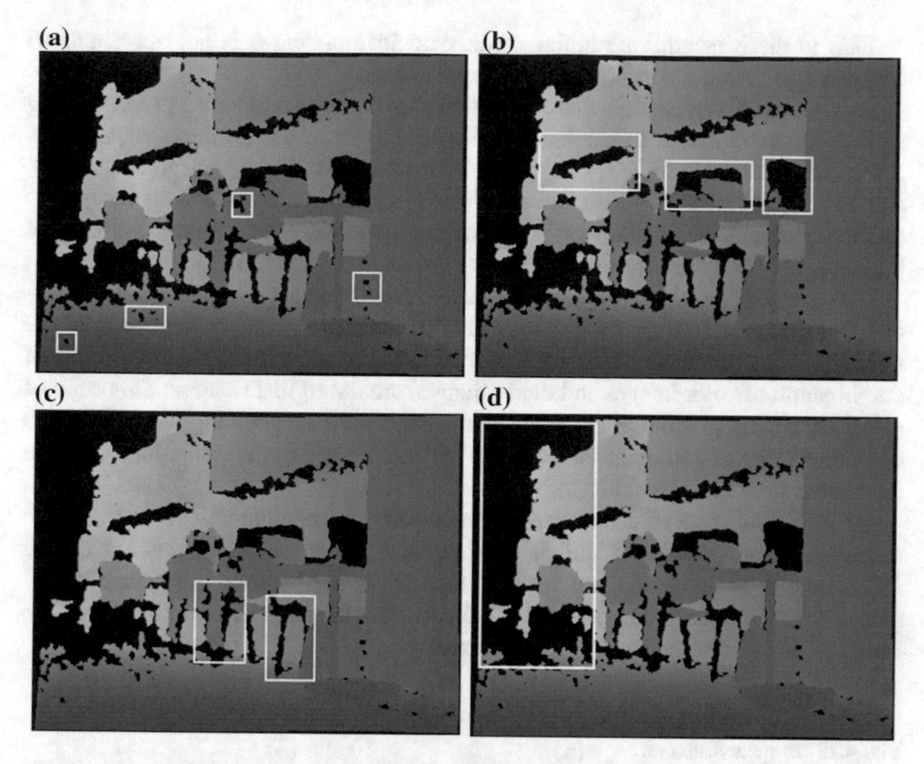

Fig. 4.14 Various types of depth maps' defects caused by: **a** light pattern recognition errors, **b** absorbing/reflecting of surfaces, **c** overlapping projected template, **d** distance

3. Noises of the projector and camera lead to the error in detection of pattern points and, consequently, to the depth map errors.
4. The triangulation method for distance calculation is characterized in that a depth measurement error increases with the distance to objects in the scene.
5. Strong distortion of the projected pattern in the surfaces is oriented along the optical axis of the camera or near this position.

Depending on the nature and causes of distortion, the defects can be classified into three different groups:

- Randomly spaced areas in the object surfaces arising from the light pattern recognition errors (Fig. 4.14a) due to the absorbing/reflecting the nature of the object surfaces (see Fig. 4.14b).
- The areas on the borders of the objects that appear because of overlapping projected template (Fig. 4.14c).
- Errors in the detection of the light pattern at a distance from the camera above the limit of the resolution of the camera sensor (Fig. 4.14d).

Due to these factors, the initial depth map in most cases is not applicable to compute the vision algorithms without pretreatment.

Analysis of the research in this area shows that the main depth map preprocessing algorithms are different interpolation methods based on the synthesis of textures, search for similar blocks [31] (Exemplar Based Method (EBM) interpolation [32]), and hybrid methods. As a rule, the existing solutions do not achieve an acceptable quality of the reconstruction of depth maps, since they lead to the blurring of brightness edges and distortion of objects borders (Fig. 4.15).

The proposed method of the depth maps' preprocessing is based on the application of the median filter for preprocessing of depth maps and modification of the method of interpolation based on the search or similar blocks with the help of combination of color images and depth maps from the RGB-D sensor. This suite of solutions allows to remove the depth map defects without causing blur effect and compensate the distortion at the edges of objects caused by interpolation based on the search or similar blocks.

At the initial stage of depth map preprocessing, the random defects in the depth maps are localized using a sliding window of a given radius. Depending on the initial quality of the depth map window size may vary between 3 × 3 and 7 × 7 pixels. Found defects are removed using the median filtering. Figure 4.16 shows the result of filtering at the stage of preprocessing.

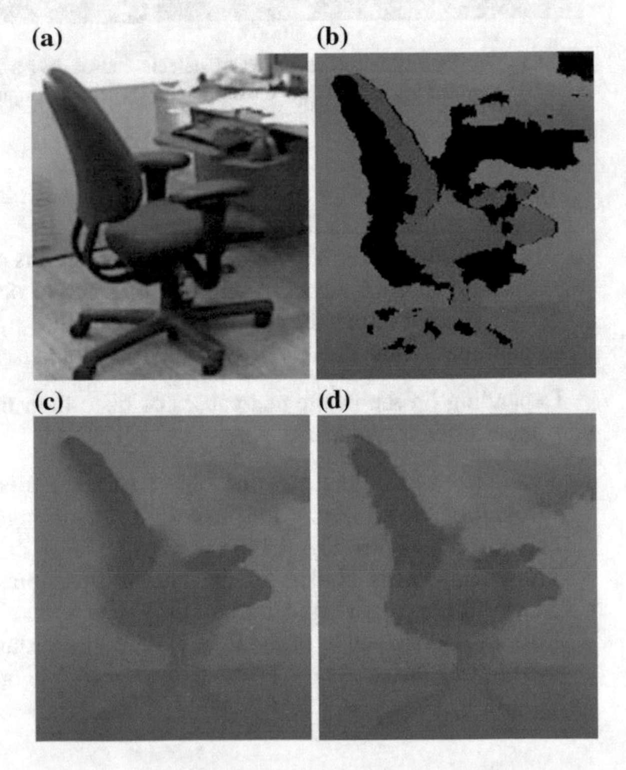

Fig. 4.15 Borders distortion that occurs, when interpolation methods are used: **a** color image in a view of grayscale ones, **b** the initial depth map, **c** method based on texture synthesis, **d** method based on similar blocks analysis

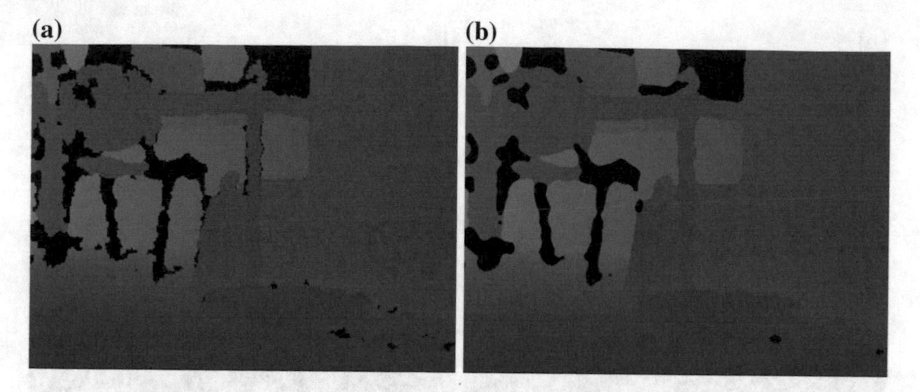

Fig. 4.16 Depth map preprocessing using median filter: **a** initial depth map, **b** the result of preprocessing

Such transformations do not add distortion to the original depth map and quite simple in comparison with the interpolation methods. After median filtering, the depth map is free from small random defects. The next step is to restore the defects caused by the overlapping and light pattern recognition errors.

Consider the principle of image reconstruction using a method similar blocks. Let the areas, in which depth values are absent, be denoted as Ω with boundary $\delta\Omega$. Point p is an element $I_{i,j}$ of an image that has a neighborhood area Ψ_p. This area is restored using Ψ_q unit of the field Φ, which serves as a source of information for the region to repair. Figure 4.17 shows the principle of reconstruction, which is as follows. For each area Ψ_p, the most similar block Ψ_q is searched using 2D least squares method. Then, intensity values Ψ_q are moved per-pixel to Ψ_p (Fig. 4.17d). Recovery order is determined by the intensity of the brightness fluctuations around point p, quantity, and reliability of available data. Areas Ω are described as a depth map areas with a zero-pixel intensity. The mask for the area intended for recovery is shown in Fig. 4.18.

An important task here is to develop an algorithm deciding defects restore order. The decisive effect should have those depths map areas, which belong to the border of the known values, the accuracy of which is the highest. For each pixel p of the border, a priority value P is calculated by Eq. 4.29, where $|\Psi_p|$ is a square of the known pixels in the window Ψ_p.

$$P(p) = \nabla I_p \frac{1}{|\Psi_p|} \sum_{q \in \Psi_p \cap \Phi} P(q) \tag{4.29}$$

The value determined by Eq. 4.29 decreases proportionally to the distance from the original boundaries of the restored area to its center. The initial values are initialized by ones in the known areas and zeros in the restored areas. Gradient intensity ∇I_p in the point p is determined by the convolution with the masks provided by Eq. 4.30.

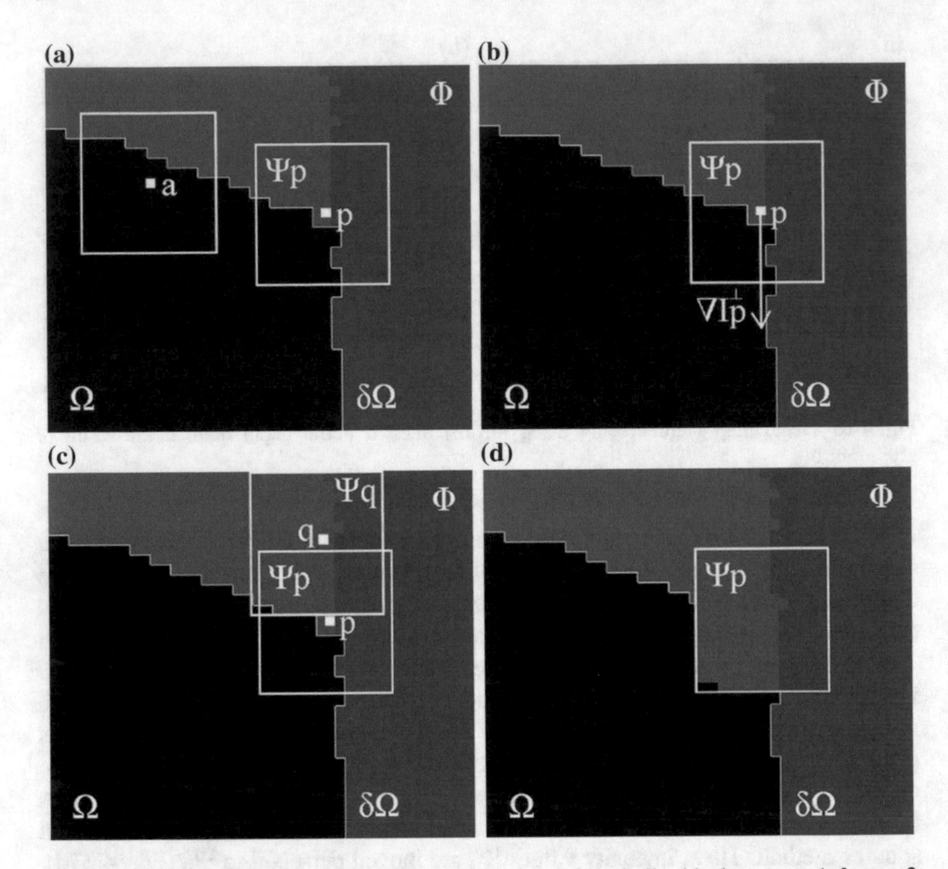

Fig. 4.17 The principle of a depth map reconstruction using similar blocks: **a** step 1, **b** step 2, **c** step 3, **d** step 4

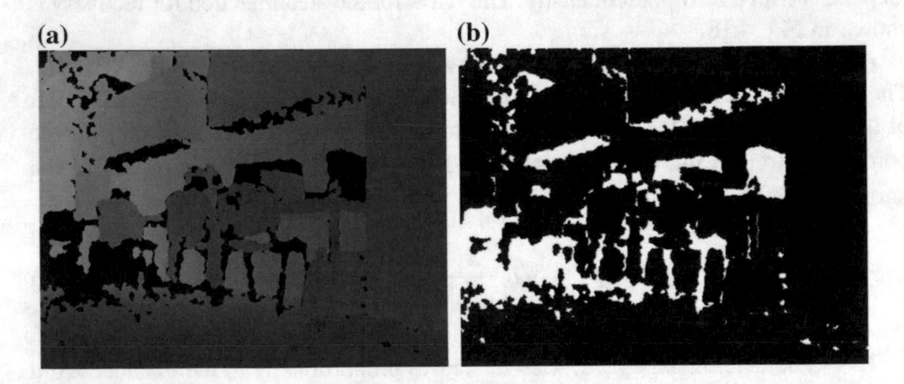

Fig. 4.18 Calculation of the defects mask: **a** initial depth map, **b** binary mask

$$G_x = \begin{bmatrix} -1 & -2 & -1 \\ 0 & 0 & 0 \\ 1 & 2 & 1 \end{bmatrix} \quad G_y = \begin{bmatrix} -1 & 0 & 1 \\ -2 & 0 & 2 \\ -1 & 0 & 1 \end{bmatrix} \tag{4.30}$$

Priority of the most authentic values of the defective area in the depth map can reduce the probability of occurrence of artifacts at the boundaries of objects. Search for the most similar block for the area with the highest priority is made by the method of least squares. Thereafter, intensity values are copied pixel by pixel into the restored area and priority values are recalculated. Thus, each work cycle includes the highest priority point selection procedure and reconstruction phase.

However, the effectiveness of the interpolation method based on similar blocks considerably limited, since the defects in the depth map is not always associated with the outlines of objects in a scene. Distortion correction at the edges of objects requires the use of methods based on a priori information about the correlation of boundaries of objects in the color image and depth map (Fig. 4.19).

To solve this problem, we can use the refinement and partitioning the defective areas using the boundaries of objects in the color image. At the beginning, it is necessary to perform an adaptive thresholding of color images. The threshold value

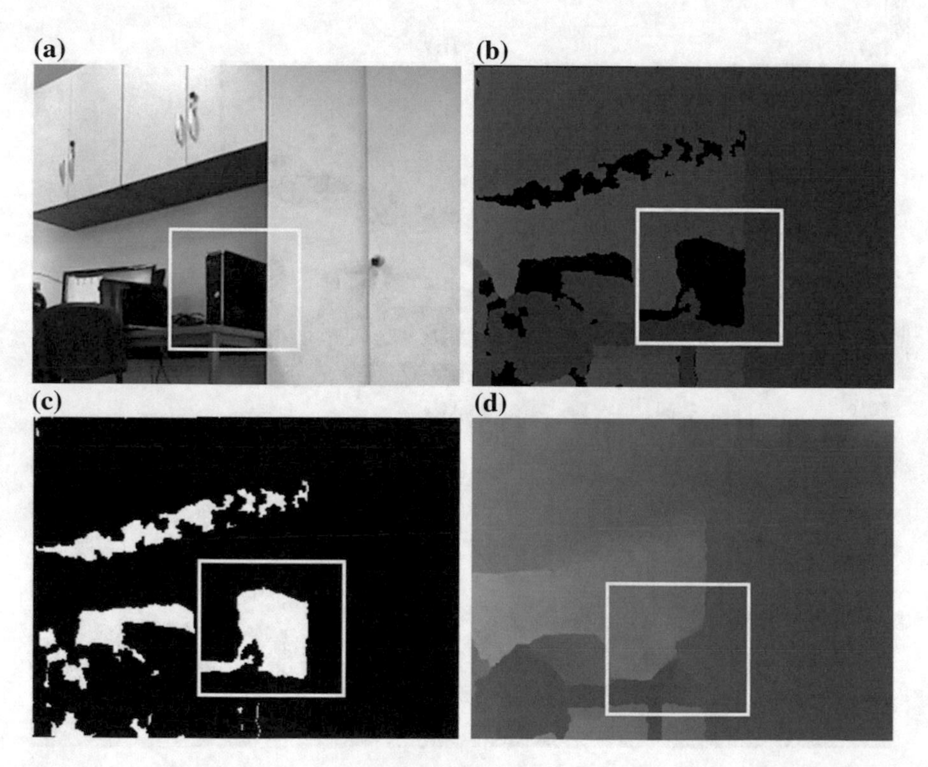

Fig. 4.19 Errors of interpolation method: **a** initial image, **b** initial depth map, **c** defects' mask, **d** reconstructed depth map

is calculated by the average intensity of the image within the defect area. Then, the edges of objects are refined using the Canny edge detector [33]. This allows to define areas of quasi-stationary, which are determined by the actual outlines of objects in the scene.

The advantage of this approach is that the boundaries of objects in the color image practically do not distorted compared with the boundaries in the depth map that allows to restore the objects' shapes in the depth map. Figure 4.20 shows the results of interpolation algorithm with specifying the boundaries of defective areas. To evaluate the effectiveness of this method, the comparative measurements of visual odometry errors during a movement of the camera were performed using the initial depth map without preprocessing and using the developed algorithm.

Numerical results show 9.8% increase in the accuracy of calculation of the camera displacement between the successive frames (the average error of 5.98 cm vs. 6.62 cm). The errors estimation of camera displacement during the calculating the transformation matrix between frames before and after restoration the depth map is depicted in Fig. 4.21. Thus, the restoration of the original depth map increases the accuracy of the SLAM algorithm.

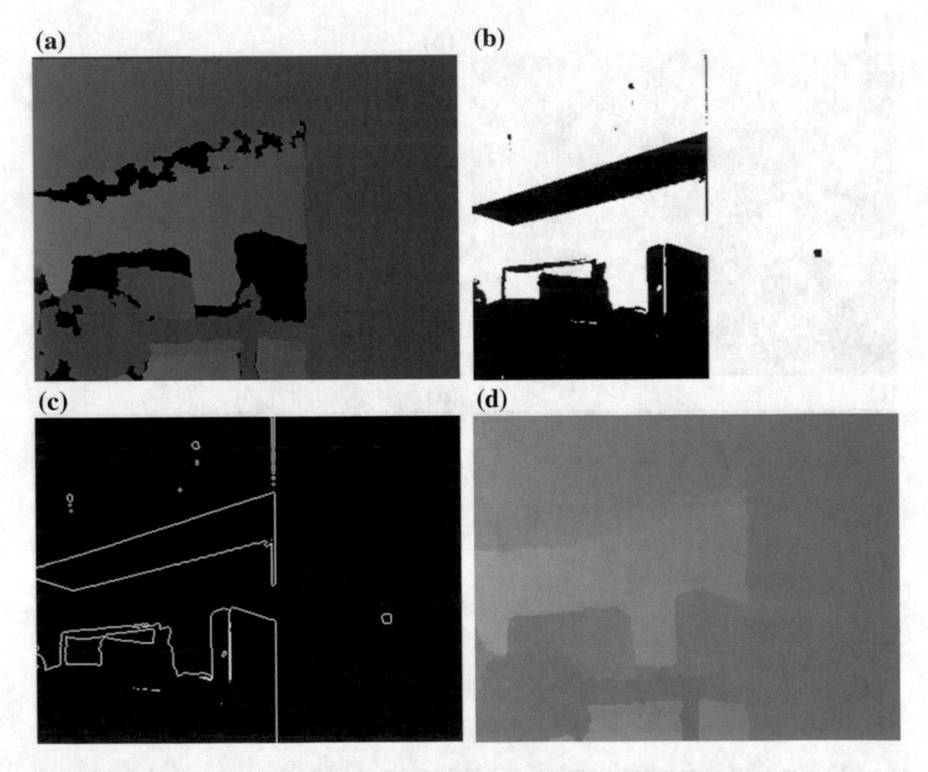

Fig. 4.20 Interpolation algorithm with boundaries refinement: **a** initial depth map, **b** the result of the thresholding, **c** objects borders, **d** reconstructed depth map

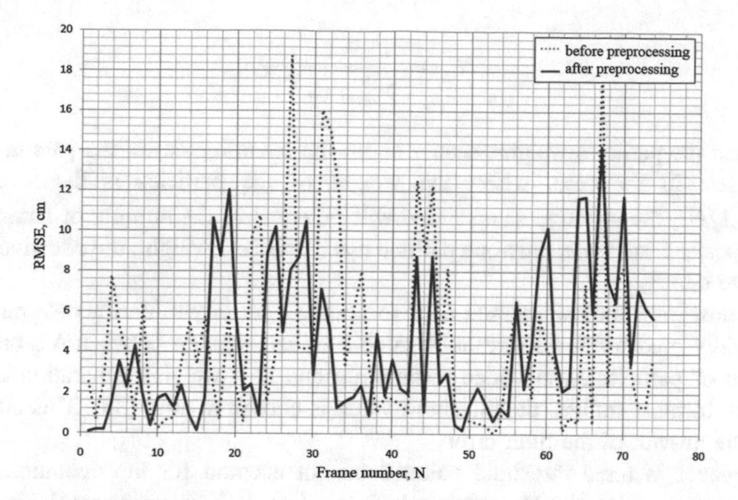

Fig. 4.21 Errors estimation of the camera displacement

After pretreatment with the help of the proposed algorithm, the depth map is free from defects caused by several factors of a different nature. Recovered region allow to obtain the depth data for an additional number of landmarks that have not been involved into the depth maps' construction without pretreatment. This eliminates the need to detect a greater number of feature points in their lack, which also increases the performance of the algorithm.

4.10 Adaptive Particles Resampling

Particle filter is a modeling method for estimating the state of a noisy system that stores the weighted normalized sample set of states called particles. In the considering problem, each particle represents by the estimated camera coordinates in space together with the state vector of the observed landmarks. During each cycle of the filter after the receipt of the vector of measurements, the new particles are created, each of which is weighted according to the Markov model of observations. After that, the weight is normalized for a new set of states.

One of the most important steps of particle filter, which has a strong impact on the performance and accuracy of the filter, is the process of updating the vector of particles. In this case, the problem is to replace the particles with smaller weights by the particles with large weights, when some values are eliminated by a certain law. In this chapter, we use a method called resampling wheel discussed in details in researches [28, 34]. Moment of the filter cycle, in which we need to update a set of particles, is determined by a quantity called effective sample size, which is defined by Eq. 4.31, where N is an overall number of particles, $\hat{w}^{[k]}$ is a normalized weight of kth particle.

$$N_{eff} = \frac{1}{\sum_{k=1}^{N} \left(\hat{w}^{[k]}\right)^2} \tag{4.31}$$

When the particles weights scatter about the mean increases, the parameter N_{eff} decreases. In the case, where the weight of the particles is the same (i.e. $\hat{w}^{[k]} = 1/N$), the effective sample size will be equal to the number of particles. In the opposite case, when only one particle has a non-zero weight, the effective size is equal to one $N_{eff} = 1$.

In most implementations, the need to update a set of particles is determined by the rigidly predetermined threshold value υ of relationship between N_{eff} and total number of particles N, at which resampling and new weights generation are performed. In most studies, the threshold value is chosen at 50% [35]. This allows to limit the growth of the filter error.

However, a fixed threshold value does not account for the evolution of the process of changing the N_{eff} value, which can lead to both over-complexity of the algorithm and degeneration of the particles sample. We can look at an example of such case, when the effective sample size is relatively small but its value oscillates around a constant value, which leads to a good level of approximation of the desired process. In this case, it is necessary to lower threshold to prevent a frequent updating of the particles sample. Also, the average effective number of particles within a specified time interval characterizes the quality of the approximation process during this interval.

The need for an adaptive threshold is a setting to update the sample particles in the characteristic of the improved SLAM algorithm, where the main method of measuring the movement of the camera is a visual odometry. The accuracy of the visual odometry algorithm depends on the number of landmarks that is constantly changed depending on the nature of the scene and related conditions. Even if the feature points' detector allows to find enough landmarks, not all of them are in the working range of the depth map. If a large object is too far or too close relative the camera, the errors of the visual odometry algorithm increase significantly. In this case, it is advisable to increase the threshold that defines the refresh rate for the particles sample.

Consider N_{eff} value change within the period $\tau - 1$. During this period, it reaches the extreme values N_{min} and N_{max}, so that $1 < N_{min} \leq N_{eff} \leq N_{max} < N$. We define a threshold value in the period $\tau - 1$ as $\upsilon_{\tau-1}$. When calculating the new value of the threshold υ_{τ}, it must be considered how fast N_{eff} value is changed within the previous period, as well as the average number of effective particles during this time. Let $k_{\tau-1}$ be a number of filter iterations for a period $\tau - 1$ and $N_{eff}(t)$ be a number of effective particles at tth iteration. Then, the values m_N and d_N characterize the average number of particles during the period and variance value determined by Eq. 4.32.

$$
\begin{aligned}
m_N &= \frac{1}{Nk_t}\sum_{t=1}^{k_{\tau-1}-1}\frac{N_{eff}(t+1)+N_{eff}(t)}{2} \\
d_N &= \frac{\left|N_{eff}(k_{\tau-1})-N_{eff}(1)\right|}{(k_{\tau-1}-1)\,(N-N_{\min})}
\end{aligned}
\tag{4.32}
$$

The threshold υ_τ is determined by the coefficients c_1 and c_2, such that c_1, $c_2 > 0$, $c_1 + c_2 = 1$, which can be set manually (Eq. 4.33).

$$
\begin{aligned}
\upsilon_\tau &= c_1\cdot m_N + c_2\cdot d_N \\
\Delta N &= N_{\max} - N_{\min} \\
N_{mean} &= \Delta N\cdot m_N + N_{\min} \\
N_{\mathrm{var}} &= \Delta N\cdot \sqrt[\Delta N]{d_N} + N_{\min}
\end{aligned}
\tag{4.33}
$$

Thus, when m_N tends to one, value N_{mean} increases linearly up to value N_{max}. Therefore, a value m_N so close to one indicates that the value N_{eff} is large relative to the total number of particles N, and the threshold should be increased.

Nonlinear function N_{var} characterizes variation of N_{eff}. On the one hand, the amount d_N close to zero means the low oscillations of coefficient N_{eff}, which indicates a high quality of approximation and needs to update the sample of particles even at a relatively low value of N_{eff}. On the other hand, if a value of d_N is significantly greater than zero, it indicates a strong oscillation of the effective sample size. In this case, the sample update should happen more often to compensate for the increased level of measurement error. The value N_{var} is dependent nonlinearly from value d_N to make it the most sensitive near zero. Visual results for the calculation of the threshold of the particle filter with a sample size of 10 at different evolution N_{eff} are shown in Figs. 4.22 and 4.23. The values of coefficients are $c_1 = 0.7$, $c_2 = 0.3$.

To estimate the effectiveness of the developed method, the comparative measurements of visual odometry errors were made in the process of movement of the camera with a the rigidly predetermined threshold of 50% and with an adaptive threshold calculation procedure (Table 4.4).

Adaptive threshold selection can significantly increase the accuracy of the camera's location. The average errors in calculating the displacement of the camera between successive frames is about 24% less than in the case of the strict threshold assignment (Table 4.4). Thus, the proposed adaptive threshold selection method allows to increase the efficiency of the filter in terms of computing power consumption and required evaluation accuracy. This method finds application for estimating camera position in the SLAM algorithm based on the video stream analysis, since the error of the measurement camera position in that case strongly varies during movement that does not usually occur with the systems based on other types of sensors, such as laser rangefinders or ultrasonic sensors.

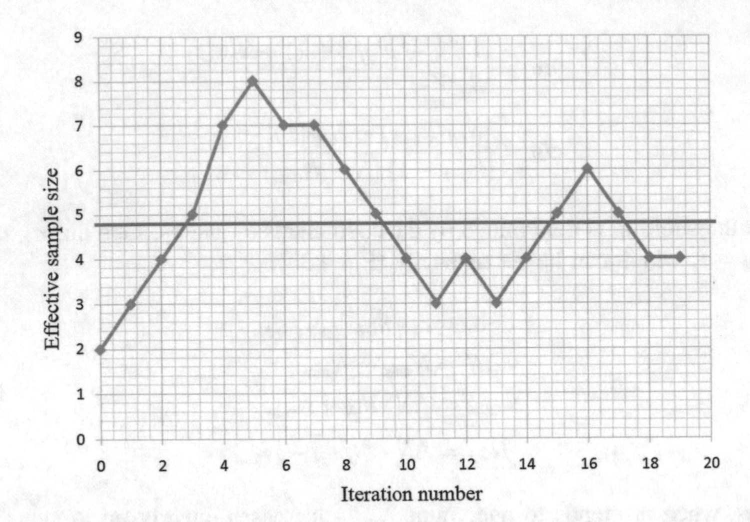

Fig. 4.22 Evolution of N_{eff} value within the period τ with oscillation factor 1.25 and the resulting adaptive threshold

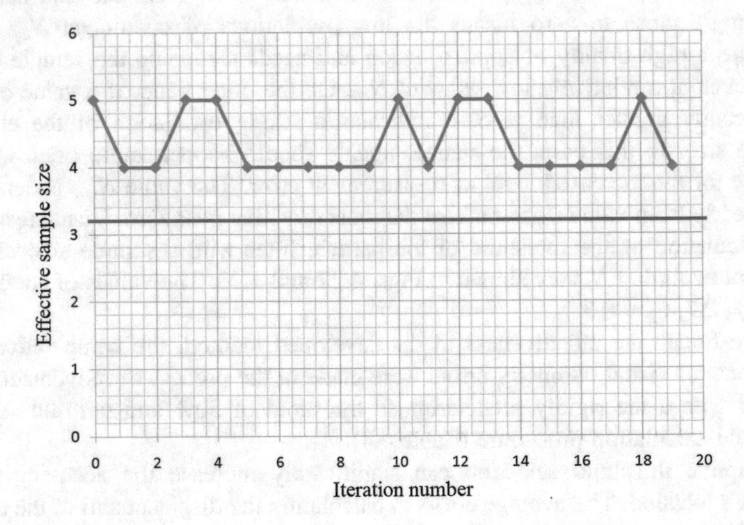

Fig. 4.23 Evolution of N_{eff} value within the period τ with oscillation factor 0.23 and the resulting adaptive threshold

Table 4.4 Accuracy of the algorithms with strict and adaptive thresholds

Threshold	Strict threshold	Adapting threshold		
	50%	$c_1 = 0.3$ $c_2 = 0.7$	$c_1 = 0.5$ $c_2 = 0.5$	$c_1 = 0.7$ $c_2 = 0.3$
RMSE (cm)	6.62	6.59	6.22	5.03

4.11 Conclusions

In this research, the classical FastSLAM algorithm was improved, i.e. the method of landmarks observations filtering with the UKF was replaced by the nonlinear models through the first order Taylor series expansion at the mean of the landmark state. Investigations of detectors and descriptors of feature points used to track spatial landmarks in RGB-D images showed the advantages of using the ORB detector-SURF descriptor pair. It combines the high detection rate inherent in the ORB method and the accuracy of the SURF descriptors. Numerical results show an average of 6% worse camera path accuracy and 34% reduction in the processing time for the ORB-SURF-pair-based algorithm in contrast to the case of using the SURF. Thus, an ORB detector can become one of the solutions allowing to increase the performance of the algorithm of visual SLAM algorithm, subject to some decrease in the accuracy of the algorithm. The use of the UKF for tracking of observing spatial landmarks has a significant advantage with respect to the EKF, especially in the cases, where a number of feature points in the images is small or for a large number of landmarks. The use of the UKF allows to reduce the root-mean-square error an average of 40% in calculating the total trajectory of the camera.

Also the chapter considers the adaptive threshold setting algorithm for updating the particle filter selection, which allows to increase significantly the accuracy of the camera localization. The accuracy of the visual odometry algorithm depends directly on the number of detected spatial feature points. Even if the detector can find a sufficient number of landmarks, not all of them are in the working range of depth maps. If a large object is too far away or too close to the camera overlapping most of the scene, the error in the visual odometry algorithm increases significantly. In this case, it is advisable to increase the threshold specifying the refresh rate of the particle sample. The use of adaptive threshold calculation algorithm shows that the average error value is up to 24% less than in the case of a hard threshold setting. Adaptive thresholding allows to increase the efficiency of the filter in terms of the required computing power and accuracy of estimation of the camera motion.

References

1. Dutta, T.: Evaluation of the Kinect sensor for 3-D kinematic measurement in the workplace. Appl. Ergon. **43**(4), 645–649 (2012)
2. Karlsson, N., di Bernardo E., Ostrowski, J., Goncalves, L., Pirjanian, P., Munich, M.: The vSLAM algorithm for robust localization and mapping. In: International Conference on Robotics and Automation (ICRA'2005), pp. 18–22 (2005)
3. Aulinas, J.: The SLAM problem: a survey. In: 2008 Conference on Artificial Intelligence Research & Development, pp. 363–371 (2015)
4. Zakaria, A.M., Said, A.M.: 3D reconstruction of a scene from multiple uncalibrated images using close range photogrammetry. In: International Symposium on Information Technology (ITSim'2010), pp. 1–5 (2010)

5. Tomasi, C., Kanade, T.: Detection and tracking of point features. Tech Report CMU-CS-91-132, Carnegie-Melon University (1992)
6. Technical description of Kinect calibration. Available from: http://wiki.ros.org/kinect_calibration/technical. Accessed 3 July 2017
7. Smisek, J., Jancosek, M., Pajdla, T.: 3D with kinect. In: Fossati, A., Gall, J., Grabner, H., Ren, X., Konolige, K. (eds.) Consumer Depth Cameras for Computer Vision, Advances in Computer Vision and Pattern Recognition, pp. 3–25. Springer, London (2013)
8. Herrera, C., Kannala, J., Heikkila, J.: Joint depth and color camera calibration with distortion correction. IEEE Trans. Pattern Anal. Mach. Intell. **34**(10), 2058–2064 (2012)
9. Wagner, D., Mulloni, A., Langlotz, T., Schmalstieg, D.: Real-time panoramic mapping and tracking on mobile phones. In: IEEE Virtual Reality Conference (VR'2010), pp. 211–218 (2010)
10. Einicke, G.A., White, L.B.: Robust extended Kalman filtering. IEEE Trans. Sig. Process. **47** (9), 2596–2599 (1999)
11. Koller, D., Montemerlo, M., Thrun, S., Wegbreit, B.: FastSLAM: A factored solution to the simultaneous localization and mapping problem. In: 18th National Conference on Artificial intelligence (AAAI'2002), pp. 593–598 (2002)
12. Murphy, K.: Bayesian map learning in dynamic environments. In: 12th Int Conf Neural Information Processing Systems (NIPS'1999), pp. 1015–1021 (1999)
13. Doucet, A., Freitas, N., Murphy, K., Russell, S.: Rao-Blackwellised particle filtering for dynamic Bayesian networks. In: 16th Conference on Uncertainty in Artificial Intelligence (UAI'2000), pp. 176–183 (2000)
14. Murphy, K., Russell, S.: Rao-blackwellized particle filtering for dynamic Bayesian networks. In: Doucet, A., de Freitas, N., Gordon, N. (eds.) Sequential MonteCarlo Methods in Practice, pp. 499–515. Springer Science+Business Media, New York (2001)
15. Szeliski, R.: Computer Vision: Algorithms and Applications. Springer, London (2010)
16. Sonka, M., Hlavac, V., Boyle, R.: Image Processing, Analysis and Machine Vision, 2nd edn. Springer Science+Business Media, London
17. Barnich, O., Droogenbroeck, M.: A universal background subtraction algorithm for video sequences. IEEE Trans. Image Process. **20**(6), 1709–1724 (2011)
18. Isard, M., Blake, A.: Condensation—conditional density propagation for visual tracking. Int. J. Comput. Vis. **29**(1), 5–28 (1998)
19. Prozorov, A., Priorov, A.: Three-dimensional reconstruction of a scene with the use of monocular vision. Meas. Tech. **57**(10), 1137–1143 (2015)
20. RGB-D SLAM dataset and benchmark. Available from: https://vision.in.tum.de/data/datasets/rgbd-dataset. Accessed 3 July 2017
21. PUT Kinect 1 & Kinect 2 data set. Available from: http://lrm.put.poznan.pl/putkk/. Accessed 3 July 2017
22. The Malaga Stereo and Laser Urban Data Set. Available from: http://www.mrpt.org/MalagaUrbanDataset. Accessed 3 July 2017
23. Sturm, J., Engelhard, N., Endres, F., Burgard, W., Cremers, D.: A benchmark for the evaluation of RGB-D SLAM systems. In: IEEE/RSJ International Conference on Intelligent Robots and Systems (IROS'2012), pp. 573–580 (2012)
24. Kuemmerle, R., Steder, B., Dornhege, C., Ruhnke, M., Grisetti, G., Stachniss, C., Kleiner, A.: On measuring the accuracy of SLAM algorithms. J Auton. Rob. **27**(4), 387–407 (2009)
25. Oxford Robotical Dataset. Available from: http://robotcar-dataset.robots.ox.ac.uk/. Accessed 3 July 2017
26. Kurt-Yavuz, Z., Yavuz, S.: A comparison of EKF, UKF, FastSLAM2.0, and UKF-based FastSLAM algorithms. In: IEEE 16th International Conference on Intelligent Engineering Systems (INES'2012), pp. 37–43 (2012)
27. Wan, E.A., Van Der Merwe, R.: The unscented Kalman filter for nonlinear estimation. In: IEEE Adaptive Systems for Signal Processing, Communications, and Control Symposium (AS-SPCC'2000), pp. 153–158 (2000)

28. Simon, J., Jeffrey, U.: Unscented filtering and nonlinear estimation. Proc. IEEE **92**(3), 401–422 (2004)
29. Wan, E.A., Van der Merwe, R.: The square-root unscented Kalman filter for state and parameter-estimation. In: IEEE International Conference on Acoustics, Speech, and Signal (ICASSP'2001), pp. 3461–3464 (2001)
30. Camera Calibration Toolbox for Matlab. Available from: http://www.vision.caltech.edu/bouguetj/calib_doc. Accessed 3 July 2017
31. Wang, H.M., Huang, C.H., Yang, J.F.: Depth maps interpolation from existing pairs of keyframes and depth maps for 3D video generation. In: IEEE International Symposium on Circuits and Systems (ISCAS'2010), pp. 3248–3251 (2010)
32. Criminisi, A., Perez, P., Toyama, K.: Region filling and object removal by exemplar-based inpainting. IEEE Trans. Image Process. **13**(9), 1200–1212 (2004)
33. Canny, J.A.: Computational approach to edge detection. IEEE Trans Pattern Anal. Mach. Intell. (PAMI) **8**(6), 679–698 (1988)
34. Blanco, J.L.: Contributions to localization, mapping and navigation in mobile robotics. Ph.D. Thesis, Universidad de Malaga
35. Douc, R., Cappe, O., Moulines, E.: Comparison of resampling schemes for particle filtering. In: 4th International Symposium on Image and Signal Processing and Analysis (ISPA'2005), pp. 64–69 (2005)

Chapter 5
Development of Fast Parallel Algorithms Based on Visual and Audio Information in Motion Control Systems of Mobile Robots

Sn. Pleshkova and Al. Bekiarski

Abstract Decision making for movement is one of the essential activities in motion control systems of mobile robots. It is based on methods and algorithms of data processing obtained from the mobile robot sensors, usually video and audio sensors, like video cameras and microphone arrays. After image processing, information about the objects and persons including their current positions in area of mobile robot observation can be obtained. The aim of methods and algorithms is to achieve the appropriate precision and effectiveness of mobile robot's visual perception, as well as the detection and tracking of objects and persons applying the mobile robot motion path planning. The precision in special cases of visual speaking person's detection and tracking can be augmented adding the information of sound arrival in order to receive and execute the voice commands. There exist algorithms using only visual perception and attention or also the joined audio perception and attention. These algorithms are usually tested in the most cases as simulations and cannot provide a real time tracking objects and people. Therefore, the goal in this chapter is to develop and test the fast parallel algorithms for decision making in the motion control systems of mobile robots. The depth analysis of the existing methods and algorithms was conducted, which provided the main ways to increase the speed of an algorithm, such as the optimization, simplification of calculations, applying high level programming languages, special libraries for image and audio signal processing based on the hybrid hardware and software implementations, using processors like Digital Signal Processor (DSP) and Field-Programmable Gate Array (FPGA). The high speed proposed algorithms were implemented in the parallel computing multiprocessor hardware structure and software platform using the well known NVIDIA GPU processor and GUDA platform, respectively. The experimental results with different parallel structures

Sn. Pleshkova (✉) · Al. Bekiarski
Technical University, St. Kl. Ohridski 8, 1000 Sofia, Bulgaria
e-mail: snegpl@tu-sofia.bg

Al. Bekiarski
e-mail: aabbv@tu-sofia.bg

M.N. Favorskaya and L.C. Jain (eds.), *Computer Vision in Control Systems-4*,
Intelligent Systems Reference Library 136,
https://doi.org/10.1007/978-3-319-67994-5_5

confirm the real time execution of algorithms for the objects and speaking person's detection and tracking using the given mobile robot construction.

Keywords Visual and audio decision making · Mobile robot · Motion control system · Visual and audio perception and attention · Parallel algorithm GPU · CUDA

5.1 Introduction

Motion is one of the main functions of mobile robots [1]. Like the human motion, every step of mobile robot motion depends from the goal of this motion. The human motion is generally defined from the human desires, for example to walk along the streets, in the rooms and corridors, to go to the work place, in the mountains, etc. The mobile robot motion can be considered in the same way but the mobile robot "desires" to go somewhere is more precise to define as the "goals". These goals are defined usually from the human and are embedded into control system of a mobile robot as algorithms for the motion to the goal or task, which the mobile robot must execute [2]. The main part of these algorithms is a decision making [3] as a mean to predict and choose the right or simple motion robot step from its current to the next position in order to prevent the crash with obstacles or go in a wrong direction. The main information for preparing decision making is based on methods and algorithms for data processing obtained from the mobile robot sensors, usually video and audio sensors, like video cameras and microphone arrays, laser range finders, etc. [4]. After processing of the images obtained from video camera and sounds received from microphone array, information about the objects and persons (perhaps, speaking to the robot) in area of mobile robot observation can be analyzed [5]. This information is used for a decision making in the motion control algorithms executed by kinematic system of a mobile robot platform [6].

In the development of methods and algorithms for a decision making, it is necessary to model and simulate the appropriate precision and effectiveness of mobile robot's audio visual perception [7] and attention [8] during the objects and person's detection and tracking, applying the mobile robot motion path planning and execution. There exist some algorithms based on the visual perception and attention [9]. Other methods can increase the precision using not only visual information but also the joined audio perception and attention [10] as the special cases of visual speaking person's detection and tracking adding the information of sound direction. The visual and audio information is used mainly in algorithms of joined visual and audio perception and attention in mobile robots tasks for detecting and tracking a speaking person, as well as receiving and execution the voice commands [11]. These algorithms are usually tested in the most cases as simulations and cannot provide a real time tracking objects and people. Therefore, the goal in this chapter is to develop, describe, and test the fast parallel algorithms for a decision making in the motion control systems of mobile robots. The simulated visual and joined audio mobile robot's perception and attention algorithms permit to achieve real time detection and tracking of the objects and speaking persons by a mobile robot. The depth analysis of the existing methods and algorithms show that the main ways to increase the speed of an algorithm deal with the optimization, simplification of calculations, applying high level programming languages, special libraries for image and audio signal processing based on the hybrid hardware and software implementations, using processors like the DSP and the FPGA. Some of the mentioned above options are applied for a comparison of high speed proposed and tested algorithms. They implemented the parallel computing multiprocessor hardware structure and software platform using the well known NVIDIA Graphics Processing Unit (GPU) processor and Compute Unified Device Architecture (CUDA) platform, respectively. The experimental results with different parallel structures confirm the real time work of algorithms for objects and speaking person's detection and tracking using for concrete mobile robot model.

This chapter is organized as follows. Section 5.2 describes briefly the related works. The development of fast parallel algorithms based on the audio and visual information for the motion control systems of mobile robots is considered in Sect. 5.3. The detailed experimental results are show in Sect. 5.4. Section 5.5 concluded the chapter.

5.2 Related Works

There are a lot of researches related to NVIDIA GPU processors and corresponding GUDA programming platform. Some of them are directed to hardware of GPUs and their internal architecture and specifications. Others consider the CUDA programming platform as a powerful mean to develop the applications for essential increasing of calculation speed and achieving a real time execution in the most cases. These applications are mainly the implementations of the GPU processors and CUDA programming platform with the time consuming mathematical operations in comparison to the same mathematical operations executed with a conventional Central Processing Unit (CPU). In order to explain more precise the related works and confirm the real time ability of GPU processors in the proposed algorithms, the internal structure of GPUs, their specifications, and CUDA programming platform capabilities to the developed mobile robot algorithms parallel executed from NVIDIA GPUs are briefly discussed.

The popular parallel NVIDIA GPU processors were firstly applied in computer games as the fast computing engines [12] and then became the serious motivation of many researches to test the ability of these NVIDIA GPU parallel processors in a combination with CUDA programming platform [13] in order to speed up the time consuming calculations in wide areas of applications. The majority of these researches and studies are directed to test and demonstrate the short time of execution of popular mathematical operations like, for example Fast Fourier Transform (FFT) [14], convolution [15], and correlation [16] in comparison to the results obtained by parallel NVIDIA GPU processors and traditional universal host computer CPU. The advantages of parallel NVIDIA GPU processors are connected with their internal structure [17] shown in Fig. 5.1.

The key features of the internal structure of NVIDIA GPU processor shown in Fig. 5.1 are the existence of many processors, memory, and interconnections. Two types of processors, such as multiprocessors and thread processors are applied. Each multiprocessor has a number of thread processors. For example, in a NVIDIA a number of multiprocessors is thirty or more and each multiprocessor has eight or more parallel thread processors that means total 240 or more thread processors. In each clock of GPU, all thread processors give the results from execution of the same operation or calculations but with different data (for example, video or audio information). Therefore, it is possible to deliver in parallel (only in one GPU clock period) the results for all different data (for example, video or audio information) processed with the same (single) operation, i.e. the thread processors works as a parallel computing structure of the type Single Instructions Multi Data (SIMD) [18]. The operations in the thread processors are managed from Thread Execution Control Unit. Two types of memory, such as Device Memory and Local Memory,

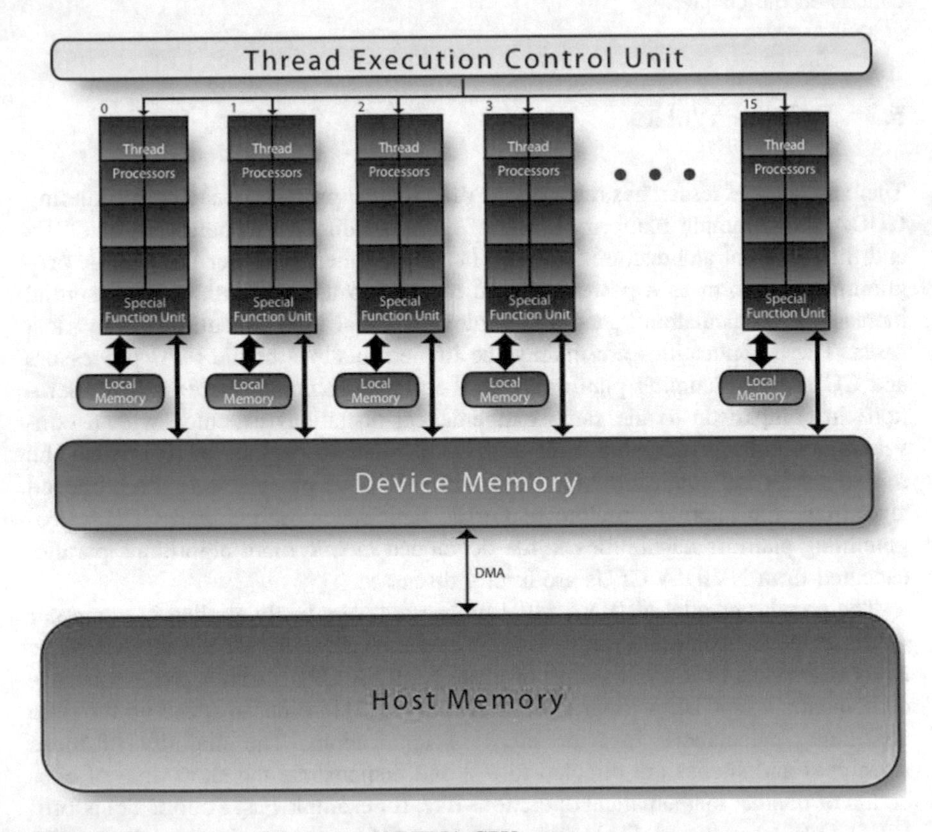

Fig. 5.1 General internal structure of NVIDIA GPU processors

are included in the internal structure of NVIDIA GPU (Fig. 5.1). For example, device memory can be 4 GB. Device memory is connected to Host Memory of the computer. The data exchange between these two memories is arranged as Direct Memory Access (DMA). The local memory is connected to each multiprocessor via Special Function Unit, which is also connected to device memory. These interconnections allow to speed up the data exchanges between the device memory and each of the local memories.

The capabilities of CUDA programming platform to developed mobile robot algorithms parallel executed from NVIDIA GPUs can be described briefly using the architecture of CUDA programming tools and facilities shown in Fig. 5.2 [19]. The CUDA architecture consists of the following several components shown in Fig. 5.2:

- CUDA Parallel Compute Engines inside NVIDIA GPUs.
- CUDA Support in OS Kernel suitable for hardware initialization, configuration, etc.
- CUDA Drivers and user-mode driver to provide a device-level API for developers.
- Parallel Thread eXecution (PTX) Instruction Set Architecture (ISA) for parallel computing kernels and functions.

These basic CUDA components can be used from developers via standard Device-level APIs such as OpenCL™ and DirectX® Compute, and via Languages integration of high level programming languages such as C/C++, Fortran, Java,

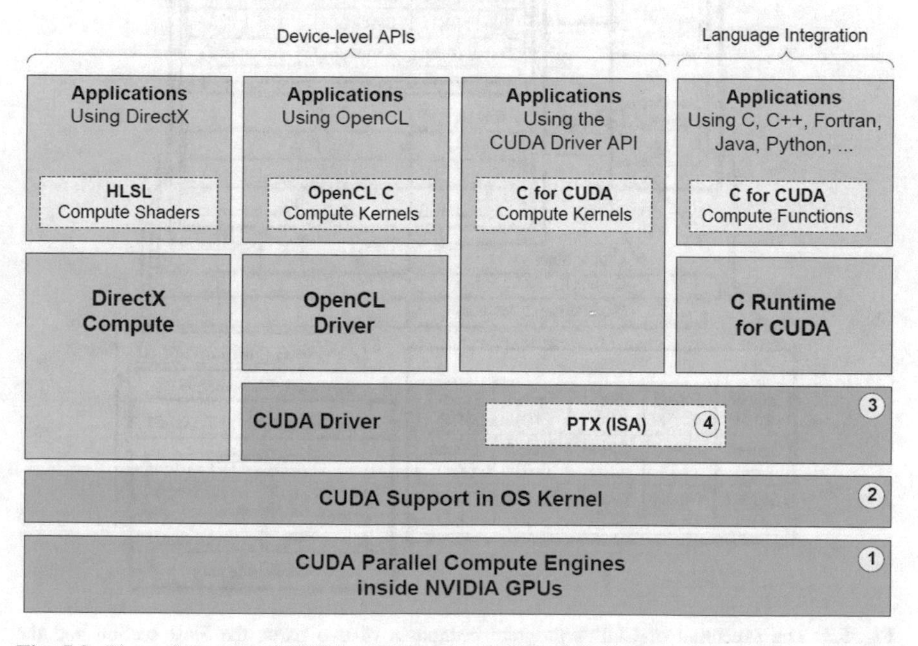

Fig. 5.2 The architecture of CUDA programming tools and facilities

Python, and the Microsoft.NET Framework. The execution of CUDA program composes of two parts [20]: the Host section and the Device section (Fig. 5.3). The Host section is executed on computer CPU while the Device section is executed on GPU. The execution of Device section on GPU is managed by kernel. The kernel handles synchronization of executing threads. It is invoked by Device call for GPU. The threads in GPU architecture can be grouped into blocks and grids as is shown in Fig. 5.3.

In GPU, a grid is with group of thread blocks and a thread block comprises the defined number of threads per block. In CUDA, the differentiation between the unique threads, thread block, and grid may be done using a set of identifiers *threadIdx*, *blockIdx* and *gridIdx* variables, respectively. Thread per block can exchange information to synchronize with each other. Per block shared memory can be used for communication between each thread within a thread block. However, there is no direct interaction or synchronization possible between the threads of different blocks.

The CUDA architecture and tools allow to develop the applications to process data and information in wide areas and fields, such as images, audio, signals, and natural resource exploration, and to achieve dramatic speedup of execution. Most of

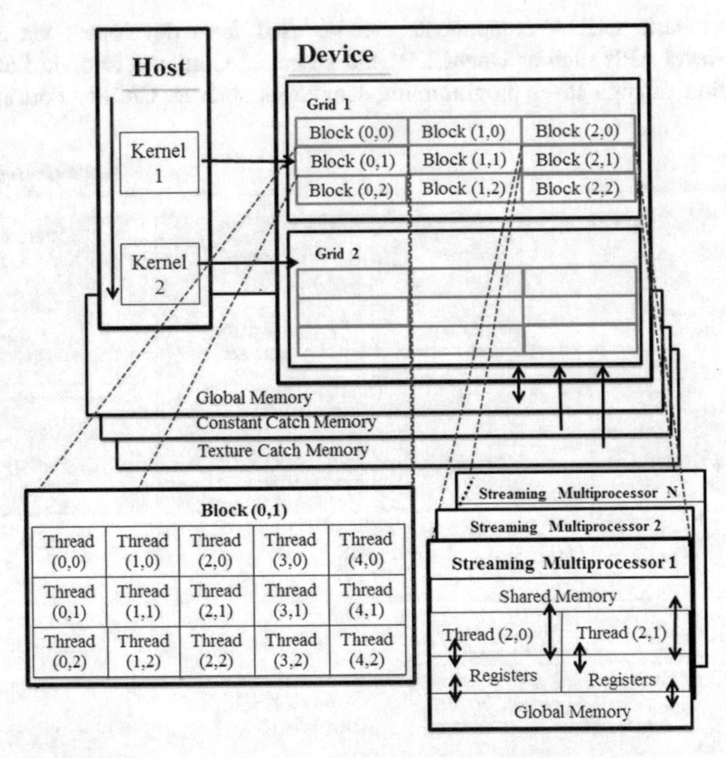

Fig. 5.3 The execution of CUDA program composes of two parts: the Host section and the Device section

these applications are examples of algorithms containing the time consuming mathematical operations. The benefit of the execution time using GPU processors and GUDA programming platform in a comparison to the conventional CPU is evident.

One of the popular mathematical operations widely used in methods and algorithms for processing information in the images, audio and signals are a matrix multiplication. There are a lot of articles [21–24], in which a matrix multiplication is the subject of descriptions and tests as mathematical operation realized using GPU processors and CUDA programming platform with the aims of demonstrate the advantages of GPU processors in comparison to conventional CPU. The plots of execution times for matrix multiplication as a function of matrix dimension using CPU or GPU are depicted in Fig. 5.4 [25].

There are also other similar mathematical operations very useful in area of methods and algorithms for processing information in the images, audio and signals like the FFT, correlation, convolution, etc. The FFT as a frequently used transformation in algorithms of images, audio, and other signal processing is a subject of many test with GPUs applying high level programming languages, such as C/C++, Fortran, Java, Python, Microsoft.NET Framework, and Matlab [26–28].

Some results published in NVIDIA documentation [29] for speed achieved, when the FFT with different transform size is performed in GPUs applying high level programming languages, for example C/C++ are represented in Fig. 5.5. The operations correlation and convolution also are in area of methods and algorithms for processing information in the images, audio, and signals. Many researches [30–33] explain the implementation using GPU processors and GUDA programming platform for testing the correlation or convolution operations and the ability to be executed in real time.

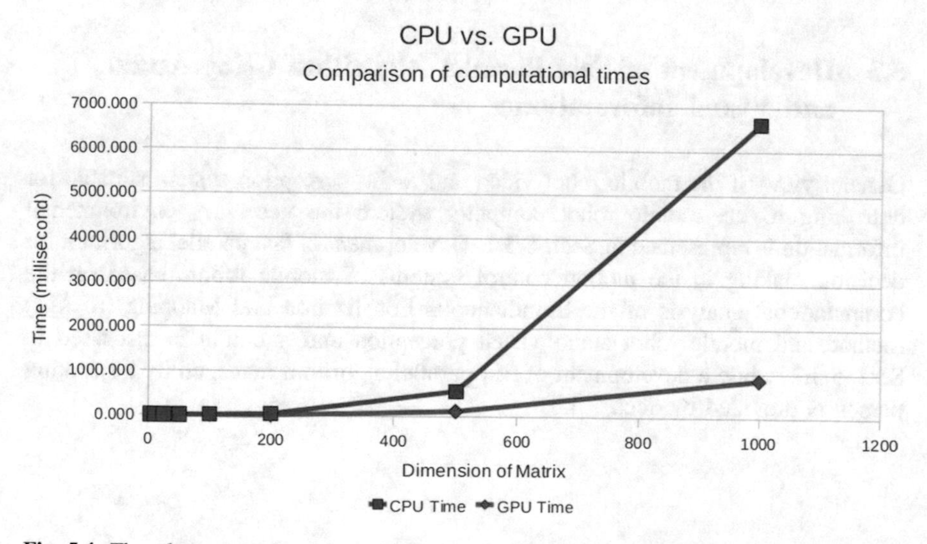

Fig. 5.4 The advantages of matrix multiplication in GPU versus CPU [25]

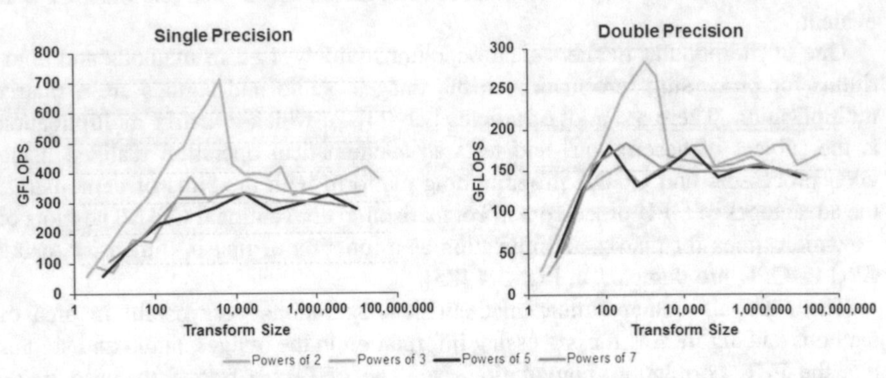

Fig. 5.5 Results published in NVIDIA documentation [29] for speed achieved, when the FFT with different Transform size performed in GPUs applying high level programming languages, for example C/C++

The briefly described GPUs internal structures, CUDA architecture, tools, and examples of implementations of time consuming mathematical operations like matrix multiplication, the FFT, correlation and convolution permit to predict the promising results in the development of fast parallel algorithms based on the joined visual and audio information for decision making in the motion control systems of mobile robots.

5.3 Development of Fast Parallel Algorithm Using Audio and Visual Information

General view of the mobile robot video and audio perception system suitable for delivering to the mobile robot computer system the necessary environmental information is represented in Sect. 5.3.1. Development of fast parallel algorithm for decision making in the motion control systems of mobile robots based on the computational analysis of the Simultaneous Localization and Mapping (SLAM) method and mobile robot audio visual perception and attention is discussed in Sect. 5.3.2, while a development of fast parallel algorithm managed by a speaking person is provided by Sect. 5.3.3.

5.3.1 *General View of Mobile Robot Video and Audio Perception System*

Decision making for motion control and navigation of a mobile robot is based on information from the environment, where the mobile robot moves. Before each step of mobile robot motion, it is necessary to capture and analyze the current information from the environment and make a decision for the right direction of motion, to prepare all suitable command for mobile robot wheels and realizing the current motion step. The possibilities of the mobile robots to capture information from the environment are similar to a human ability of perceiving this environmental information mainly like visual information and also as audio information. The general view of the video and audio perception system of a mobile robot suitable for delivering the necessary environmental information in the mobile robot computer system is presented in Fig. 5.6.

As it is seen from Fig. 5.6, the mobile robot perceives visual and audio information from its environment through a video camera and microphone array. Also a possibility to use the supplementary information for distance through a laser rangefinder is available. The presented in Fig. 5.6 Video Camera, Microphone Array, and Laser Range Finder can be considered as mobile robot visual, audio, and depth

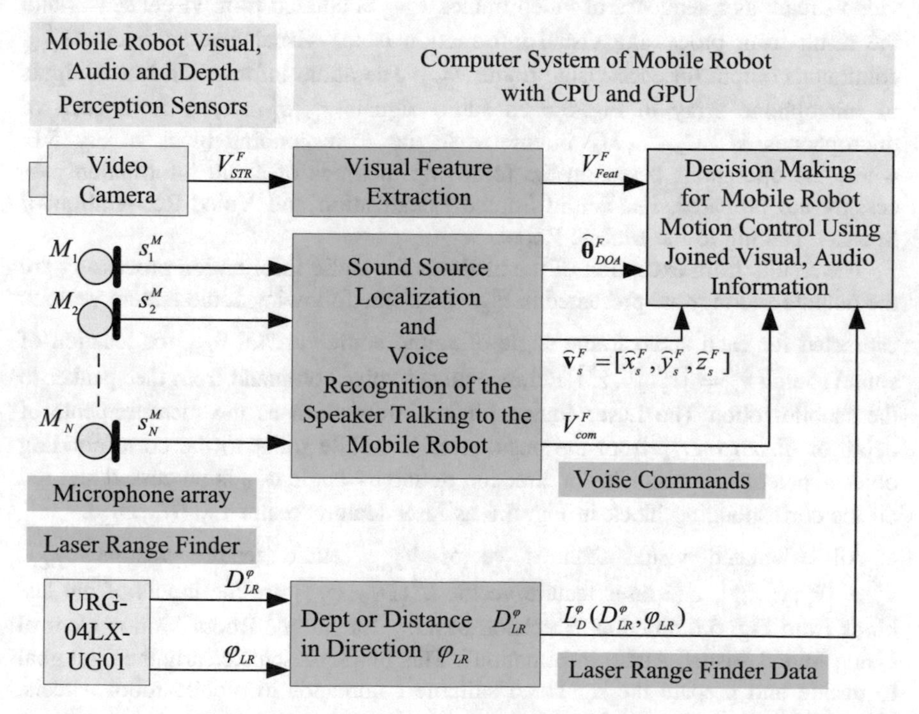

Fig. 5.6 The general view of the mobile robot video and audio perception system suitable for delivering to the mobile robot computer system the necessary environmental information

perceptual sensors, respectively. The practical implementation of mobile robot visual and depth perceptual sensors is well-known and popular as RGB-D Depth Camera [34]. The visual information from a visual sensor (output of Video Camera in Fig. 5.6) enters in form of a continuous video stream as a sequence of video frames V_{STR}^{F}. The audio information obtained from an audio sensor (outputs of Microphone Array in Fig. 5.6) consists of the audio signals $s_M^1, s_M^2, \ldots, s_M^N$ from the microphones $M_1, M_2, \ldots, M_N$. The information from Laser Range Finder (Type URG-04LX-UG01 [35] in Fig. 5.6) represent the measurements in direction define by angle φ_{LR} of Depth or Distance D_{LR}^{φ} from the position of the mobile robot to the objects, persons, or obstacles and perceived from mobile robot visual, audio, and depth perceptual sensors. The information from outputs of visual, audio and laser sensors is entered in the correspondent blocks of mobile robot computer system, where it can be processed using traditional universal host computer CPU or with parallel NVIDIA GPU processor.

The methods and algorithms for processing information received from visual, audio and laser sensors are different but all of them are dedicated to satisfy the goal of decision making for the motion control and navigation of a mobile robot based on information from the environment, where the mobile robot moves. The general method for processing visual information is implemented in the block Visual Feature Extraction in Fig. 5.6, where the visual information in a form of continuous video stream as a sequence of video frames V_{STR}^{F} is entered from visual sensor and the result from processing visual information is the visual features vector V_{Feat}^{F} continuous output for each visual frame V_{STR}^{F}. The audio information from outputs of microphone array in Fig. 5.6 as audio signals $s_M^1, s_M^2, \ldots, s_M^N$ from each of microphones $M_1, M_2, \ldots, M_N$ is entered in the corresponding block in Fig. 5.6, where the operations based on the following methods of audio information processing are prepared, i.e. Sound Source Localization and Voice Recognition of Speaker Talking to the Mobile Robot.

The results from execution of the methods for audio information processing are the outputs and they are presented in Fig. 5.6 as the following audio feature vectors: estimated for each audio frame angle of sound source arrival $\hat{\theta}_{DOA}^{F}$ or location of sound source $\hat{\mathbf{v}}_s^F = \left[\hat{x}_s^F, \hat{y}_s^F, \hat{z}_s^F\right]$ and recognized voice command from the speaker to the mobile robot. The Laser Range Finder information as the measurements of depth or distance D_{LR}^{φ} from the position of a mobile robot to the corresponding objects, persons, or obstacles in direction define by angle φ_{LR} is processed and fed on the corresponding block in Fig. 5.6 as laser feature vector $L_D^{\varphi}(D_{LR}^{\varphi}, \varphi_{LR})$.

All calculated visual features vector V_{Feat}^{F}, audio feature vectors $\hat{\theta}_{DOA}^{F}$, $\hat{\mathbf{v}}_s^F = \left[\hat{x}_s^F, \hat{y}_s^F, \hat{z}_s^F\right]$, and laser feature vector $L_D^{\varphi}(D_{LR}^{\varphi}, \varphi_{LR})$ are the inputs of the last block from Fig. 5.6. titled as “Decision Making for Mobile Robot Motion Control Using Joined Visual, Audio Information”. This block presents clearly the final goal to decide and prepare the right and suitable commands to mobile robot wheels, using which a mobile robot can make the next current motion step.

Following sections of this chapter present briefly the chosen methods to realize the blocks shown in Fig. 5.6 for processing the information from visual, audio, and laser sensors in order to calculate the visual features vector V_{Fea}^{F}, audio feature vectors $\hat{\theta}_{DOA}^{F}$, $\hat{\mathbf{v}}_{s}^{F} = \left[\hat{x}_{s}^{F}, \hat{y}_{s}^{F}, \hat{z}_{s}^{F}\right]$, and laser feature vector $L_{D}^{\varphi}(D_{LR}^{\varphi}, \varphi_{LR})$ and to implement these methods as algorithms by the traditional universal host computer CPU and parallel NVIDIA GPU processor.

5.3.2 *Development of Fast Parallel Algorithm Based on Simultaneous Localization and Mapping Method and Mobile Robot Audio Visual Perception and Attention*

There exist many different methods and algorithms applied in the decision making tasks for motion control of mobile robots based on the visual information perceived from mobile robot [36]. The SLAM method is one the actual and wide spread method [37] based mainly on the visual mobile robot perception and attention [38], using also information from the audio sensors (microphone array) [39] and laser range finder mounted on a mobile robot platform [40]. In this chapter, the SLAM method was chosen in sense of its ability to work not only as simulation but in real time as parallel algorithms for decision making in motion control system of a mobile robot using CUDA programming platform embedded into GPU devices. Consider briefly the SLAM method in a view of the general mathematical equation.

The SLAM method is based on the continuous sequential probabilistic processing and estimation of observations captured from mobile robot video, audio sensors, laser distance measurement sensors, and mobile robot wheel motion sensors [41]. The current information of observations from the mobile robot sensors consist usually from the spatial positions of the objects, persons, and obstacles in a mobile robot motion environment and also from natural defined or artificial added "landmarks" suitable to compose and continuous update a map of area of observation leading to improve the precision of decision making for right prediction and execution of each next motion step of mobile robot (Fig. 5.6). Therefore, for each k time step of mobile robot motion the current information of observations from mobile robot sensors are presented as the observation vector or matrix $\mathbf{z}_{obs}^{k}$ provided by Eq. 5.1, where $\mathbf{z}_{vo}^{k}, \mathbf{z}_{ao}^{k}, \mathbf{z}_{lo}^{k}, \mathbf{z}_{mo}^{k}$ are the corresponding vectors or matrices of observations from video $\mathbf{z}_{vo}^{k}$, audio $\mathbf{z}_{ao}^{k}$, and laser $\mathbf{z}_{lo}^{k}$ sensors, and natural or artificial defined "landmarks" $\mathbf{z}_{mo}^{k}$ taken in k time step of mobile robot motion, respectively.

$$\mathbf{z}_{obs}^{k} = \left\{\mathbf{z}_{vo}^{k}, \mathbf{z}_{ao}^{k}, \mathbf{z}_{lo}^{k}, \mathbf{z}_{mo}^{k}\right\} \tag{5.1}$$

Except of the presented with Eq. 5.1 observation vector or matrix $\mathbf{z}_{obs}^{k}$ in the SLAM method, it is necessary to initiate, define, and use also the following information in a form of vectors or matrices:

- The current mobile robot position as vector $\mathbf{p}_{mr}^{k}$ in area of observation in a form of coordinates $x_{mr}^{k}, y_{mr}^{k}, z_{mr}^{k}$ (Eq. 5.2).

$$\mathbf{p}_{mr}^{k} = \left\{x_{mr}^{k}, y_{mr}^{k}, z_{mr}^{k}\right\} \tag{5.2}$$

- The current mobile robot control vector $\mathbf{c}_{mr}^{k}$ in area of observation containing odometry information from mobile robot wheel motion sensors for the linear v_{mrl}^{k} and angular v_{mra}^{k} current speed of mobile robot moving from previous position $\mathbf{p}_{mr}^{k-1}$ to current position $\mathbf{p}_{mr}^{k}$ (Eq. 5.3).

$$\mathbf{c}_{mr}^{k} = \left\{v_{mrl}^{k}, v_{mra}^{k}\right\} \tag{5.3}$$

- Vector $\mathbf{p}_{lm}^{n}$ containing the true position for each n defined natural or artificial landmark from the set of $n = 1, 2\ldots, N$ landmark in area of mobile robot observation supposing that the true position of all landmarks rest constant (Eq. 5.4).

$$\mathbf{p}_{lm}^{n} = \left\{x_{lm}^{n}, y_{lm}^{n}, z_{lm}^{n}\right\} \tag{5.4}$$

- The set of positions of all landmarks $n = 1, 2\ldots, N$ forming the landmark map $\mathbf{M}_{lm}$ in mobile robot environment (Eq. 5.5).

$$\mathbf{M}_{lm} = \left\{\mathbf{p}_{lm}^{1}, \mathbf{p}_{lm}^{2}, \ldots, \mathbf{p}_{lm}^{n}, \ldots \mathbf{p}_{lm}^{N}\right\} \tag{5.5}$$

The simultaneous principle of the SLAM method require to collect and keep the information for all previous 0, 1, 2, …, $k-1$ steps of execution for a decision making for a mobile robot. To satisfy this requirement, it is defined the following vectors or matrices to hold the information for all previous 0, 1, 2, …, $k-1$ steps and the information for the current step k:

- The previous information of observations from mobile robot sensors $\mathbf{Z}_{obs}^{0,k-1}$ and the current observation vector $\mathbf{z}_{obs}^{k}$ using Eq. 5.1:

$$\mathbf{Z}_{obs}^{0,k} = \left\{\mathbf{z}_{obs}^{0}, \mathbf{z}_{obs}^{1}, \mathbf{z}_{obs}^{2}, \ldots, \mathbf{z}_{obs}^{k-1}, \mathbf{z}_{obs}^{k}\right\} = \left\{\mathbf{Z}_{obs}^{0,k-1}, \mathbf{z}_{obs}^{k}\right\}. \tag{5.6}$$

- The previous mobile robot positions $\mathbf{p}_{mr}^{0,k-1}$ and the current position $\mathbf{p}_{mr}^{k}$ using Eq. 5.2:

$$\mathbf{P}_{mr}^{0,k} = \left\{\mathbf{p}_{mr}^{0}, \mathbf{p}_{mr}^{1}, \mathbf{p}_{mr,}^{2} \ldots, \mathbf{p}_{mr}^{k-1}, \mathbf{p}_{mr}^{k}\right\} = \{\mathbf{P}_{mr}^{0,k-1}, \mathbf{p}_{mr}^{k}\}. \tag{5.7}$$

- The previous of mobile robot control vector $\mathbf{C}_{mr}^{0,k-1}$ and the current control vector $\mathbf{c}_{mr}^{k}$ using Eq. 5.3:

$$\mathbf{C}_{mr}^{0,k} = \left\{\mathbf{c}_{mr}^{0}, \mathbf{c}_{mr}^{1}, \mathbf{c}_{mr,}^{2} \ldots, \mathbf{c}_{mr}^{k-1}, \mathbf{c}_{mr}^{k}\right\} = \{\mathbf{C}_{mr}^{0,k-1}, \mathbf{c}_{mr}^{k}\}. \tag{5.8}$$

The probabilistic representation of the SLAM method can be described with the following definitions:

- The joint posterior density $P_{mr,lm}^{k}$ between the mobile robot position $\mathbf{p}_{mr}^{k}$ in k time step (Eq. 5.2) and positions of all landmarks defined in the landmark map $\mathbf{M}_{lm}$ (Eqs. 5.4–5.5), depending from information of all observations $\mathbf{Z}_{obs}^{0,k}$ (Eq. 5.6), all collected mobile robot control vectors $\mathbf{C}_{mr}^{0,k}$ (Eq. 5.8), and initial mobile robot position $\mathbf{p}_{mr}^{0}$ derived from Eq. 5.6:

$$P_{mr,lm}^{k} = P\left(\mathbf{p}_{mr}^{k}, \mathbf{M}_{lm} | \mathbf{Z}_{obs}^{0,k}, \mathbf{C}_{mr}^{0,k}, \mathbf{p}_{mr}^{0}\right). \tag{5.9}$$

- The model of observation is presented as the probability P_{obs}^{k} in order to accomplish the observation $\mathbf{z}_{obs}^{k}$ if the mobile robot position $\mathbf{p}_{mr}^{k}$ (Eq. 5.2) and positions of all $n = 1, 2\ldots, N$ landmark in the map $\mathbf{M}_{lm}$ (Eq. 5.5) are known:

$$P_{obs}^{k} = P(\mathbf{z}_{obs}^{k}|, \mathbf{p}_{mr}^{k}, \mathbf{M}_{lm}). \tag{5.10}$$

- The mobile robot model of motion presented as the probability distribution P_{mt}^{k} of mobile robot motion to position $\mathbf{p}_{mr}^{k}$ if the previous position is $\mathbf{p}_{mr}^{k-1}$ and the current mobile robot control vector is $\mathbf{c}_{mr}^{k}$:

$$P_{mt}^{k} = P(\mathbf{p}_{mr}^{k}|, \mathbf{p}_{mr}^{k-1}, \mathbf{c}_{mr}^{k}). \tag{5.11}$$

The represented with Eqs. 5.9–5.11 probabilistic model of the SLAM method is suitable for implementation as an algorithm consisting from the following two recursively executed operations:

- The calculations of the joint posterior density $P_{mr,lm}^{k}$ (Eq. 5.9) as a time update (tu) function F_{tu}^{k} in each new k time step using the model of mobile robot motion presented as probability distribution P_{mt}^{k} (Eq. 5.11) and the joint posterior density $P_{mr,lm}^{k}$ (Eq. 5.9) from previous $k - 1$ time step is provided by Eq. 5.12.

$$P^k_{mr,lm} = F^k_{tu}\left(P\left(\mathbf{p}^k_{mr}|\mathbf{p}^{k-1}_{mr}, \mathbf{c}^k_{mr}\right), P\left(\mathbf{p}^{k-1}_{mr}, \mathbf{M}_{lm}|\mathbf{Z}^{0,k-1}_{obs}, \mathbf{C}^{0,k-1}_{mr}, \mathbf{p}^0_{mr}\right)\right) \quad (5.12)$$

- The calculations of the joint posterior density $P^k_{mr,lm}$ (Eq. 5.9) as a measurement update (mu) function F^k_{mu} in each new k time step using the model of mobile robot motion presented as probability distribution P^k_{obs} (Eq. 5.10) and the joint posterior density $P^k_{mr,lm}$ (Eq. 5.9) from previous $k-1$ time step is represented by Eq. 5.13.

$$P^k_{mr,lm} = F^k_{mu}\left(P\left(\mathbf{z}^k_{obs}|\mathbf{p}^k_{mr}, \mathbf{M}_{lm}\right), P\left(\mathbf{p}^k_{mr}, \mathbf{M}_{lm}|\mathbf{Z}^{0,k-1}_{obs}, \mathbf{C}^{0,k}_{mr}, \mathbf{p}^0_{mr}\right), P\left(\mathbf{z}^k_{obs}|\mathbf{Z}^{0,k-1}_{obs}, \mathbf{C}^{0,k}_{mr}\right)\right) \quad (5.13)$$

From Eqs. 5.12–5.13 one can conclude about the following two possible solutions:

- The building the map $\mathbf{M}_{lm}$ calculating the conditional density P_{lm} under assumption that the current mobile robot position as vector $\mathbf{p}^k_{mr}$ is known (Eq. 5.14).

$$P_{lm} = P\left(\mathbf{M}_{lm}|\mathbf{P}^{0,k}_{mr}, \mathbf{Z}^{0,k}_{obs}, \mathbf{C}^{0,k}_{mr}\right) \quad (5.14)$$

- The detection of the mobile robot position $\mathbf{p}^k_{mr}$ calculating the conditional density P_{mr} under assumption that the landmark locations in the map $\mathbf{M}_{lm}$ are known (Eq. 5.15).

$$P^k_{lm} = P\left(\mathbf{p}^k_{mr}|\mathbf{Z}^{0,k}_{obs}, \mathbf{C}^{0,k}_{mr}, \mathbf{M}_{lm}\right) \quad (5.15)$$

Equations 5.9–5.15 describe the SLAM method as general probabilistic model. There are many possible solutions to realize this general probabilistic model as the working algorithm applying Bayesian filters particle filters, the Extended Kalman Filter SLAM (EKF SLAM), etc. [42]. The EKF SLAM method is the most popular method from these listed solutions. There exist realizations of the EKF SLAM method as concrete algorithm in motion control using the perceived audio visual perception and attention information from mobile robot audio and visual sensors [7, 8].

The EKF SLAM algorithm is presented in Table 5.1 (as the Algorithm 1) and described briefly only to prepare the necessary computational analysis in order to define a possibility of its realization as a fast parallel algorithm based on mobile robot visual perception and attention using NVIDIA GPU processor and CUDA platform for decision making in motion control system of a mobile robot. The Algorithm 1 shown in Table 5.1 begins with a definition of initial position $\mathbf{p}^0_{mr}$ if a mobile robot is placed in an unknown environment (Step 1, Table 5.1). This algorithm is usually executed from a host CPU of computer system of a mobile robot.

After definition of initial position $\mathbf{p}_{mr}^{0}$, the mobile robot begin to execute the main part of the EKF SLAM algorithm (Step 2 from Table 5.1) as a sequence of operations executed from mobile robot computer system. First, it is necessary to input the current information of observations from the mobile robot audio, visual, and laser sensors (Step 3 from Table 5.1) presented as the observation vector or matrix $\mathbf{z}_{obs}^{k}$ (Eq. 5.1) in each k time step of mobile robot motion. Second, the next steps are for concurrently build the map $\mathbf{M}_{lm}$ of mobile robot surrounding environment while localizing the mobile robot current position $\mathbf{p}_{mr}^{k}$ within these environments performing the SLAM method. The presented above general probabilistic definition of the SLAM method needs at each discrete k time step to calculate the joint posterior density of the robot state and the landmark locations having the distribution provided by Eq. 5.16.

$$P\left(\mathbf{p}_{mr}^{k}, \mathbf{M}_{lm} \middle| \mathbf{z}_{obs}^{k}, \mathbf{c}_{mr}^{k}\right) \quad (5.16)$$

Table 5.1 The necessary steps of the EKF SLAM algorithm for motion control using the perceived audio visual perception and attention information from the mobile robot audio and visual sensors (Algorithm 1)

Step	Step description	Execution by CPU or GPU
1.	**Begin with definition** of initial position $\mathbf{p}_{mr}^{0}$, when a mobile robot is placed in an unknown environment	CPU
2.	**Begin recursive execution** of the EKF SLAM algorithm in each current discrete k time step	CPU
3.	**Input** in each k time step of mobile robot motion the current information of observations from mobile robot sensors presented as the observation vector or matrix $\mathbf{z}_{obs}^{k}$ (Eq. 5.1)	CPU or GPU
4.	**State estimation** or prediction in the current discrete k time step	CPU or GPU
4.1.	**Estimate state vector** $\mathbf{p}_{mr}^{k}$ and covariance matrix $\sum_{\mathbf{k}}$ (Eqs. 5.18, 5.21–5.23)	CPU or GPU
4.2.	**Update for correction** the observation vector $\mathbf{z}_{obs}^{k}$ (Eq. 5.1) using extracted in each k time step of mobile robot motion the observed information as the following observation features (Fig. 5.6): – The audio features as an angle of sound source arrival $\hat{\theta}_{DOA}^{F}$ or location of sound source $\hat{\mathbf{v}}_{s}^{F} = \left[\hat{x}_{s}^{F}, \hat{y}_{s}^{F}, \hat{z}_{s}^{F}\right]$ – The visual features as a vector V_{Fea}^{F} – The laser range finder features as a laser feature vector including depth or distance D_{LR}^{φ} in direction define by angle φ_{LR} from the position of the mobile robot to the corresponding objects, persons, or obstacles	CPU or GPU
5.	**Go to** Step 2 or Stop execution of algorithm	CPU

In Eq. 5.16, the control information obtained from mobile robot wheel motion sensors described by vector $\mathbf{c}_{mr}^{k}$ (Eq. 5.3) and also a set of observations from mobile robot video, audio, and laser sensors, i.e. the current observation vector $\mathbf{z}_{obs}^{k}$ from Eq. 5.1 are considered. If it is used the Bayesian rule, then the posterior distribution from Eq. 5.16 can be written in a view of Eq. 5.17.

$$P\left(\mathbf{p}_{mr}^{k}, \mathbf{M}_{lm} | \mathbf{z}_{obs}^{k}, \mathbf{c}_{mr}^{k}\right) = \eta P\left(\mathbf{z}_{obs}^{k} | \mathbf{p}_{mr}^{k}, \mathbf{M}_{lm}\right) P\left(\mathbf{p}_{mr}^{k}, \mathbf{M}_{lm} | \mathbf{z}_{obs}^{k-1}, \mathbf{c}_{mr}^{k}\right) \tag{5.17}$$

The common popular and wide spread estimation of the posterior distribution (Eqs. 5.16–5.17) in the SLAM method is the EKF [43]. The EKF represents the posterior distribution in the SLAM method as a high-dimensional, multivariate Gaussian parameterized by a mean $\boldsymbol{\mu}_{p,m}^{k}$ and a covariance matrix $\sum_{p,m}^{k}$ [4, 5] of mobile robot system state in the current discrete k time step (Eq. 5.18).

$$\boldsymbol{\mu}_{p,m}^{k} = \begin{bmatrix} \boldsymbol{\mu}_{p}^{k} \\ \boldsymbol{\mu}_{m}^{k} \end{bmatrix} \quad \sum_{p,m}^{k} = \begin{bmatrix} \sum_{p}^{k} & \sum_{p,m}^{k} \\ \sum_{m,p}^{k} & \sum_{m}^{k} \end{bmatrix} \tag{5.18}$$

The mobile robot position $\mathbf{p}_{mr}^{k}$ in the current discrete k time step can be represented as a non-linear motion model, in which $\mathbf{w}^{k}$ is the zero-mean Gaussian system noise with the covariance $\mathbf{Q}^{k}$ (Eq. 5.19).

$$\mathbf{p}_{mr}^{k} = f\left(\mathbf{p}_{mr}^{k-1}, \mathbf{c}_{mr}^{k}\right) + \mathbf{w}^{k} \tag{5.19}$$

The mobile robot observations $\mathbf{z}_{obs}^{k}$ in the current discrete k time step can be represented as a model, in which $\mathbf{r}^{k}$ is defined by the Gaussian measurement errors from visual, audio, and laser mobile robot sensors with the covariance $\mathbf{R}^{k}$ (Eq. 5.20).

$$\mathbf{z}_{obs}^{k} = h\left(\mathbf{p}_{mr}^{k-1}\right) + \mathbf{r}^{k} \quad \mathbf{R}^{k} = \begin{bmatrix} \sigma_{r}^{2} & 0 \\ 0 & \sigma_{\varphi}^{2} \end{bmatrix} \tag{5.20}$$

The nonlinear models of mobile robot position $\mathbf{p}_{mr}^{k}$ and observations $\mathbf{z}_{obs}^{k}$ presented by Eqs. 5.19–5.20 can be approximated as linear around the most likely system state $\boldsymbol{\mu}_{p,m}^{k}$, using Taylor expansion: $g \approx g + g'$ where g' is the Jacobian [44] of the function with respect to its variables.

With the above presented assumptions, the EKF-SLAM algorithm is executed as a recursive algorithm (Step 4 from Table 5.1) divided into two main steps: state estimation or prediction and state update or correction.

In the estimation or prediction state (Step 4.1 from Table 5.1) of the EKF-SLAM algorithm, the following operation are included and executed (choosing host CPU or GPU of mobile robot computer system): the estimated state vector $\bar{\mathbf{p}}_{mr}^{k}$ and covariance matrix calculated from the previous state and covariance and the control input $\sum_{\mathbf{k}}$ (Step 4.1 in Table 5.1):

$$\bar{\mathbf{p}}_{mr}^{k} = f\left(\mathbf{p}_{mr}^{k-1}, \mathbf{c}_{mr}^{k}\right) \quad \sum_{\mathbf{k}} = \mathbf{F}_{p,k} \sum_{k-1} \mathbf{F}_{p,k}^{T} + \mathbf{F}_{c,k} Q^{k} \mathbf{F}_{c,k}^{T}, \tag{5.21}$$

where

$$\mathbf{F}_{p,k} = \frac{\partial f}{\partial \boldsymbol{\mu}_{p,m}^{k-1}} \tag{5.22}$$

is the Jacobian of the state transition function f with respect to the mobile robot state model (Eq. 5.18), assuming that the estimated and updated map features regarding to the world reference frame are the same and the only time variant part of the mobile robot state is the robot state, i.e. Eq. 5.18 can be simplified as:

$$\boldsymbol{\mu}_{p,m}^{k} = \bar{\mu}_{m}^{k} \quad \sum_{p,m^{k}} = \sum_{m^{k}}. \tag{5.23}$$

In the state update or correction (Step 4.2 from Table 5.1) of the EKF-SLAM algorithm, the operations for processing the input in Step 3 of Table 5.1 are included and executed (choosing host CPU or GPU of mobile robot computer system). These operations include current information about the observations from audio, visual, and laser sensors presented as the observation vector or matrix $\mathbf{z}_{obs}^{k}$ (Eq. 5.1). Depending on the specifics of each of the audio, visual and laser sensors, the observed information is a subject of different processing methods. Nevertheless, the results of all them are intended to extract, estimate, and update for correction in each k time step (Step 4.2 from Table 5.1) and represented the following observation features (Fig. 5.6):

- The audio features as an angle of sound source arrival $\hat{\theta}_{DOA}^{F}$ or location of a sound source $\hat{\mathbf{v}}_{s}^{F} = \left[\hat{x}_{s}^{F}, \hat{y}_{s}^{F}, \hat{z}_{s}^{F}\right]$.
- The visual features as vector V_{Fea}^{F}.
- The laser range finder features as laser feature vector including depth or distance D_{LR}^{φ} in direction defined by angle φ_{LR} from the position of the mobile robot to the corresponding objects, persons, or obstacles.

After execution the Step 4.1 and Step 4.2 for estimation and correction mobile robot motion state as two main steps of the EKF-SLAM algorithm it is necessary to execute the final Step 5 in Algorithm 1 in Table 5.1 with two possibilities: to go on Step 2 for in loop or recursive execution the each next time step or to Stop algorithm execution.

The Algorithm 1 presented in Table 5.1 is the general description of the proposed fast parallel algorithm for decision making in the motion control systems of the mobile robots based on a computational analysis of the SLAM method. Each step of Algorithm 1 is analyzed for computational difficulties in order to decide, whether it is more convenient to execute the operations in corresponding step by CPU or GPU of mobile robot computer system (the last column on Table 5.1). Two possibilities for execution shown in the last column in Table 5.1, such as the execution only with CPU (without using GPU) or execution using GPU, are applied in the experiments in order to test and compare the execution of the developed fast algorithm for decision making in motion control system of a mobile robot based on audio visual information and using in calculations only with CPU or also NIVIDA GPU and CUDA platform.

5.3.3 Development of Fast Parallel Algorithm Based on Mobile Robot Audio Perception and Attention Managed by Speaking Person

In the Step 3 of Algorithm 1 (Table 5.1) for each k time step of mobile robot motion, the current information of observations obtained from mobile robot sensors and presented as the observation vector or matrix $\mathbf{z}^k_{obs}$ (Eq. 5.1) is used. One of the current observation from vector or matrix $\mathbf{z}^k_{obs}$ is the audio information described as a vector $\mathbf{z}^k_{ao}$. This information is very important for decision making in motion control system of a mobile robot especially in mobile robot navigation under the voice commands from a speaking person. Therefore, this section presents the description of developed fast parallel algorithm based on the mobile robot audio perception and attention for a decision making in motion control system managed by a speaking person. This algorithm can be considered as a special case of processing the observed in the Step 3 of Algorithm 1. The current audio features as an angle of sound source arrival $\hat{\boldsymbol{\theta}}^F_{DOA}$ or location of sound source $\hat{\mathbf{v}}^F_s = [\hat{x}^F_s, \hat{y}^F_s, \hat{z}^F_s]$ extracted in Step 4.2 of Algorithm 1 is necessary for the developed fast decision making algorithm.

The decision making in the motion control systems of mobile robots using the audio perception and attention can be considered as a very simple and raw model of human audio perception and attention [8]. The main principle in mobile robot audio perception and attention is based on calculation of sound directions of arrivals from the audio sensors or microphone arrays mounted on mobile robot moving platform [10]. First, the information of sound perception is analyzed. Second, the mobile robot audio attention is concentrated in the determined direction of a sound arrival, for example to follow the voice commands send from a person to the robot [11]. Therefore, the development an efficient real time working algorithm of mobile robot audio perception and attention requires to consider and to analyze the time consuming calculations mainly of the algorithms for precise determination of direction of sound arrival or sound source localization.

The methods and algorithms directing the sound arrival or sound source localization include Direction Of Arrival (DOA) [45], Time Delay Estimation (TDE) [46] also called Time Difference Of Arrival (TDOA) [47] or Inter aural Time Difference (ITD) [48], Generalized Cross Correlation (GCC) [49], and Generalized

Cross Correlation with PHAse Transform (GCC-PHAT) [50]. These methods are the subject of intensive research and applications not only in area of mobile robot audio perception and attention but also in other wide spread applications like audio visual conference systems, sound enhancements, sound recognition and identification, and sound masking. Their characteristics, especially their precision and accuracy in the determination of a direction of arrival or sound source localization, are studied and tested with the appropriate algorithms and simulations in all of the above mentioned applications including in mobile robot perception, attention and mobile robot motion control and navigation. The goal is to choose and analyze some of these methods in sense of their ability to work not only as simulation but in real time as parallel working algorithms for a decision making in mobile robot motion control using CUDA programming platform embedded in GPU devices. From the numerous listed methods, the most popular GCC and its modification called as the Steered Response Power (SRP)-PHAT [51] using the inverse Fourier transform of the cross-power spectral density function multiplied by a proper weighting were chosen. The description of the chosen for analysis and test the SRP-PHAT method is presented briefly. Let $s_1^M, s_2^M, \ldots, s_i^M, \ldots s_N^M$ be the input audio signals from N numbers of microphones M_1, M_2, …, M_i, …, M_N included in a microphone array and served as mobile robot audio sensors mounted on a mobile robot platform (Fig. 5.7).

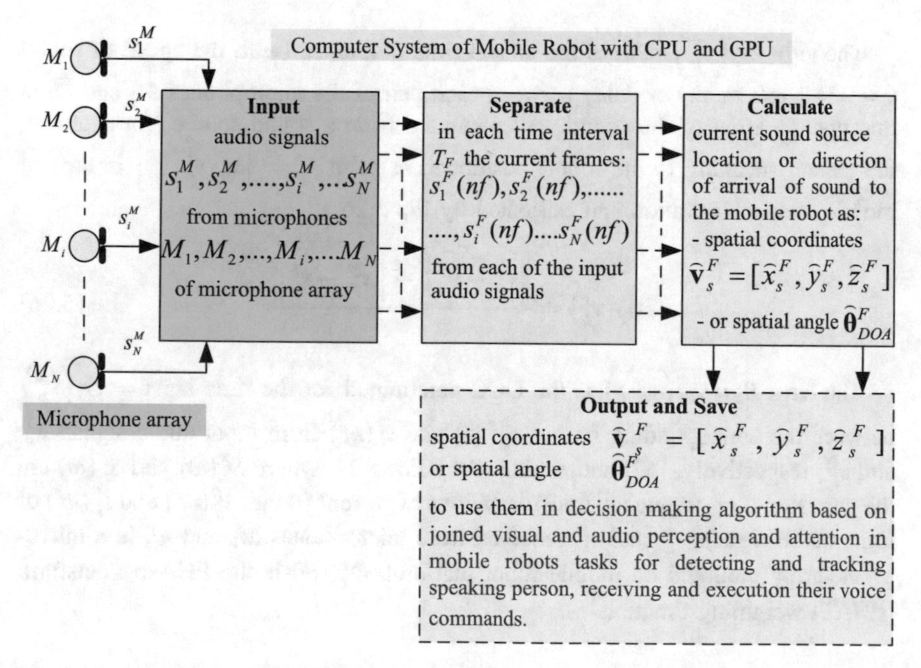

Fig. 5.7 Microphone array consisting from M_1, M_2, …, M_i, …, M_N microphones as the mobile robot audio sensors mounted on a mobile robot platform

In each time interval T_F, the input audio signals $s_1^M, s_2^M, \ldots, s_i^M, \ldots s_N^M$ are separated in the nfth current frames $s_1^F(nf), s_2^F(nf), \ldots, s_i^F(nf) \ldots s_N^F(nf)$. For each of these nfth current frames for a given point with spatial coordinates as vector $\mathbf{v}_p^F = \left[x_p^F, y_p^F, z_p^F\right]$ in area of mobile robot observation, the SRP $P_{nf}^{SR}(\mathbf{v}_p^F)$ can be determined using Eq. 5.24, where w_i^F is the weight suitable chosen for the corresponding $s_i^F(nf)$ frame from input audio signal s_i^M, $\tau\left(\mathbf{v}_p^F, i\right)$ is an arrival time of the sound waves direct from point $\mathbf{v}_p^F = \left[x_p^F, y_p^F, z_p^F\right]$ to the microphone M_i.

$$P_{nf}^{SR}\left(\mathbf{v}_p^F\right) = \int_{nfT}^{(nf+1)T} \left| \sum_{i=1}^{N} \left| w_i^F s_i^F(nf)\left(t - \tau\left(\mathbf{v}_p^F, i\right)\right) \right| \right|^2 dt \quad (5.24)$$

Equation 5.24 can be simplified by Eq. 5.25, where $\tau_{k,i}\left(\mathbf{v}_p^F\right)$ is an arrival time difference, $R_{k,i}\left(\tau_{k,i}\left(\mathbf{v}_p^F\right)\right)$ is the GCC determined for the time lag $\tau = \tau_{k,i}\left(\mathbf{v}_p^F\right)$.

$$P_{nf}^{SR}\left(\mathbf{v}_p^F\right) = \sum_{i=1}^{N} \sum_{k=i+1}^{N} R_{k,i}\left(\tau_{k,i}\left(\mathbf{v}_p^F\right)\right) \quad (5.25)$$

The term $\tau_{k,i}\left(\mathbf{v}_p^F\right)$ defines the arrival time difference (with the speed of sound $c = 384.2$ m/s in air) or delay between each pair of the microphones M_k and M_i in microphone array of the sound waves emitted from a sound source (for example, the person speaking to the robot) positioned at point $\mathbf{v}_p^F = \left[x_p^F, y_p^F, z_p^F\right]$ in area of mobile robot observation and estimated by Eq. 5.26.

$$\tau_{k,i}(\mathbf{v}_p^F) = \frac{\left\| \mathbf{v}_p^F - \mathbf{v}_k^F \right\| - \left\| \mathbf{v}_p^F - \mathbf{v}_i^F \right\|}{c} \quad (5.26)$$

The term $R_{k,i}\left(\tau_{k,i}\left(\mathbf{v}_p^F\right)\right)$ is the GCC determined for the time lag $\tau = \tau_{k,i}\left(\mathbf{v}_p^F\right)$ between the corresponding frames $s_k^F(nf)$ and $s_i^F(nf)$ from input audio signals s_k^M and s_i^M, respectively, [52] and evaluated by Eq. 5.27, where $S_k^F(\omega)$ and $S_i^F(\omega)$ are the corresponding frequency transformations of current frames $s_k^F(nf)$ and $s_i^F(nf)$ of input audio signals s_k^M and s_i^M received from microphones M_k and M_i in a microphone array mounted on mobile robot platform, $\Phi_{k,i}^F(\omega)$ is the PHAse Transform (PHAT) weighting function.

$$R_{k,i}(\tau) = \int_{-\infty}^{+\infty} \Phi_{k,i}^{F}(\omega) S_k^F(\omega) \left(S_i^F(\omega)\right)^* e^{j\omega\tau} d\omega \tag{5.27}$$

The PHAT weighting function is very effective for time delay estimation in sound source localization algorithms based on the GCC if in the area of mobile robot observation, when the reflected sound waves and reverberations exist. The PHAT weighting function is defined in frequency domain by Eq. 5.28.

$$\Phi_{k,i}^{F}(\omega) = \frac{1}{\left|S_k^F(\omega)\left(S_i^F(\omega)\right)^*\right|} \tag{5.28}$$

The calculated from Eq. 5.25 the SRP $P_{nf}^{SR}\left(\mathbf{v}_p^F\right)$ is used to determine a possible or approximately sound source location $\widehat{\mathbf{v}}_s^F = \left[\widehat{x}_s^F, \widehat{y}_s^F, \widehat{z}_s^F\right]$ as the maximum value of $P_{nf}^{SR}\left(\mathbf{v}_p^F\right)$ for the current time interval T_F provided by Eq. 5.29.

$$\widehat{\mathbf{v}}_s^F = \arg\max\left(P_{nf}^{SR}\left(\mathbf{v}_p^F\right)\right) \tag{5.29}$$

Equation 5.29 is useful information for possible absolute spatial coordinates $\widehat{\mathbf{v}}_s^F = \left[\widehat{x}_s^F, \widehat{y}_s^F, \widehat{z}_s^F\right]$ of sound source located on the corresponding distance from a mobile robot but more often the purpose of sound source localization is to determine only the direction, from which the sound from the source arrives to the robot. Therefore, in these cases, it is sufficient (instead of sound source absolute position $\widehat{\mathbf{v}}_s^F = \left[\widehat{x}_s^F, \widehat{y}_s^F, \widehat{z}_s^F\right]$) to determine only a vector of spatial angle $\boldsymbol{\theta}_s^F$ to this position, representing a direction from a sound source of arrival and consisting from two components: angle of azimuth $\theta_{s,az}^F$ and angle of elevation $\theta_{s,el}^F$:

$$\boldsymbol{\theta}_s^F = \left[\theta_{s,az}^F, \theta_{s,el}^F\right]. \tag{5.30}$$

Therefore, instead of definition a given point with the spatial coordinates as vector $\mathbf{v}_p^F = [x_p^F, y_p^F, z_p^F]$ in area of mobile robot observation it is possible to consider only the direction from mobile robot place to this given point $\mathbf{v}_p^F = [x_p^F, y_p^F, z_p^F]$ as a vector of the spatial angle $\boldsymbol{\theta}_p^F$ composed from two components: angle of azimuth $\theta_{p,az}^F$ and angle of elevation $\theta_{p,el}^F$ and define it using Eq. 5.31.

$$\boldsymbol{\theta}_p^F = \left[\theta_{p,az}^F, \theta_{p,el}^F\right] \tag{5.31}$$

Using a definition of a given point with spatial coordinates as vector $\mathbf{v}_p^F = \left[x_p^F, y_p^F, z_p^F\right]$ in area of mobile robot observation provided by Eq. 5.31 as a vector $\boldsymbol{\theta}_p^F$, it is necessary to modify the defined with the Eq. 5.26 delay $\tau_{k,i}\left(\mathbf{v}_p^F\right)$ between each pair of the microphones M_k and M_i in microphone array in a view of Eq. 5.32.

$$\tau_{k,i}(\boldsymbol{\theta}_p^F) = \frac{(\mathbf{v}_k^F - \mathbf{v}_i^F)}{c}\boldsymbol{\theta}_p^F \tag{5.32}$$

This leads also to modification of Eq. 5.25 for the SRP $P_{nf}^{SR}\left(\mathbf{v}_p^F\right)$ in the way provided by Eq. 5.33.

$$P_{nf}^{SR}\left(\boldsymbol{\theta}_p^F\right) = \sum_{i=1}^{N}\sum_{k=i+1}^{N} R_{k,i}\left(\tau_{k,i}\left(\boldsymbol{\theta}_p^F\right)\right) \tag{5.33}$$

The calculation the SRP as $P_{nf}^{SR}\left(\boldsymbol{\theta}_p^F\right)$ (Eq. 5.33) instead of $P_{nf}^{SR}\left(\mathbf{v}_p^F\right)$ (Eq. 5.17) require to change a determination of possible sound source location $\widehat{\mathbf{v}}_s^F = \left[\widehat{x}_s^F, \widehat{y}_s^F, \widehat{z}_s^F\right]$. Such sound source location ought to be calculated as the maximum value of $P_{nf}^{SR}\left(\boldsymbol{\theta}_p^F\right)$ for a possible spatial angle $\widehat{\theta}_s^F$, indicating only a direction of a sound source arrival from the place of sound source to the mobile robot position, using Eq. 5.34.

$$\widehat{\theta}_s^F = \arg\max\left(P_{nf}^{SR}\left(\theta_p^F\right)\right) \tag{5.34}$$

The mentioned above Eqs. 5.24–5.34 describe briefly the method SRP-PHAT. These equations are used to development the appropriate sound source localization algorithm as a real time execution with GPU processor and GUDA platform.

As it is seen from Fig. 5.7 this algorithm is implemented in computer system of mobile robot, which receive the input audio signals $s_1^M, s_2^M, \ldots, s_i^M, \ldots s_N^M$ from N numbers of microphones M_1, M_2, …, M_i, …, M_N included in microphone array mounted on a mobile robot platform. The actual description of the proposed sound source localization algorithm, its analysis and test for real time execution with GPU processor and GUDA platform must be preceded by some necessary general and usual steps as input, manipulations, and output audio signals and also results from sound source localization. These necessary steps are presented in Table 5.2 (Algorithm 2).

In Table 5.2, the proposition from the analysis how to estimate the advisability and effectiveness of the execution of the corresponding steps of algorithm by CPU or GPU is also presented. Therefore, Algorithm 2 helps to increase the speed of algorithm and achieve of source localization in a real time. From the presented in Algorithm 2 results, one can be concluded that the Step 5 is the basic operation in a

Table 5.2 The necessary general and usual steps as input, manipulations and output audio signals and also results from sound source localization (Algorithm 2)

Step	Step description	Execution by CPU or GPU
1.	**Initialize geometrical characteristics** of microphone array mounted on mobile robot platform and served as mobile robot audio sensors (Fig. 5.7): – **type of microphone array** (linear, circular, spherical, etc.); – **number of microphones** N in microphone array; distance $d_{i,j}$ between microphones M_i and M_j; – **local coordinate system** of microphone array with origin $\mathbf{O}_M$ as vector $\mathbf{O}_M = \left[x_M^o, y_M^o, z_M^o\right]$	CPU
2.	**Initialize time interval** T_F defined as time duration of the audio frames separate from each of input audio signals $s_1^M, s_2^M, \ldots, s_i^M, \ldots s_N^M$ captured from M_1, M_2, …, M_i, …, M_N microphones and processed as frame sequence to calculate the current direction of arrival of sound from a person speaking to a mobile robot	CPU
3.	**Input** continuous and synchronously in computer system of mobile robot (Fig. 5.1) all audio signals $s_1^M, s_2^M, \ldots, s_i^M, \ldots s_N^M$ from M_1, M_2, …, M_i, …, M_N microphones of microphone array	CPU
4.	**Separate** in each time interval T_F the current *nf*th frame $s_1^F(nf), s_2^F(nf), \ldots, s_i^F(nf), \ldots, s_N^F(nf)$ from each of the input audio signals $s_1^M, s_2^M, \ldots, s_i^M, \ldots s_N^M$ with the length of each frame $L_F = T_F/T_S$, where T_S is the sample time $T_S = 1/F_S$ and F_S is sample frequency of input audio signals	CPU
5.	**Calculate** *nf*th current direction of a sound arrival to a mobile robot (possible sound source location as spatial coordinates $\widehat{\mathbf{v}}_s^F = \left[\widehat{x}_s^F, \widehat{y}_s^F, \widehat{z}_s^F\right]$ or spatial angle $\widehat{\theta}_{DOA}^F$) as the result of processing (with GPU or with CPU algorithm) the corresponding current *nf*th frames $s_1^F(nf), s_2^F(nf), \ldots, s_i^F(nf), \ldots, s_N^F(nf)$ separated from input audio signals $s_1^M, s_2^M, \ldots, s_i^M, \ldots s_N^M$ captured from M_1, M_2, …, M_i, …, M_N microphones of microphone array	CPU or GPU
6.	**Output and Save** the results from calculation of *nf*th current direction of a sound arrival to the mobile robot (possible sound source location as the spatial coordinates $\widehat{\mathbf{v}}_s^F = \left[\widehat{x}_s^F, \widehat{y}_s^F, \widehat{z}_s^F\right]$ or spatial angle $\widehat{\theta}_{DOA}^F$) to analyze the real time capability and also use these results as suitable audio information in decision making algorithm based on the joined visual and audio perception and attention in mobile robots tasks for detecting and tracking speaking person, as well as receiving and execution the voice commands	CPU
7.	**Go to Step 4 or Stop** execution of algorithm	CPU

sound source localization. It is significantly more computing power consuming, therefore it is relevant and appropriate to execute and test using GPU and to compare with the execution time using CPU. The other steps (initializations, audio signals input, segmentations and output) are closely related with the CPU of mobile robot computer system and the microphone array connecter to CPU, therefore their execution is more appropriate by CPU. In the developed Algorithm 3 (Table 5.3), more precise the operations needed for realizing the Step 5 from Algorithm 2 (Table 5.2) are presented. These operations correspond to the interpretation in Algorithm 3 (Table 5.3) of the described briefly the method SRP-PHAT using the Eqs. 5.24–5.34.

The developed algorithms, such as Algorithms 1 EKF SLAM algorithm for motion control using the perceived audio visual perception and attention information from mobile robot audio and visual sensors, Algorithms 2 for sound source localization, and Algorithms 3 realizing the SRP-PHAT method for sound source localization, that are presented in Tables 5.1 and 5.3, respectively, are the subject of experiments as a practical realization and execution in two ways in mobile robot computer system: only with CPU, without using GPU, or using GPU of corresponding operations.

5.4 Experimental Results

The main goal of the developed Algorithm 1, Algorithm 2, and Algorithm 3 is to realize the task of decision making in mobile robot motion control with the corresponding precision using audio visual information. This goal is a subject of the previous researches [7, 8]. Therefore, the experimental results from these researches mainly as the precision of decision making in mobile robot motion control using audio and visual information are not presented here. These results are used in this chapter only as the base in the goal to increase the speed developing the fast variants of these algorithms applying NVIDIA GPU's and CUDA programming platform. The experiments in this chapter are directed to test and analyze the speed of the developed fast variants of Algorithm 1, Algorithm 2, and Algorithm 3 for the tasks of decision making in motion control system of a mobile robot using audio visual information applying only CPU or GPU's and CUDA programming platform.

The experiments are based on the presented in Fig. 5.6 the mobile robot video and audio perception system suitable for delivering the necessary environmental information to the mobile robot computer system. For experiments, the mobile robot platform of type Surveyor SRV-1 [53], which is equipped with the Blackfin camera [54] used as visual sensor, microphone evaluation board STEVAL-MKI126V2 based on STA321MPL and MP34DB01 microphones from MEMS type [55] used as audio sensor, and Hokuyo URG-04LX-UG01 scanning laser rangefinder [56] were utilized. Some of the experiments, requiring the general mobile robot algorithms, are based on the previous research and developments [7, 8, 11]. Also, two types of GPU processors, such as NVIDIA GeForce GTX560 [57] and NVIDIA GeForce

Table 5.3 The precise description of operations of the Step 5 in Algorithm 2 (Table 5.2) for realization of proposed with Eqs. 5.24–5.34 method the SRP-PHAT of sound source localization (Algorithm 3)

Step	Step description	Execution by CPU or GPU
1.	Input from Step 4 of Algorithm 1 (Table 5.1) the current frames $s_1^F(nf), s_2^F(nf), \ldots, s_i^F(nf), \ldots, s_N^F(nf)$ as array (with size $L_F \times N$, where L_F is length of frame, N is numbers of microphones) in CPU $\mathbf{s}_{mem}^{CPU}$: $\mathbf{s}_{mem}^{CPU}(nf) = s_1^F(nf), s_2^F(nf), \ldots, s_i^F(nf), \ldots, s_N^F(nf)$ Transfer array $\mathbf{s}_{mem}^{CPU}(nf)$ in GPU: $\mathbf{s}_{mem}^{CPU}(nf) \rightarrow \mathbf{s}_{mem}^{GPU}(nf)\,\mathbf{s}_{mem}^{CPU}(nf)$	CPU CPU to GPU
2.	Calculate the spectrums $s_k^F(\omega)$ and $s_i^F(\omega)$ of all current frames $s_k^F(nf)$ and $s_i^F(nf)$ contained in CPU $\mathbf{s}_{mem}^{CPU}(nf)$ or GPU $\mathbf{s}_{mem}^{GPU}(nf)$ arrays, respectively, applying the FFT and put the results in CPU $\mathbf{S}_{mem}^{CPU}(nf)$ or GPU $\mathbf{S}_{mem}^{GPU}(nf)$ arrays, respectively: $\mathbf{S}_{mem}^{CPU}(nf) = FFT\left(\mathbf{s}_{mem}^{CPU}(nf)\right)$, $\mathbf{S}_{mem}^{GPU}(nf) = FFT\left(\mathbf{s}_{mem}^{GPU}(nf)\right)$	CPU or GPU
3.	Calculate the PHAT weighting function $\Phi_{k,i}^F(\omega)$ applying Eq. 5.20 and using spectrums $S_k^F(\omega)$ and $S_i^F(\omega)$ contained in CPU $\mathbf{S}_{mem}^{CPU}(nf)$ or GPU $\mathbf{S}_{mem}^{GPU}(nf)$ arrays, respectively, and put the results in CPU $\mathbf{\Phi}_{mem}^{CPU}(nf)$ or GPU $\mathbf{\Phi}_{mem}^{GPU}(nf)$ arrays, respectively: $\mathbf{\Phi}_{mem}^{CPU}(nf) = \frac{1}{\left\lvert \mathbf{S}_{mem}^{CPU}(\omega)\,\left(\mathbf{S}_{mem}^{CPU}(\omega)\right)^{*}\right\rvert}$, $\mathbf{\Phi}_{mem}^{GPU}(nf) = \frac{1}{\left\lvert \mathbf{S}_{mem}^{GPU}(\omega)\,\left(\mathbf{S}_{mem}^{GPU}(\omega)\right)^{*}\right\rvert}$	CPU GPU
4.	Calculate the arrival time difference as a delay $\tau_{k,i}\left(\mathbf{v}_p^F\right)$ or a delay $\tau_{k,i}\left(\mathbf{\theta}_p^F\right)$ between each pair of the microphones M_k and M_i using Eq. 5.18 or Eq. 5.25, respectively: $\tau_{k,i}\left(\mathbf{v}_p^F\right) = \frac{\lVert \mathbf{v}_p^F - \mathbf{v}_k^F \rVert - \lVert \mathbf{v}_p^F - \mathbf{v}_i^F \rVert}{c}$, $\tau_{k,i}\left(\mathbf{\theta}_p^F\right) = \frac{\left(\mathbf{v}_k^F - \mathbf{v}_i^F\right)}{c}\mathbf{\theta}_p^F$	CPU or GPU

(continued)

Table 5.3 (continued)

Step	Step description	Execution by CPU or GPU
5.	Calculate in frequency domain the GCC $R_{k,i}(\tau)$ from Eq. 5.19 using time lag $\tau = \tau_{k,i}\left(\mathbf{v}_p^F\right)$ or $\tau = \tau_{k,i}\left(\boldsymbol{\theta}_p^F\right)$	CPU or GPU
6.	Transform the calculated in above step the GCC $R_{k,i}(\tau)$ from frequency to time domain using the IFFT	CPU or GPU
7.	Calculate the SRP using Eq. 5.25 or Eq. 5.33: $P_{nf}^{SR}\left(\mathbf{v}_p^F\right) = \sum_{i=1}^{N}\sum_{k=i+1}^{N} R_{k,i}\left(\tau_{k,i}\left(\mathbf{v}_p^F\right)\right)$ or $P_{nf}^{SR}\left(\boldsymbol{\theta}_p^F\right) = \sum_{i=1}^{N}\sum_{k=i+1}^{N} R_{k,i}\left(\tau_{k,i}\left(\boldsymbol{\theta}_p^F\right)\right)$	CPU or GPU
8.	**Calculate** the possible sound source location $\overset{\frown}{\mathbf{v}}_s^F = \left[\overset{\frown}{x}_s^F, \overset{\frown}{y}_s^F, \overset{\frown}{z}_s^F\right]$ as the maximum value of $P_{nf}^{SR}\left(\mathbf{v}_p^F\right)$ using Eq. 5.21 $\overset{\frown}{\mathbf{v}}_s^F = \arg\max\left(P_{nf}^{SR}\left(\mathbf{v}_p^F\right)\right)$ or possible spatial angle $\overset{\frown}{\theta}_{DOA}^F$ using Eq. 5.27 $\overset{\frown}{\theta}_s^F = \arg\max\left(P_{nf}^{SR}\left(\boldsymbol{\theta}_p^F\right)\right)$	CPU or GPU
9.	**Output** the following data to Step 6 of Algorithm 1 (Table 5.1): Calculated sound source location as: – spatial coordinates $\overset{\frown}{\mathbf{v}}_s^F = \left[\overset{\frown}{x}_s^F, \overset{\frown}{y}_s^F, \overset{\frown}{z}_s^F\right]$; – or spatial angle $\overset{\frown}{\theta}_{DOA}^F$	CPU
10.	**Return to** Step 6 of Algorithm 1 (Table 5.1)	CPU

GTX920 M [58], are applied. The GPU programming platform is based on the latest version CUDA Toolkit v8.0 [59].

The program implementation of the developed fast variants of Algorithm 1, Algorithm 2, and Algorithm 3 is based on programming language Microsoft Visual Studio Professional Version 2015 [60] and Matlab Release 2016b programming system [61]. The CUDA v8.0 programming model is completed with the software environment that allows to use C or C++ high-level programming language [62] or MATLAB GPU Computing Support for NVIDIA CUDA-Enabled GPUs [63] included in Parallel Computing Toolbox [64].

The experiments include the speed analysis of the developed fast variants of Algorithm 1, Algorithm 2, and Algorithm 3. For such analysis, a suitable methodology of estimating the execution time based on a comparison the computational costs was developed. This methodology provides a decision making in motion control system of mobile robot using audio visual information applying only CPU of mobile robot computer system or GPU's and CUDA programming platform. The proposed methodology includes all three developed Algorithm 1, Algorithm 2, and Algorithm 3 but for the briefness it is demonstrated only for the Algorithm 2 and Algorithm 3 in the following way.

From Eqs. 5.24–5.34 Algorithm 2 (Table 5.2) and Algorithm 3 (Table 5.3) one can see that a computation of a sound source location as position $\mathbf{v}_p^F = \left[x_p^F, y_p^F, z_p^F\right]$ or as angle $\widehat{\theta}_s^F$ using direction of arrival requires to process the pairs of the audio signals from the microphones in mobile robot microphone array. If the number of microphones in a microphone array is N, then the number of microphone pairs N_{MP} between each M_k, M_i microphone is estimated by Eq. 5.35.

$$N_{MP} = \frac{1}{2}N(N-1) \tag{5.35}$$

In the Step 2 of Algorithm 3 (Table 5.3), the spectrums $S_k^F(\omega)$ and $S_i^F(\omega)$ of all current frames $s_k^F(nf)$ and $s_i^F(nf)$ applying the FFT to the appropriate CPU or GPU arrays $\mathbf{s}_{mem}^{CPU}(nf)$ and $\mathbf{s}_{mem}^{GPU}(nf)$ are calculated and the results are put in CPU or GPU arrays $\mathbf{S}_{mem}^{CPU}(nf)$ and $\mathbf{S}_{mem}^{GPU}(nf)$, respectively. For number of microphones N in a microphone array, the arrays $\mathbf{s}_{mem}^{CPU}(nf)$ and $\mathbf{s}_{mem}^{GPU}(nf)$ contain N current frames, respectively, and, therefore, it is necessary to apply N number of the FFT to the current frames in the arrays $\mathbf{s}_{mem}^{CPU}(nf)$ and $\mathbf{s}_{mem}^{GPU}(nf)$. For the FFT of single frames $s_k^F(nf)$ or $s_i^F(nf)$ with the length of each frame $L_F = T_F/T_S$, where T_S is the sample time, $T = 1/F_S$, F_S is a sample frequency of input audio signals, the number of arithmetic operations N_{FFT}^{SF} are usually calculated using Eq. 5.36.

$$N_{FFT}^{SF} = 5L_F \log_2 L_F \tag{5.36}$$

This means that the FFT for all frames $s_k^F(nf)$ or $s_i^F(nf)$ in the arrays $\mathbf{s}_{mem}^{CPU}(nf)$ and $\mathbf{s}_{mem}^{GPU}(nf)$ requires the number of arithmetic operations N_{FFT}^{ALLF} provided by Eq. 5.37.

$$N_{FFT}^{ALLF} = NN_{FFT}^{SF} = N5L_F \log_2 L_F \tag{5.37}$$

The execution of the GCC $R_{k,i}(\tau)$ in the Step 5 of Algorithm 3 (Table 5.3) using time lag $\tau = \tau_{k,i}\left(\mathbf{v}_p^F\right)$ or $\tau = \tau_{k,i}\left(\boldsymbol{\theta}_p^F\right)$ is required for N_{MP} pairs spectrums $S_k^F(\omega)$ and $S_i^F(\omega)$ of corresponding N_{MP} pairs of current frames $s_k^F(nf)$ and $s_i^F(nf)$ of input audio signals from N_{MP} pairs between each M_k, M_i microphones. Let us denote this parameter as N_{GCC}^{ALLP} arithmetic operations under assumption that for the GCC $R_{k,i}(\tau)$ of a single frame pair the $N_{GCC}^{SP} = 6L_F$ arithmetic operations are necessary:

$$N_{GCC}^{ALLP} = N_{MP}N_{GCC}^{SP} = N_{MP}6L_F. \tag{5.38}$$

The calculation of the PHAT weighting function $\Phi_{k,i}^F(\omega)$ in CPU or GPU arrays $\mathbf{S}_{mem}^{CPU}(nf)$ and $\mathbf{S}_{mem}^{GPU}(nf)$ is pair wise operation related with the number of pairs N_{MP} between each M_k, M_i microphones in a microphone array and also depends from the length of each frame L_F. Therefore, the number N_{PHAT}^{ALL} for all pair's wise operations of the PHAT weighting function $\Phi_{k,i}^F(\omega)$ can be calculated using Eq. 5.39.

$$N_{PHAT}^{ALL} = N_{MP}L_F \tag{5.39}$$

The transformation of the GCC $R_{k,i}(\tau)$ from frequency to time domain using the IFFT is prepared for N_{MP} number of pairs and in this case each pair requires the following N_{IFFT}^{SF} arithmetic operations like the number of arithmetic operation N_{FFT}^{SF} for the FFT defined with Eq. 5.36:

$$N_{IFFT}^{SF} = 5L_F \log_2 L_F. \tag{5.40}$$

Therefore, for the IFFT of all number of pairs N_{MP} the number of arithmetic operations N_{IFFT}^{ALLF} can be calculated by Eq. 5.41.

$$N_{IFFT}^{ALLF} = N_{MP}N_{IFFT}^{SF} = N_{MP}5L_F \log_2 L_F \tag{5.41}$$

It can be determined that for calculation of a single value of $\mathbf{v}_p^F$ or $\boldsymbol{\theta}_p^F$ applied in Eq. 5.29 or Eq. 5.34, respectively, for calculations of the arrival time difference or delay $\tau_{k,i}\left(\mathbf{v}_p^F\right)$ or $\tau_{k,i}\left(\boldsymbol{\theta}_p^F\right)$ using Eq. 5.26 or Eq. 5.32, respectively, it is necessary $N_{PT}^{SV} = 20$ arithmetic operations. The number of calculations N_{PT}^{ALLV} for all values of $\mathbf{v}_p^F$ or $\boldsymbol{\theta}_p^F$ depends from the number of microphones N and can be defined by Eq. 5.42.

$$N_{PT}^{ALL} = N_{PT}^{SV} N = 20N. \tag{5.42}$$

If the calculation of values $\mathbf{v}_p^F$ or $\boldsymbol{\theta}_p^F$ in Eq. 5.29 or Eq. 5.34 for computing one single arrival time difference or delay $\tau_{k,i}\left(\mathbf{v}_p^F\right)$ or $\tau_{k,i}\left(\boldsymbol{\theta}_p^F\right)$ requires $N_{TD}^{SV} = 2$ arithmetic operations, then for all N_{MP} microphone pairs a total number of arithmetic operations can be estimated using Eq. 5.43.

$$N_{TD}^{ALLV} = N_{TD}^{SV} N_{MP} = 20N_{MP} \tag{5.43}$$

The number of arithmetic operations for a single SRP using Eq. 5.25 equals $N_{SR}^{SV} = 5$. Therefore, for all N_{MP} microphone pairs the required number of arithmetic operations can be calculated by Eq. 5.44.

$$N_{SR}^{ALLV} = N_{SR}^{SV} N_{MP} = 5N_{MP} \tag{5.44}$$

Using Eqs. 5.35–5.44, it is possible to calculate the total number of arithmetic operation N_{SL}^{T} as a sum of all arithmetic operations for sound source location using Algorithm 2 (Table 5.2) and Algorithm 3 (Table 5.3) provided by Eqs. 5.45–5.46.

$$N_{SL}^{T} = N_{FFT}^{ALLF} + N_{GCC}^{ALLP} + N_{PHAT}^{ALL} + N_{IFFT}^{ALLF} + N_{PT}^{ALLV} + N_{TD}^{ALLV} + N_{SR}^{ALLV} \tag{5.45}$$

$$\begin{aligned} N_{SL}^{T} = {} & N5L_F \log_2 L_F + N_{MP} 6L_F + N_{MP} L_F + N_{MP} 5L_F \log_2 L_F \\ & + 20N + 2N_{MP} + 5N_{MP} \end{aligned} \tag{5.46}$$

Equations 5.45 and 5.46 are used as a mean of calculating the number of arithmetic operations in practical implementation of the proposed Algorithm 2 (Table 5.2) and Algorithm 3 (Table 5.3) in sound source localization for mobile robot decision making. In a similar way, the equations (not presented here for briefness) for calculating the number of arithmetic operations in practical implementation of the proposed Algorithm 1 (Table 5.1) EKF SLAM for motion control using the perceived audio visual perception and attention information from mobile robot audio and visual sensors are obtained. The results from these calculations are used in comparison of effectiveness in real time implementation of these algorithms using GPU and CUDA platform instead of CPU.

The results achieved from the numerous experiments testing fast variants of Algorithm 1, Algorithm 2, and Algorithm 3 are summarized and briefly presented in this chapter in a comparative form (Tables 5.4 and 5.5) to demonstrate the effectiveness of using NIVIDA GPU and CUDA platform in time consuming operations of the developed fast algorithms.

It is seen from Tables 5.4 and 5.5 that the achieved in the experiments values for the execution time of Algorithm 1, Algorithm 2, and Algorithm 3 demonstrate clearly the advantages and real time execution of using NIVIDA GPU and CUDA

Table 5.4 The execution time comparison of Algorithm 1, Algorithm 2, and Algorithm 3 using NIVIDA GPU and CUDA platform or only using CPU of mobile robot computer system and applying C++ programming language

Execution time, s	Using CPU	Using GPU GeForce GTX560	Using GPU GeForce GTX920 M
Algorithm 1	0.0380	0.0060	0.0045
Algorithm 2	0.0157	0.0028	0.0024
Algorithm 3	0.0128	0.0021	0.0016

Table 5.5 The execution time comparison of Algorithm 1, Algorithm 2 and Algorithm 3 using NIVIDA GPU and CUDA platform or only using CPU of mobile robot computer system and applying Matlab programming system

Execution time, s	Using CPU	Using GPU GeForce GTX560	Using GPU GeForce GTX920 M
Algorithm 1	2.580	0.8560	0.6045
Algorithm 2	1.2318	0.4670	0.3590
Algorithm 3	1.0610	0.3789	0.2213

platform instead of CPU in the developed fast algorithm for decision making in mobile robot motion control based on audio visual information.

5.5 Conclusions

The precise analysis of the measured execution time of Algorithm 1, Algorithm 2, and Algorithm 3 in the proposed and developed fast parallel algorithm based on mobile robot audio visual perception and attention using NVIDIA GPU processor and CUDA platform in decision making for mobile robot motion control leads to the following conclusions:

- The execution time of Algorithm 1, Algorithm 2, and Algorithm 3 using NIVIDA GPU and CUDA platform is less than the execution time using CPU of mobile robot system.
- The execution time of Algorithm 1, Algorithm 2, and Algorithm 3 applying C++ programming language in NIVIDA GPU and CUDA platform is less than the execution time using the ability of Matlab programming system to work with NIVIDA GPU and CUDA platform.
- The execution time of Algorithm 1, Algorithm 2, and Algorithm 3 using GeForce GTX920 M NIVIDA GPU and CUDA platform is less than the execution time if the GeForce GTX560 NIVIDA GPU and CUDA platform is applied.
- For Algorithm 1 (which include the execution also of the Algorithm 2 and Algorithm 3) in Table 5.4, the values of execution time $T_{ecex}^{GPU} = 0.0060$ s using

GeForce GTX560 NIVIDA GPU and $T_{exec}^{GPU} = 0.0045$ s using GeForce GTX920 M NIVIDA GPU are less than the standard time T_{im}^{F} of an image frame provided by Eq. 5.47.

$$T_{exec} < T_{im}^{F} \tag{5.47}$$

- Equation 5.47 indicates the ability of real time execution of Algorithm 1, Algorithm 2, and Algorithm 3, when NIVIDA GPU's and CUDA platform are used.

Considering the summarized in Tables 5.4 and 5.5 experimental results, one can concluded that the achieved in the experiments values for the execution time Algorithm 1, Algorithm 2, and Algorithm 3 demonstrate clearly the advantages and real time execution of using NIVIDA GPU and CUDA platform instead of CPU in the developed fast algorithm for a decision making in motion control system of a mobile robot based on audio visual information.

The future research can be directed to improve the achieved effectiveness and possibility of real time execution of NIVIDA GPU and CUDA platform in the developed fast algorithm for a decision making in motion control system of a mobile robot based on audio visual information by the optimization of proposed algorithms and applying new autonomous and more suitable for mobile robot applications NVIDIA GPU and also new releases of CUDA programming with the extended facilities.

References

1. Siegwart, R., Nourbakhsh, I.R.: Introduction to Autonomous Mobile Robots. The MIT Press (2004)
2. Erdmann, M., Hsu, D., Overmars, M., Van der Stappen, A.F. (eds.): Algorithmic Foundations of Robotics VI. Springer, Heidelberg (2005)
3. Siciliano, Br., Khatib, O. (eds.): Springer Handbook of Robotics. Springer, Heidelberg (2008)
4. Dehkharghani, S.S., Bekiarski, A., Pleshkova, S.: Application of probabilistic methods in mobile robots audio visual motion control combined with laser range finder distance measurements. In: 11th WSEAS International Conference on Circuits, Systems, Electronics, Control & Signal Processing (CSECS'2012), pp. 91–98 (2012)
5. Dehkharghani, S.S., Bekiarski, A., Pleshkova, S.: Method and algorithm for precise estimation of joined audio visual robot control. In: Iran's 3rd International Conference on Industrial Automation (2013)
6. Venkov, P., Bekiarski, Al., Dehkharghani, S.S., Pleshkova, Sn.: Search and tracking of targets with mobile robot by using audio-visual information. In: International Conference on Automation and Informatics (CAI'2010), pp. 463–469 (2010)
7. Al, Bekiarski: Visual mobile robots perception for motion control. In: Kountchev, R., Nakamatsu, K. (eds.) Advances in Reasoning-Based Image Processing Intelligent Systems, ISRL, vol. 29, pp. 173–209. Springer, Berlin (2012)
8. Pleshkova, Sn., Bekiarski, Al.: Audio visual attention models in the mobile robots navigation. In: Kountchev, R., Nakamatsu, K. (eds.) New Approaches in Intelligent Image Analysis, ISRL, vol. 108, pp. 253–294. Springer International Publishing, Switzerland (2016)

9. Dehkharghani, S.S., Pleshkova, S.: Geometric thermal infrared camera calibration for target tracking by a mobile robot. Comptes rendus de l'Academie bulgare des Sciences **67**(1), 109–114 (2014)
10. Bekiarski, Al., Pleshkova, Sn.: Microphone array beamforming for mobile robot. In: 8th WSEAS International Conference on Circuits, Systems, Electronics, Control & Signal Processing (CSECS'2009), pp. 146–149 (2009)
11. Dehkharghani, S.S.: Development of methods and algorithms for audio-visual mobile robot motion control. Doctoral Thesis, Technical University, Sofia, Bulgaria (2013)
12. Cebenoyan, C.: GPU Computing for Games. Available from: https://developer.nvidia.com/sites/default/files/akamai/gameworks/CN/computing_cgdc_en.pdf. Accessed 6 May 2017
13. NVIDIA CUDA Compute Unified Device Architecture. Programming Guide. Version 1.0. Available from: http://developer.download.nvidia.com/compute/cuda/1.0/NVIDIA_CUDA_Programming_Guide_1.0.pdf. Accessed 6 May 2017
14. Nukada, A., Matsuoka, S.: Auto-tuning 3-D FFT library for CUDA GPUs. In: Conference for High Performance Computing Networking, Storage and Analysis, pp. 30.1–30.10 (2009)
15. Kijsipongse, E., U-ruekolan, S., Ngamphiw, C., Tongsima, S.: Efficient large Pearson correlation matrix computing using hybrid MPI/CUDA. In: Computer Science and Software Engineering (JCSSE'2011), pp. 237–241 (2011)
16. Podlozhnyuk, V.: Image Convolution with CUDA. NVIDIA. Available from: http://igm.univ-mlv.fr/~biri/Enseignement/MII2/Donnees/convolutionSeparable.pdf. Accessed 6 May 2017
17. HPC wire. Available from: https://www.hpcwire.com/2008/09/10/compilers_and_more_gpu_architecture_applications/. Accessed 6 May 2017
18. Cypher, R., Sanz, J.L.C.: The SIMD Model of Parallel Computation. Springer Publishing Company, Incorporated (2011)
19. NVIDIA CUDA Architecture. Introduction & Overview. Version 1.1 Available from: http://developer.download.nvidia.com/compute/cuda/docs/CUDA_Architecture_Overview.pdf. Accessed 6 May 2017
20. Sinha, R.S., Singh, S., Singh, S., Banga, V.K.: Accelerating genetic algorithm using general purpose GPU and CUDA. International Journal of Computer Graphics **7**(1), 17–30 (2016)
21. NVIDIA Developer CUDA Toolkit Documentation. Available from: http://docs.nvidia.com/cuda/cuda-samples/index.html#matrix-multiplication. Accessed 6 May 2017
22. Ding, C., Karlsson, C., Liu, H., Davies, T., Chen, Z.: Matrix multiplication on GPUs with on-line fault tolerance. In: 9th IEEE Int Symposium Parallel and Distributed Processing with Applications, IEEE Compute Society, pp. 311–317 (2011)
23. Restocchi, V.: Matrix-matrix multiplication on the GPU with NVIDIA CUDA. Available from: https://www.quantstart.com/articles/Matrix-Matrix-Multiplication-on-the-GPU-with-Nvidia-CUDA (2015). Accessed 6 May 2017
24. Fatahalian, K., Sugerman, J., Hanrahan, P.: Understanding the Efficiency of GPU algorithms for matrix-matrix multiplication. In: Akenine-Möller, T., McCool, M. (eds.) Graphics Hardware: The Eurographics Association, Stanford University, pp. 1–5 (2004)
25. CUDA—Matrix Multiplication. Available from: https://rohitnarurkar.wordpress.com/2013/11/02/cuda-matrix-multiplication/ Accessed 6 May 2017
26. Ramey, W.: Languages, APIs and Development Tools for GPU Computing. NVIDIA Corporation. Available from: http://www.nvidia.com/content/GTC/documents/SC09_Languages_DevTools_Ramey.pdf. Accessed 6 May 2017
27. Java bindings for CUDA. Available from: http://jcuda.org/ Accessed 6 May 2017
28. GPU Programming in MATLAB. https://www.mathworks.com/company/newsletters/articles/gpu-programming-in-matlab.html. Accessed 6 May 2017
29. NVIDIA Developer CUDA Toolkit Documentation. http://docs.nvidia.com/cuda/cuda-samples/index.html#simple-cufft. Accessed 6 May 2017
30. Gembris, D., Neeb, M., Gipp, M., Kugel, A., Männer, R.: Correlation analysis on GPU systems using NVIDIA's CUDA. J. Real-Time Image Proc. **6**(4), 275–280 (2011)
31. Papamakariosa, G., Rizosa, G., Pitsianisab, N.P., Sunb, X.: Fast computation of local correlation coefficients on graphics processing units. In: Luk, F.T., Schmalz, M.S., Ritter, G.

X., Barrera, J., Astola, J.T. (eds.) Mathematics for Signal and Information Processing. Proceedings of SPIE, vol. 7444, pp. 12.1–12.8 (2010)
32. Podlozhnyuk V. Image Convolution with CUDA. Available from: http://igm.univ-mlv.fr/~biri/Enseignement/MII2/Donnees/convolutionSeparable.pdf. Accessed 6 May 2017
33. Iandola, F., Sheffield, D., Anderson, M., Phothilimthana, P.M., Keutzer, K.: Communication-minimizing 2D convolution in GPU registers. In: IEEE International Conference on Image Processing (ICIP'2013), pp. 2116–2120 (2013)
34. Hizook (Robotics news for Academics and professionals). Available from: http://www.hizook.com/blog/2010/03/28/low-cost-depth-cameras-aka-ranging-cameras-or-rgb-d-cameras-emerge-2010. Accessed 6 May 2017
35. Hokuyo. Available from: https://www.hokuyo-aut.jp/02sensor/07scanner/urg_04lx_ug01.html. Accessed 6 May 2017
36. Pomerleau, D.A. Neural Nertwork perception for mobile robot guidance. Springer Science +Business Media LLC (1993)
37. Wang, Zh., Dissanayake, G., Huang, Sh.: Simultaneous Localization and Mapping: Exactly Sparse Information Filters. World Scientific Publishing Co. Pte. Ltd (2011)
38. Liénard, J.S.: Variability, ambiguity and attention: a perception model based on analog induction. In: Cantoni, V., di Gesù, V., Setti, A., Tegolo, D. (eds.) Attention in Human and Machine Perception 2: Emergence, Attention, and Creativity. Springer Science+Business Media LLC, pp. 87–99 (1999)
39. Levorato, R., Pagello, E.: DOA acoustic source localization in mobile robot sensor networks. IEEE International Conference on Autonomous Robot Systems and Competitions (ICARSC'2015), pp. 71–76 (2015)
40. Chou, Y., Liu, J.: A robotic indoor 3D mapping system using a 2D laser range finder mounted on a rotating four-bar linkage of a mobile platform. Int. J. Adv. Rob. Syst. **10**, 257–271 (2013)
41. Martínez-Gómez, J., Fernández-Caballero, A., García-Varea, I., Rodríguez, L., Romero-González, Cr.: A taxonomy of vision systems for ground mobile robots. Int. J. Adv. Robot. Syst. 11.1–11.11 (2014)
42. Naminski, M.R.: An analysis of simultaneous localization and mapping (SLAM) algorithms. Macalester College, Honors Projects Mathematics, Statistics, and Computer Science (2013)
43. Thrun, S., Burgar, W., Fox, D.: Probabilistic Robotics. Massachusetts Institute of Technology (2006)
44. Michiel, H.: Jacobian. Encyclopedia of Mathematics. Springer, Berlin (2001)
45. Dmochowski, Benesty, J.: Direction of arrival estimation using the parameterized spatial correlation matrix. IEEE Trans Audio Speech Lang. Process. **15**(4), 1327–1339 (2007)
46. Chen, J., Huang, Y., Benesty, J.: Time delay estimation. Audio signal processing for next-generation multimedia communication systems. In: Huang, Y., Benesty, J. (eds.) Audio Signal Processing for Next-Generation Multimedia Communication Systems, pp. 197–227. Kluwer Academic Publishers (2004)
47. Dvorkind, T., Gannot, S.: The time difference of arrival estimation of speech source in a noisy and reverberant environment. Sig. Process. **85**(1), 177–204 (2005)
48. Vonderschen, K., Wagner, H.: Detecting interaural time differences and remodeling their representation. Trends Neurosci. **37**(5), 289–300 (2014)
49. Tashev, I.: Sound Capture and Processing. Wiley (2009)
50. Dhull, S., Arya, S., Sahu, O.P.: Comparison of time-delay estimation techniques in acoustic environment. Int. J. Comput. Appl. **8**(9), 29–31 (2010)
51. Dmochowski, J., Benesty, J.: A generalized steered response power method for computationally viable source localization. Trans. Audio Speech Lang. Process. **15**(8), 2510–2526 (2007)
52. DiBiase, J.H.: A high-accuracy, low-latency technique for talker localization in reverberant environments using microphone arrays. Ph.D. thesis, Brown University (2000)
53. Surveyor SRV-1 Blackfin Robot Review. Available from: http://www.robotreviews.com/reviews/surveyor-srv-1-blackfin-robot-review. Accessed 6 May 2017

54. Surveyor SRV-1 Blackfin Camera Board. Available from: http://makezine.com/2007/10/26/surveyor-srv1-blackfin-ca/. Accessed 6 May 2017
55. ST. http://www.st.com/content/st_com/en.html. Accessed 6 May 2017
56. Scanning range finder URG-04LX-UG01. Available from: http://www.hokuyo-aut.jp/02sensor/07scanner/urg_04lx_ug01.html. Accessed 6 May 2017
57. Geforce. Available from: http://www.geforce.com/hardware/desktop-gpus/geforce-gtx-560. Accessed 6 May 2017
58. NVIDIA GeForce 920 M. Available from: https://www.notebookcheck.com/NVIDIA-GeForce-920M.138762.0.html..Accessed 6 May 2017
59. CUDA Toolkit 8.0. Available from: https://developer.nvidia.com/cuda-downloads/ Accessed 6 May 2017
60. Microsoft Visual Studio Professional Version 2015. Available from: https://www.visualstudio.com/downloads/. Accessed 6 May 2017
61. Matlab Release 2016b. Available from: https://www.mathworks.com/products/new_products/latest_features.html. Accessed 6 May 2017
62. CUDA C Programming Guide. Available from: https://docs.nvidia.com/cuda/cuda-c-programming-guide/index.html. Accessed 6 May 2017
63. MATLAB GPU Computing Support for NVIDIA CUDA-Enabled GPUs. Available from: https://www.mathworks.com/discovery/matlab-gpu.html. Accessed 6 May 2017
64. Parallel Computing Toolbox. Available from: https://www.mathworks.com/products/parallel-computing/. Accessed 6 May 2017

Chapter 6
Methods and Algorithms of Audio-Video Signal Processing for Analysis of Indoor Human Activity

Irina V. Vatamaniuk, Victor Yu Budkov, Irina S. Kipyatkova and Alexey A. Karpov

Abstract In this chapter, the methods and algorithms of audio and video signal processing for analysis of indoor human activity are presented. The concept of ambient intelligent space and several implementations are discussed. The main idea is the development of proactive information services based on the analysis of user behavior and environment. The methods of image processing, such as illumination normalization and blur estimation based on focus estimation methods and image quality assessment, are described afterwards. A short overview of face recognition methods including the principal component analysis, Fisher linear discriminate analysis, and local binary patterns is presented. Their efficiency was subjected to comparative analysis both in terms of the processing speed and precision under several variants of the area selected for image analysis, including the procedure seeking face in the image and limitation of the size of the zone of interest. Several approaches to audiovisual monitoring of meeting room participants are discussed. The main goal of the described system is to identify events in the smart conference room, such as the time, when a new user enters the room, a speech begins, or an audience member is given the floor. The participant registration system including face recognition accuracy assessment and recording system involving assessment results of the pointing of the camera, when photographing participants are presented.

I.V. Vatamaniuk (✉) · V.Y. Budkov I.S. Kipyatkova · A.A. Karpov
Saint-Petersburg Institute for Informatics and Automation of the Russian Academy of Sciences, 14-ya liniya V.O., 39, St. Petersburg 199178, Russian Federation
e-mail: vatamaniuk@iis.spb.su

V.Y. Budkov
e-mail: budkov@iis.spb.su

I.S. Kipyatkova
e-mail: kipyatkova@iis.spb.su

A.A. Karpov
e-mail: karpov@iis.spb.su

M.N. Favorskaya and L.C. Jain (eds.), *Computer Vision in Control Systems-4*,
Intelligent Systems Reference Library 136,
https://doi.org/10.1007/978-3-319-67994-5_6

Keywords Audio-video signal processing · Indoor human activity
Face recognition · Audiovisual monitoring · Speech recognition
Image blur

6.1 Introduction

Proactive information management services based on the analysis of user behavior and environment is the main idea of the concept of ambient intelligence. One example of such a smart space is an intelligent meeting room, which is equipped with a network of software modules, activation devices, and multimedia and audiovisual sensors. Awareness of the conference room about the spatial position of the participants, their current activities, the role of this event, and their preferences are provided by intelligent control of the integrated equipment. In practice, such smart conference rooms often operate in automatic mode with the support of expert operators. The mathematical and software support of the processing of audiovisual data during the monitoring of events in the conference room makes it possible to determine the current position of the participants during their presentations and stages of events, as well as to automate the personalized recording (journaling) of multimedia data.

One of the goals of audiovisual monitoring is to determine the time points (events), when the state of the participants of activities (their position in space and behavior) changes. These changes can occur, when a new participant enters the meeting room, the presentation begins, and the audience is given the floor. For small meetings, when all the participants are at the same table, personal or panoramic cameras are effective, but an increase in the number of participants leads to the expansion of space and increase of the cost of the equipment needed to carry out monitoring. Another goal of audiovisual monitoring is the automatic recording and analysis of audiovisual data reflecting events taking place in the room in order to determine the current situation in the intelligent meeting room. The automatic analysis of multimodal data collected is complicated by the fact that the participants may randomly change the position of their bodies, heads, and directions of their gaze. In order to ensure the capture and tracking of participants of the activities, Pan/Tilt/Zoom (PTZ) cameras and arrays of cameras and microphones are used. The location of the sound source using a microphone array is effective only in small lecture or conference rooms. Personal microphones for all participants or a system of microphone arrays distributed throughout the audience are used for recording in large rooms. The joint analysis of audio and video data recorded during monitoring in the conference room meetings is being studied in a number of current Russian and international research projects.

Section 6.2 briefly reviews the existing technologies and their implementations in various meeting rooms. Also, a basic sketch of the intelligent meeting room is provided. Section 6.3 presents modern technologies of digital signal processing and pattern recognition. We consider in more detail the problem of image blur and the methods enabling one to determine the quality reduction due to the image blur. Section 6.4 describes approaches to audiovisual monitoring of meeting room participants. The experiment results are presented in Sect. 6.5. Section 6.6 concludes the chapter.

6.2 Technologies and Frameworks of Smart Meeting Rooms

The research work of recent years in the development of prototypes of intelligent spaces (in most cases, intelligent meeting rooms) for scientific projects listed in Table 6.1 has been analyzed. Consider the functionality of these prototypes of intelligent meeting rooms. By functionality, we understand their equipment and methods for the realization of information support services and the automation of the support of activities. The sensory equipment of the intelligent meeting room can be divided into two types: the primary equipment for the recording of events in the entire space of the conference room and supporting equipment for a more detailed analysis of specific areas of interest identified by the primary equipment. A ceiling panoramic camera and a microphone array are usually used in intelligent meeting rooms as primary sensory equipment for audiovisual monitoring.

Auxiliary equipment includes a separate microphone and a high-resolution camera, covering different areas of the room, a PTZ camera used for keeping track

Table 6.1 List of organizations that are investigating prototypes of intelligent meeting rooms

No.	Organization	City, country	Abbreviation	URL
1	Polytechnic University of Catalonia	Barcelona, Spain	UPC	http://imatge.upc.edu/web/?q=node/106
2	University of Trento	Trento, Italy	ITC	http://www.fbk.eu/
3	Netherlands Organisation for Applied Scientific Research	Delft, The Netherlands	TNO	http://www.amiproject.org/
4	Information Technology Services Center of the University of Hong Kong	Hong Kong, China	ITSC	http://www.ust.hk/itsc/
5	Idiap Research Institute	Martigny, Switzerland	Idiap	http://www.idiap.ch/scientificresearch/smartmeetingroom
6	National Institute of Standards and Technology	Gaithersburg, United States	NIST	http://www.nist.gov/smartspace
7	Petrozavodsk State University	Petrozavodsk, Russia	PetrSU	http://petrsu.ru/
8	St. Petersburg Institute for Informatics and Automation of the Russian Academy of Sciences	St. Petersburg, Russia	SPIIRAS	http://www.spiiras.nw.ru/

of faces of certain participants in the room and recording of their talks, the built-in microphone of the personal web camera mounted on the conference table, etc. Only a single item of each of the above types of equipment is listed to show the necessary minimum configuration of the audiovisual monitoring system of the intelligent meeting room. The scalability of hardware and software makes it possible to adjust the entire range of equipment under specific operating conditions in the case of rooms of larger sizes, greater numbers of participants and observation areas, and higher requirements for the recording of activities.

Most intelligent meeting rooms are equipped with audio and video recording equipment, as well as with advanced I/O multimedia devices. Methods of image analysis (the recognition and identification of faces, detection and tracking of participants, and identification of positions), sound analysis (voice activity detection, localization and separation of sound sources, identification of speakers, and recognition of speech and non-speech acoustic events), and combined methods (estimation of the position and orientation of the speaker's head and multimodal recognition of emotions) are used in order to process the audiovisual signals recorded by equipment located inside the room. Now let us consider the equipment of rooms with some examples.

In UPC, ITC, and NIST [1, 2] two different types of cameras (fixed and PTZ cameras) are installed. In a UPC intelligent meeting room an optional ceiling panoramic camera is mounted. In addition, the listed rooms are equipped with the personal microphones and two different microphone arrays, namely T-shaped and Mark III [3]. TNO and Idiap intelligent meeting rooms are equipped with the stationary cameras and circular microphone arrays. The software based on the methods of identification of text notes on the board, as well as the identification and annotating of activities of participants and recognition of gestures and emotions of participants are used [2, 4]. The ITSC room is intended for the educational activities. There is only one stationary camera, and the methods for determining and tracking changes on the projection screen, identification and tracking of the movement of the lecturer, and the identification of their face are applied [5]. A prototype of the intelligent meeting room of the Petrozavodsk State University has been under development since 2012. Currently, it has equipment for presentations, audio and video recording, and climate sensors. Methods of personalized service for participants in research and education activities, who can use their mobile devices to control presentation equipment, are used [6, 7].

A multimodal database PROMETHEUS, containing data from heterogeneous sensors, is presented in [8]. It serves for human behavior analysis (both indoor and outdoor), contains multi-person scenarios and group interaction, and can be applied in the smart meeting room development.

When designing the intelligent meeting room, ergonomic aspects of the location of the multimedia, audio, and video recording equipment were taken into account in order to provide coverage and service for the largest possible number of participants. The sensory equipment of the SPIIRAS intelligent meeting room (Fig. 6.1) consists of four types of video cameras (two fixed cameras, two PTZ cameras, one ceiling panoramic camera, and ten personal webcams) and two types of audio

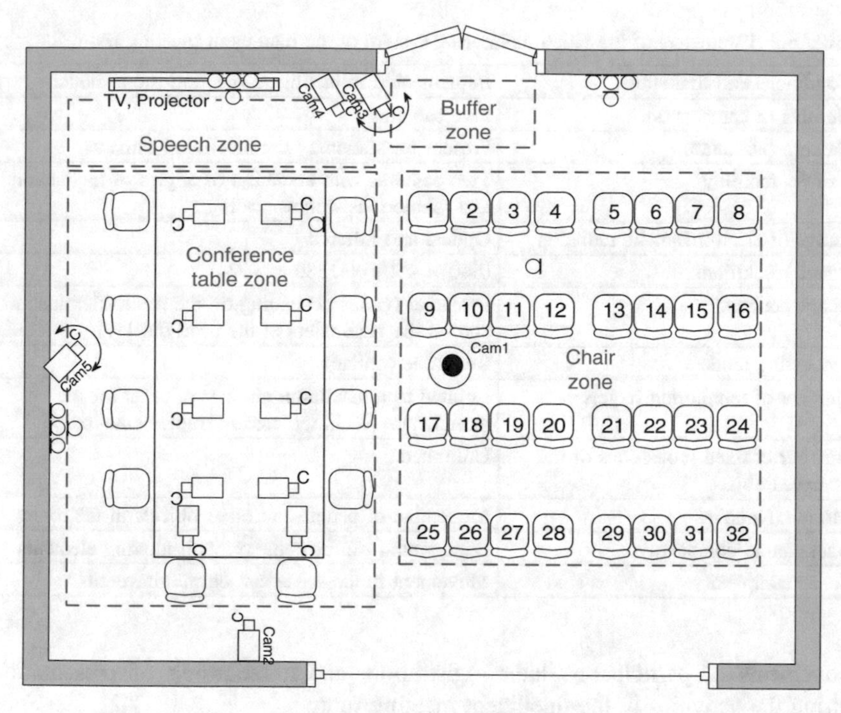

Fig. 6.1 Scheme of the smart room with an arrangement of cameras

recording equipment (two separate ceiling microphones and three microphone arrays with T-shape configuration). The system for monitoring participants in the intelligent meeting room was implemented based on the distributed processing of audiovisual signals. Both existing methods of digital data processing (image segmentation, calculation and comparison of color distribution histograms, etc.) and new methods developed by the authors were used, in particular, a method for the registration of participants in activities and a method for the audiovisual recording of their presentations [9]. The parameters of the video monitoring system of the SPIIRAS intelligent meeting room are presented in Table 6.2.

The method of the automation of the registration of participants using a multi-stage bimodal analysis of room areas makes it possible to detect the participants and obtain the audiovisual data necessary to identify those present. The method of audiovisual recording of the presentations of participants in the intelligent meeting room uses a bimodal approach (for audio and video) for the detection of the active speaker, which provides timely automatic capture and synchronization of audio and video data streams from the speaker. A software system for audiovisual monitoring and assessment of the situation in the room was developed based on the proposed methods. The method provides the processing of audiovisual data streams from a set of network cameras and microphone arrays for the detection and tracking of

Table 6.2 Parameters of the video monitoring system of the intelligent meeting room

Conditions and limitations	Implemented in intelligent meeting room model
Number of cameras	Five cameras
Camera calibration	Present for stationary unidirectional cameras
Camera mobility	Two cameras with alteration of angles of inclination and rotation, as well as scaling
Range of electromagnetic radiation	Optical and infrared
Image resolution	(640 × 480) − (1280 × 1024)
Image contrast	Dynamic (varies depending on the room illumination and on the parameters of the used display)
Processing mode	Real time (online)
Number of monitoring targets	Limited to a maximum number of seats for the participants (42 in the current implementation)
Number of fixed projections of the observed object	Unlimited
Motion dynamics	Movement of people and other objects in the room
Background complexity	Background can be complex with moving elements
Scene complexity	Movement in the scene can be multilayered

movements of participants, their registration, and a recording of presentations during the activities in the intelligent meeting room.

The following Sect. 6.3 describes some state-of-the-art image processing techniques for analysis of indoor human activity.

6.3 Image Processing Methods for Analysis of Indoor Human Activity

Image processing methods for analysis of indoor human activity include the image normalization methods presented in Sect. 6.3.1 and face recognition methods discussed in Sect. 6.3.2.

6.3.1 Image Normalization Methods

The automatic image processing usually lies in the frame preprocessing, including removal of low-quality frames, illumination normalization, cleaning of digital noise, etc. and extraction of image characteristics, segmentation and recognition of the images that are of interest in the given applied field. A need often occurs to select from a set of images the best frames meeting the requirements of a particular technical problem. This job makes no problem for an expert, but if the volume of data is great, then at the stage of preprocessing it is advisable to computerize the process of image quality estimation and screening of frames carrying no reliable data.

The methods for increasing image contrast, which improve the visual discernibility of the fragments by transforming the brightness scale and using the adaptive techniques of parameter adjustment, and application of these methods in the systems of digital processing of video information were analyzed in [10]. The methods of a priori estimation of image blur were proposed in [11]. The blur window is estimated for the case of Gaussian and uniform blur of the original monochromatic gray-scale image.

Three image components, such as contour, gradient, and texture ones, which are semantically valuable for perception by the human visual system, are considered in [12]. Consideration is also given to the image defects, such as blur of the light and color boundaries, presence of foreign patterns in the form of moire fringes caused by the noise of spatial discretization, presence of false contours due to an insufficient number of the quantization levels, and so on. The method of image restoration [12] relies on increasing sharpness on the boundaries and areas of the gradient variation of the intensity and noise filtration in the gradient areas. The algorithms to estimate the image quality are required practically in all fields of use of the computer vision technologies, such as the system of photo and video observation and registration, process control systems (industrial robots, self-sufficient transportation facilities), systems modeling objects and environment, systems of man-computer interaction, and so on [10, 13, 14].

Human perceives any image integrally, but judges its quality by the main Parameters, such as the brightness, contrast, prevailing tone, sharpness, and noisiness. In the case of computer-aided image processing, these parameters can be estimated either separately or integrally.

In the cases, where an array of similar images arrives to the input,—for example, at photo registration—not all quality indices are varied. At photographing moving objects by a fixed camera, the brightness, contrast, and noisiness may vary insignificantly, whereas estimation of the image sharpness comes to the foreground. Variations of sharpness may result from the errors of both the recording and translation hardware and the software for digital image processing. Image sharpness can be enhanced by focusing the camera on the object and using additional methods of digital processing of the original image. High-quality video hardware must reproduce the maximum-contrast brightness transition without distortions.

Image blur results from the diffuse mixture of two contrast colors. Human perceives and estimates sharpness by the presence of the contour of contrast brightness or tone transition between the neighboring areas in the image. If during photographing the camera was focused incorrectly or the object was moving, then the resulting image is partially or completely blurred.

The paper [15] presents a framework for solving the illumination problem in the face recognition tasks. Features extracted from face images may be divided into two categories: small-scale and large-scale; features may also be called as small and large intrinsic details, respectively. The first aim in [16] was the decomposition of face image into two features images by implementation of Logarithm Total

Variation (LTV) method with interpoint Second-Order Cone Program (SOCP) algorithm for approximate estimation of the LTV model [17]. The second aim was processing of decomposed face image elements for smoothing of small-scale and illumination normalization of large-scale features. For illumination normalization, the methods of Non-Point Light Quotient Image (NPL-QI) relighting and truncating Discrete Cosine Transform coefficients in logarithm domain (LOG-DCT) were separately employed and estimated [15]. All processed features are used for the reconstruction of a normalized face image. It shows [18] that implementation of the NPL-QI allows one to extend the illumination estimation from single light source to any type illumination conditions and simulate face images under arbitrary conditions. The LOG-DCT is not suitable for image normalization because it loses some low-frequency information during image restoration.

The paper [16] presents a method of Illumination Transition of an Image (ITI) for illumination estimation and image re-rendering. This method is based on the calculation of face similarity factor: the comparison of a more similar image with pre-referenced personal face image having a larger weight allows generating a person-specific reference face $T_{x,z}$:

$$T_{x,z} = \frac{\sum_i k_i A_i z + \varepsilon}{\sum_i k_i A_i x + \varepsilon}, \tag{6.1}$$

where $k = (k_1, k_2, \ldots, k_N)^{\mathrm{T}}$ is the weight coefficient, A_i is an array of images of subject i, x and z are current and expected illumination parameters, respectively, ε is a small constant ($\varepsilon = 1$), which is used in the case of being divided by zero [16].

The experimental results [16] show that current method may be used for illumination normalization of facial images as preprocessing in face recognition methods.

In continuation of works [15, 16, 18], the paper [19] presents the implementation of Empirical Mode Decomposition (EMD) [20] for illumination preprocessing of facial images. This method can adaptively decompose a complex signal into Intrinsic Mode Functions (IMFs), which are relevant to intrinsic physical significances. A face image may be represented as Lambertian reflectance assumption [15]:

$$I(x, y) = R(x, y)L(x, y), \tag{6.2}$$

where x, y are the coordinates of pixels in the image, $I(x, y)$ is the observed intensity, $R(x, y)$ is the reflectance (albedo), $L(x, y)$ is the shading or illumination component.

The right side of Eq. 6.2 is multiplicative and the EMD decomposes signal into the IMFs in an additive form. Thus, a factorization in the logarithm domain of the EMD (LEMD) was used in [16] for conversion of multiplicative model into additive:

$$f = \log I = \log R + \log L. \tag{6.3}$$

Because the EMD can decompose a signal into a set of the IMFs with different frequencies, the R and L components may be estimated according to Eq. 6.4, where d_k are the IMFs with different frequencies k, r is residue, K is the number of frequencies, K_0 is a settable parameter, in [19] $K_0 = 2$.

$$R = \exp\left(\sum_{k=1}^{K_0} d_k\right) \quad L = \exp\left(\sum_{k=K_0}^{K} d_k + r\right) \tag{6.4}$$

The complete algorithm for illumination normalization in facial images presented in [19] is mentioned below:

1. Computation of the logarithm of image.
2. Perform the 1-D EMD of gathered logarithm (experiment result in [19] shows that 1-D EMD method has better recognition performances).
3. Detection of shadow regions. In this stage, each IMF has been analyzed. There is shadow in a IMF if $D \cdot e^{m_{IMF}^2} > \theta_1$ is satisfied the binarization operator with threshold $\theta_1 = m_{IMF}/2$ is applied to analyzed IMF or residue, where D and m_{IMF} are variance and mean pixel values recently, as well as the threshold $\theta_1 = 0.1$ [19], and then each connected black area is marked as a shadow region.
4. Grayscale adjustment. Gray level substitution of each detected region by average gray level (with the use of Gaussian weighting) of its neighboring region.
5. Restoration of the image, using processed IMFs and residue, and its conversion to original image space.

For experiments, only frontal-face images from same databases as in work [18] were carried out.

The limited resolution of the forming system, defocusing, distorting environment (atmosphere, for example), motion of camera relative to the registered object, and so on are the main causes of the distortions leading to reduced sharpness [21]. The mathematical model of image formation is given by Eq. 6.5, where $z(x, y)$ is the output of the linear system of image generation, $n(x, y)$ is the two-dimensional additive noise.

$$s(x, y) = z(x, y) + n(x, y) \tag{6.5}$$

The image obtained by linear distortion of the original image in the absence of noise obeys the integral of the convolution:

$$z(x, y) = h(x, y) * u(x, y) = \iint_{\upsilon, v} h(x - \upsilon, y - v) u(\upsilon, v) d\upsilon\, dv, \tag{6.6}$$

where * is the symbol of 2D convolution, $h(x, y)$ is the two-dimensional pulse characteristic or Point Dissipation Function (PDF) of the linear distorting system.

Therefore, the value of the brightness function $u(\upsilon, \nu)$ of the original image at the point with the coordinates (υ, ν) is "smeared" depending on the form of the PDF $h(x, y)$ and distorted by the additive noise.

Reproduction of minor details characterizes the image sharpness, which is defined by the resolution of the forming system. At defocusing, a point is reproduced as a spot, and in the observed image two neighboring points merge into a single point. The blur value depends on the objective's focal distance, as well as on the distances from the objective to the object and the plane of the formed image. The discrete image is sharp (focused) if the blur diameter does not exceed the step of discretization of the observed image. Otherwise, the linear distortions become noticeable [21].

Therefore, one can judge the image blur by estimating its focus. Additionally, the complex quality estimation also enables one to reveal the blurred images. To estimate the blur, it is advisable to transform the color images into the achromatic ones and consider the rectangular matrix of the values of brightness of the image pixels. The size of this matrix can coincide with the image size in pixels or correspond to the size of the extracted image area.

The methods for image focus estimation can be categorized as follows [22]:

1. Methods based on the gradient estimation, where the gradient or the first image derivative characterizes the sharper brightness transitions at the boundaries of the objects in the image.
2. Methods based on Laplacian estimation, where the Laplacian or the second image derivative can be used to determine the number of sharp boundaries in the image.
3. Methods based on the wavelet transformations, where variation of the coefficients of the discrete wavelet transformation is used to describe the frequency and spatial image areas.
4. Methods based on considering the statistical image characteristics and estimating blur as a deviation from the normal distribution followed usually by the undistorted images.
5. Methods based on the discrete cosine transformations, where the coefficients of the discrete cosine transformation that are similar to the wavelet transformations estimate the image focus in the frequency domain.
6. Methods of integral estimation.

The aforementioned types of methods are distinguished for their complexity and image processing time. Therefore, the optimal method for image processing is selected depending on the knowledge domain and allocated computing resources and time. Now briefly consider four of the above methods, which have demonstrated the best experimental results at estimation of image blur and enable real time processing.

The Tenengrad method [23, 24] belongs to the category, where the image gradient is estimated. It relies on the estimation of the mean square of the brightness gradient of the pixels of a monochrome image carried out by the Sobel operator determining at each point an approximate vector of the brightness gradient [25]. The approximate values of the derivatives at each point are calculated using convolution of two rotation masks, orthogonal 3 × 3 matrices with the initial image in horizontal and vertical position, provided by Eq. 6.7, where $A(x, y)$ is the original image, $G_x(x, y)$ and $G_y(x, y)$ are the masks based on the Sobel operator and * is the symbol of 2D convolution.

$$G_x(x,y) = \begin{bmatrix} 1 & 0 & -1 \\ 2 & 0 & -2 \\ 1 & 0 & -1 \end{bmatrix} * A(x,y) \quad G_y(x,y) = \begin{bmatrix} 1 & 2 & 1 \\ 0 & 0 & 0 \\ -1 & -2 & -1 \end{bmatrix} * A(x,y). \tag{6.7}$$

The approximate values of the mean value of the squared gradient G_{TENG} over the processed image are given by Eq. 6.8.

$$G_{TENG} = \sum\sum\left(G_x(x,y)^2 + G_y(x,y)^2\right) \tag{6.8}$$

Increased mean value of the gradient square implies that there are sharp brightness transitions in the image and, consequently, higher sharpness of the boundaries. Relatively small amount of computations and high processing speed deserve mentioning among the advantages of the present method. It works best over a small image area where contrast transitions exist.

Now consider the method of image blur estimation based on measuring the local brightness contrast [26]. This measurement is carried out by calculating the ratio of intensity of each pixel of the monochrome image to the mean level of grey in the neighborhood of the given pixel provided by Eq. 6.9, where $I(x, y)$ is the considered pixel and $\mu(x, y)$ is the mean value of intensity of the pixels in its neighborhood.

$$R(x,y) = \begin{cases} \frac{\mu(x,y)}{I(x,y)} & I(x,y) \le \mu(x,y) \\ \frac{I(x,y)}{\mu(x,y)} & I(x,y) > \mu(x,y) \end{cases} \tag{6.9}$$

The size to the considered neighborhood centered at the point (x, y) is determined heuristically. The blur coefficient is equal to the sum of values of $R(x, y)$ over the entire image or over the considered image area.

Now briefly consider a method based on measuring blur by the curvature [26, 27]. The brightness matrix of the pixels of the monochrome image is represented as a 3D second-order surface, each point of which has two coordinates of the corresponding pixel and the value of its brightness. The curvature of this surface corresponds to the brightness transitions between the pixels and is approximated by Eq. 6.10.

$$f(x,y) = ax + by + cx^2 + dy^2 \tag{6.10}$$

The higher the value of curvature of the 3D surface, the more focused is the image under consideration. The coefficients a, b, c, and d are calculated approximately by the least-squares method through convolution of the original image with the matrices M_1 and M_2 provided by Eq. 6.11, where M'_1 and M'_2 are the transposed matrices M_1 and M_2, I is the initial image.

$$\begin{gathered} M_1 = \begin{pmatrix} -1 & 0 & 1 \\ -1 & 0 & 1 \\ -1 & 0 & 1 \end{pmatrix} \quad M_2 = \begin{pmatrix} 1 & 0 & 1 \\ 1 & 0 & 1 \\ 1 & 0 & 1 \end{pmatrix} \\ a = \frac{M_1 * I}{6} \quad b = \frac{M_1' * I}{6} \\ c = \frac{3}{10} M_2 * I - \frac{1}{5} M_2' * I \quad d = \frac{3}{10} M_2' * I - \frac{1}{5} M_2 * I \end{gathered} \tag{6.11}$$

The sum of the magnitudes of the coefficients estimates the image blur by Eq. 6.12.

$$G_c(x,y) = |a| + |b| + |c| + |d| \tag{6.12}$$

This method is not suited for uniform images with smooth transitions of brightness such as cloudless sky because in this case the curvature of the 3D surface varies insignificantly and cannot be an adequate mark of blur.

The last method is the Natural Image Quality Evaluator (NIQE) [28]. It is based on the statistical study of the natural images. By the natural are meant the images obtained with the use of photography, frame capture from video, and so on, that is, images that are not generated artificially and distorted by artificial noise. In the natural monochrome images, the matrix of normalized pixel brightness coefficients tends to the normal distribution. Any noise, blur including, leads to a deviation from the normal distribution. The concept underlying the method lies in comparing two multidimensional Gaussian models of attributes, the model of the considered image and that constructed on the basis of a set of images prepared in advance.

To calculate the attributes necessary to construct the model, the image pixel brightness coefficients are normalized by subtracting the local mean from the initial matrix of the brightness coefficients of the monochrome image followed by the division by the Root-Mean-Square Deviation (RMSD) in a view of Eq. 6.13, where $x \in \{1, 2,\ldots, M\}$ and $y \in \{1, 2,\ldots, N\}$ are the spatial indices, M and N are the image dimensions, $\mu(x, y)$ is the expectation, $\sigma(x, y)$ is the RMSD.

$$\hat{I}(x,y) = \frac{I(x,y) - \mu(x,y)}{\sigma(x,y) + 1} \tag{6.13}$$

The unit constant is introduced in the denominator to avoid the division by zero at zero value of the RMSD. The expectation and the RMSD are given by Eq. 6.14, where $w = \{w_{k,l}|k = -K,\ldots, K, l = -L,\ldots, L\}$ is the 2D circularly symmetric weight function, which is the discrete representation of the 2D Gauss function equal to $K = L = 3$ in the case at hand [25].

$$\begin{aligned}\mu(x,y) &= \sum_{k=-K}^{K}\sum_{l=-L}^{L} w_{k.l} I(x+k, y+l)\\ \sigma(x,y) &= \sqrt{\sum\nolimits_{k=-K}^{K}\sum\nolimits_{l=-L}^{L} w_{k.l}[I(x+k,y+l)-\mu(x,y)]^2}\end{aligned} \tag{6.14}$$

This normalization allows one to reduce substantially the dependence between the brightness coefficients of the neighboring pixels by reducing them to a form convenient for construction of the multidimensional Gaussian model. Since the clearness of the entire image is often limited by the hardware focus depth, it is advisable to decompose the considered image into areas of $P \times P$ pixels and then, upon estimating the local sharpness of each area b, choose the sharpest ones for further analysis. The local sharpness can be calculated in terms of the variance $\sigma(x, y)$:

$$\delta(b) = \sum_i \sum_j \sigma(x,y). \tag{6.15}$$

Experts determine the threshold of sharpness, with respect to which the local sharpness of the area is estimated. The selected areas overcoming the threshold are described by the generalized normal distribution with zero mean provided by Eq. 6.16:

$$f(x;\alpha,\beta) = \frac{\alpha}{2\beta\Gamma(1/\alpha)}\exp\left(-\left(\frac{|x|}{\beta}\right)^{\alpha}\right), \tag{6.16}$$

where $\Gamma(\cdot)$ is the gamma function

$$\Gamma(\alpha) = \int_0^{\infty} t^{\alpha-1}e^{-t}dt \quad \alpha > 0. \tag{6.17}$$

The parameters α and β can be estimated using the method of moments described in [29].

The deviation of the image model from the generalized normal distribution can be revealed by analyzing the products of pairs of neighboring coefficients of the normalized pixel brightness $\hat{I}(x, y)\hat{I}(x, y + 1)$, $\hat{I}(x, y)\hat{I}(x + 1, y)$, $\hat{I}(x, y)\hat{I}(x + 1, y + 1)$, $\hat{I}(x, y)\hat{I}(x + 1, y - 1)$, where $x \in \{1, 2,\ldots, M\}$, $y \in \{1, 2,\ldots, N\}$, along four directions—horizontal, vertical, and main and secondary diagonals. These parameters obey the generalized asymmetrical normal distribution provided by Eq. 6.18.

$$f(x;\gamma,\beta_l,\beta_r)=\begin{cases}\frac{\gamma}{(\beta_l+\beta_r)\Gamma\left(\frac{1}{\gamma}\right)}\exp\left(-\left(\frac{-x}{\beta_l}\right)^{\gamma}\right) & \forall x<0\\ \frac{\gamma}{(\beta_l+\beta_r)\Gamma\left(\frac{1}{\gamma}\right)}\exp\left(-\left(\frac{x}{\beta_r}\right)^{\gamma}\right) & \forall x\geq 0\end{cases} \quad (6.18)$$

The parameter γ controls the form of the distribution curve, β_l and β_r are the parameters controlling the left and right scatters, respectively, the coefficients γ, β_l, and β_r can be estimated efficiently using the method of moments. The mean value of distribution is also one of the attributes allowed for at model construction provided by Eq. 6.19.

$$\eta=(\beta_r-\beta_l)\frac{\Gamma\left(\frac{2}{\gamma}\right)}{\Gamma\left(\frac{1}{\gamma}\right)} \quad (6.19)$$

As a result of the above computations, we obtain a set of attributes that is compared with the multidimensional Gaussian model based on a set of different images of a certain quality in a view of Eq. 6.20, where $(x_1,\ldots, x_k)$ is a set of computed attribute, ν and Σ are the mean value and covariation of the matrix of the multidimensional Gaussian model calculated by the maximum likelihood method.

$$f(x_1,\ldots,x_k)=\frac{1}{(2\pi)^{k/2}|\Sigma|^{k/2}}\exp\left(-\frac{1}{2}(x-\nu)'\Sigma^{-1}(x-\nu)\right) \quad (6.20)$$

The image quality coefficient $D(\cdot)$ is given by Eq. 6.21, where ν_1 and ν_2 are the vectors of the mean values of the multidimensional reference Gaussian model and the model constructed for the tested image, respectively, Σ_1 and Σ_2 are the respective covariance matrices of these models.

$$D(\nu_1\nu_2,\Sigma_1,\Sigma_2)=\sqrt{(\nu_1-\nu_2)^{\mathrm{T}}\left(\frac{\Sigma_1+\Sigma_2}{2}\right)^{-1}(\nu_1-\nu_2)} \quad (6.21)$$

The coefficient $D(\cdot)$ indicates the mismatch between the models. The smaller $D(\cdot)$ the closer the distribution of the image at hand to the normal one; otherwise, there exists noise pollution of the image, which can be blur as well.

6.3.2 *Face Recognition Methods*

The majority of the face recognition methods include two basic steps, such as the determination of the position of user face in the image with simple or complex background [30] and analysis of the facial features enabling one to identify the user. The availability and position of the user face are determined by analyzing the pixel membership to the image foreground (area of face) and background [31]. Determination of the face area presents no difficulties with the images having pure background that means a uniform and single-color background. However, if the image consists of more than one layer with other objects, then the problem complicates. The face area is usually determined using methods based on determining the key face points, such as the eyes, nose, lips, or analyzing the image color space, as well as the methods using other characteristic facial features.

Segmentation of the face area is followed by the parameter normalization in the sizes, orientation, illumination, and other characteristics. To carry out normalization, it is important to determine the key face points, with respect to which parameters are corrected. The procedure of feature calculation and generation of the biometrical personal pattern to be stored in the face database can be executed only upon completion of the normalization procedure.

An interesting approach to face recognition is discussed in [32]. It represents an improvement of the non-negative matrix factorization method [33] by deriving a model preserving the local topology structure. This information serves to reveal latent manifold structure of face patterns. The main disadvantage of this approach is the computation complexity and high computing time. Despite the topology preserving, the non-negative matrix factorization method of face recognition is of some interest. However, it will not be discussed in the experimental part due to the fact that it cannot be applied in the real-time applications.

The Principal Components Analysis (PCA) [34], Linear Discriminant Analysis (LDA) [35], and Local Binary Pattern (LBP) analysis [36] represent the most popular current methods of face recognition. Now consider in more detail all three mentioned above methods.

The purpose of face recognition system is the division of the input signals (image files) into several classes (users) [37]. The application of such systems is relevant for a wide range of tasks: images and movies processing, human-computer interaction, identification of criminals, etc. Input signals may contain a lot of noise due to various Conditions, such as lighting, users pose, their emotions, different hairstyles. However, the input signals are not completely random and even more common features are partly present in each incoming signal. Among the common features in input data the following objects can be observed: the eyes, mouth, nose, and the relative distances between these objects. Such common features in the fields of research on face recognition are called eigenfaces [34] or principal components. The subspace of eigen features in the image is calculated according to Eq. 6.22, where Φ is an input image arranged as a vector, w_i are weight functions, u_i are eigenfaces.

$$\Psi = \sum_{i=1}^{K} w_i u_i \left(w_i = u_i^{\mathrm{T}} \Phi\right) \tag{6.22}$$

After that, the resulting matrix Ψ is converted into the eigenvectors of the covariance matrix C corresponding to the original face images:

$$\Omega = \begin{bmatrix} w_1 \\ w_2 \\ \dots \\ w_n \end{bmatrix} \quad C = \frac{1}{M}\sum_{n=1}^{M} \Phi_n \Phi_n^{\mathrm{T}}. \tag{6.23}$$

Further the Mahalanobis distance is calculated according to Eq. 6.24, where k is the number of used eigenfaces, λ is a scalar of eigenvalues.

$$e^r = \Omega - \Omega^k = \sum_{i=1}^{M} \frac{1}{\lambda_i}\left(w_i - w_i^k\right)^2 \tag{6.24}$$

After the calculation of e^r value, its comparison with the pre-selected threshold T^r is performed for the belongingness definition of the analyzed face to users' faces, which are added to the training database. For more details see [34].

The main idea of the LDA is to find such a linear transformation to separate the features into clusters after transformation [38], which is achieved due to the scattered matrix analysis [39]. For the problem of M-class scatter matrix S_b and S_w between-classes and within-classes are defined by Eq. 6.25:

$$\begin{aligned} S_b &= \sum_{i=1}^{M} Pr(C_i)(\mu_i - \mu)(\mu_i - \mu)^{\mathrm{T}} = \Phi_b \Phi_b^{\mathrm{T}}, \\ S_w &= \sum_{i=1}^{M} Pr(C_i)\Sigma_i = \Phi_w \Phi_w^{\mathrm{T}}, \end{aligned} \tag{6.25}$$

where $Pr(C_i)$ is the priori probability of class C_i, which takes the value $1/M$ with the assumption of equal priori probabilities, μ is the overall average vector, Σ_i is an average scatter of sample vectors of different classes C_i around their representation in the form of the mean vector μ_i:

$$\sum_i = E\left[(x - \mu_i)(x - \mu_i)^{\mathrm{T}} | C = C_c|\right]. \tag{6.26}$$

Distribution of class features can be calculated using a ratio of the scatter matrices S_b and S_w determinants provided by Eq. 6.27, where A is a matrix with size $m \times n$, $m \leq n$.

$$J(A) = \arg\max_{A} \frac{|AS_bA^{\mathrm{T}}|}{|AS_wA^{\mathrm{T}}|} \tag{6.27}$$

For optimization of Eq. 6.27, the approach described in the paper [40] is used. Thus, Eq. 6.27 becomes Eq. 6.28 [38].

$$S_b A^* = \lambda S_w A^* \tag{6.28}$$

Solution of the aforementioned Eq. 6.28 is to calculate the inverse matrix S_w. Solution of the eigenvalue problem for the matrix $S_w^{-1} S_b$ is described in [40]. However, this method is numerically unstable since it performs a direct appeal to the high dimension matrix of probabilities. In practice, the most commonly used LDA algorithm is based on finding matrix A, which can simultaneously diagonalize matrix S_b and S_w in a view of Eq. 6.29, where Λ is a diagonal matrix with elements sorted in descending order.

$$AS_wA^{\mathrm{T}} = I \ AS_bA^{\mathrm{T}} = \Lambda \tag{6.29}$$

For the classification problem, linear discriminate functions look as Eq. 6.30.

$$D_i(X) = A^{\mathrm{T}}(X - \mu_i) \quad i = 1, 2, \ldots, m, \tag{6.30}$$

More details are presented in [38].

Consider the application of local binary patterns in the face recognition problem [36]. The original LBP operator described in the paper [41] is intended to describe the structures due to marking the image pixels, which is based on the analysis of neighboring pixels relative to central pixel in 3 × 3 region size and presenting the analysis result of the region as a binary number. At the end of processing the whole image, a histogram can be formed on resulting mask, which is further used to describe the texture. Later the operator was extended to use neighborhoods of different sizes [42]. The application of neighborhoods with a circular shapes and bilinear interpolation of pixel values allows using any value of the radius R and the number of pixels P in the neighborhood (P, R) [36]. The histogram of image mask $f_l(x, y)$ can be calculated according to Eq. 6.31:

$$H_i = \sum_{x,y} I\{f_l(x, y) = i\} \quad i = 0, \ldots, n - 1, \tag{6.31}$$

where n is the number of different markers obtained during the application of the operator LBP, and

$$I\{A\} = \begin{cases} 1 & A \text{ is true,} \\ 0 & A \text{ is false.} \end{cases} \tag{6.32}$$

This histogram contains information about the local micro patterns, such as the edges, spots, and plane over the whole image. For effective face representation, a spatial information is used. For this, image is divided into regions $R_0, R_1, \ldots, R_{m-1}$ and the resulting histogram is defined by Eq. 6.33.

$$H_{i,j} = \sum_{x,y} I\{f_l(x,y) = i\} I\{(x,y) \in R_j\} \quad i = 0,\ldots,n-1 \quad j = 0,\ldots,m-1 \tag{6.33}$$

During the separation of the image, the analysis of information is performed, which is contained in each of the resultant field [43, 44], and in consequence of that, each area is assigned a weight; in this case, the chi-square [36] is calculated using Eq. 6.34, where w_j is weight value of region j.

$$X_w^2(S,M) = \sum_{i,j} w_j \frac{\left(S_{i,j} - M_{i,j}\right)^2}{S_{i,j} + M_{i,j}} \tag{6.34}$$

For more details see [36].

6.4 Approaches to Audiovisual Monitoring of Meeting Room Participants

In this Sect. 6.4, the method of participant localization, tracking, and registration (Sect. 6.4.1) and the method of audiovisual recording of participant activity (Sect. 6.4.2) are discussed in order to provide an audiovisual monitoring of meeting room participants.

6.4.1 Method of Participant Localization, Tracking, and Registration

Consider the method of the automation of registration of participants of activities $M_{auto}^{pre\_rec}$, in which three cameras (Cam1, Cam3, Cam4) are connected in series for the processing of each sitting participant (Fig. 6.1). Cameras Cam2 and Cam5 are intended for a speaker's image processing and are not involved in the registration process. There are three stages of the registration algorithm shown in Fig. 6.2. This method includes a plurality of video stream processing methods.

In the operation of the method of determining occupied chairs M_{v_c} the following methods are involved: the method of cropping the frame area M_{cut}, method of the creation of color distribution histograms M_{h_cre}, and method of comparing histograms M_{h_comp} in the form of Eq. 6.35.

$$M_{auto}^{par\_reg} = \left\langle M_{cut}, M_{h\_cre}, M_{h\_comp} \right\rangle. \tag{6.35}$$

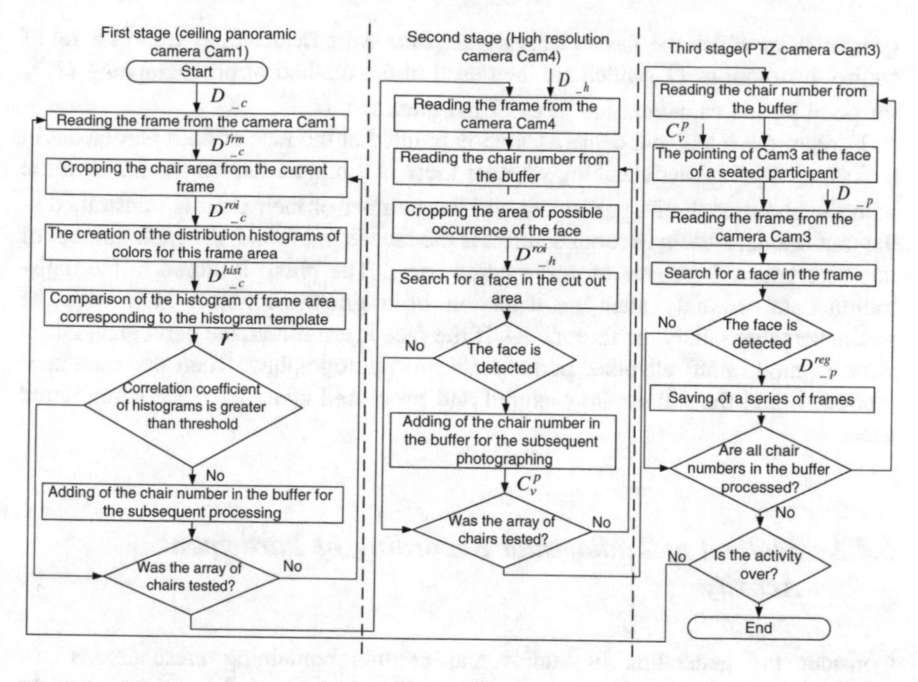

Fig. 6.2 Diagram of the participant registration automation

At the beginning of the first stage, a frame from the ceiling panoramic camera Cam1 is read out, then in a circuit according to the number of chairs in the conference room the chair area $D^{roi}_{v\_c}$ with predetermined sizes and coordinates is cropped from the frame $D^{frm}_{v\_c}$ of the video signal D_{v_c} using the method M_{cut}. After that, the method M_{h_cre} creates the color distribution histogram $D^{hist}_{v\_c}$ by the obtained field of the frame $D^{roi}_{v\_c}$. Next, using the method M_{h_comp}, the created histogram $D^{hist}_{v\_c}$ is compared with the prearranged reference histogram $D^{hist\_temp}_{v\_c}$ for the considered chair with the calculation of the correlation coefficient r provided by Eq. 6.36, where $H'_k(i) = H_k(i) - (1/Q)\left(\sum_j H_k(j)\right)$, Q is the number of histogram cells, k is the serial number of the compared histogram, i and j are numbers of the line and column of the histogram.

$$r = M_{h\_comp}\left(H_1 = D^{hist}_{v\_c}, H_2 = D^{hist\_temp}_{v\_c}\right) = \frac{\sum_i H'_1(i) * H'_2(i)}{\sqrt{\sum_i H'^2_1(i) * \sum_i H'^2_2(i)}} \qquad (6.36)$$

At the second stage, after the processing of all chairs is finished, the frame is read from the high resolution camera Cam4, in which using the method $M^{zpf}_{v\_h}$ the faces of participants are searched for in the areas of possible occurrence of faces $M^{roi}_{v\_h}$ corresponding to occupied chairs and the list of numbers of chairs $M^{zpf}_{v\_h}$ is generated, in which the faces of the participants were detected, as well as a set of control instructions C_v^p, which are then used in the method of photographing $M_{v\_p}^{reg}$ for pointing the camera at the faces of the participants.

Further, the intelligent camera Cam3 is pointed at the face of each participant in a close up $D_{v\_p}^{reg}$. After checking whether there is a human face in the frame, a file with the photograph of a participant and the number of their chair is transmitted in the multimodal system of room control. If the face is not found, the chair number of this user enters the queue of unregistered users. The photo is stored in the registration database only after the detection of a graphical object in it with the parameters that satisfy the face model. If the face is not found, the participant enters standby mode until all other participants are photographed. Then the camera is repointed, and the frames are captured and processed again even for unregistered users.

6.4.2 Method of Audiovisual Recording of Participant Activity

Consider the generation of audiovisual records containing presentations and statements of active participants in the discussion. As a result of the operation of the system of audiovisual monitoring of meetings, six databases are created that contain audio and video data of the event and its participants at the conference table, who are seated in rows of chairs arranged in the intelligent meeting room. The resulting databases are used for logging of the progress of activities and generation of multimedia content from the web system of teleconferencing support [45].

Next, consider an algorithm for pointing a camera at a current speaker or an active participant and the subsequent recording of their presentation. This algorithm uses a Multifunction System of Video Monitoring (MSVM) and a Multichannel System of Audio Localization (MSAL) that detects the position of the sound source based on the evaluation of the phase difference of signals recorded by pairs of microphones from four arrays [14, 46]. The MSVM searches for participants and tracks their movements inside the conference room. It also determines occupied chairs. Based on these data, the MSAL can assume that active participants are in these chairs during the discussions. During the activities in the intelligent meeting room, the MSAL and the MSVM systems work together. The diagram of interaction of multichannel audiovisual processing systems for recording of speeches of participants, who are sitting in the audience, is shown in Fig. 6.3.

When audio appears in the chair area, a detection of the position of the sound source is initiated, and a request for the verification of the presence of the participant in the chair closest to the coordinates of the detected sound source is sent to the MSVM (event E_1). If the chair is occupied, then the appropriate response is sent to the MSAL. At the same time, a unit for participant detection points the camera that serves the chair area at the active participant. If two persons are speaking

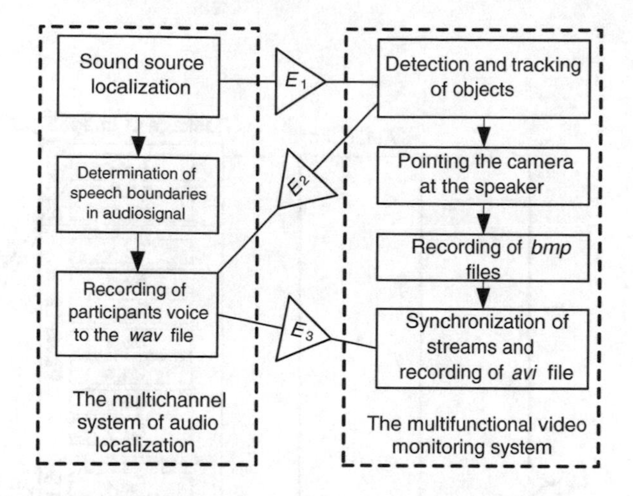

Fig. 6.3 Diagram of interaction of multichannel audiovisual processing systems in speech recording

simultaneously, the preference is given to the speaker recording. If there is no speaker, the preference is given to the person, who began to speak earlier.

In order to avoid missing participant statements, the audio signal is checked for speech every 20 ms. Short phrases and noise less than half a second are not considered. Thus, false statements are eliminated. The camera that records speech from the chair area has a recording speed of up to 30 frames per second. Frames from the camera are recorded in *bmp* files. The camera positioning takes a few seconds that are necessary to stabilize images after mechanical changes of the inclination, rotation, and scaling parameters. Therefore, frames are recorded in *bmp* files only after the camera is pointed properly. For this reason, in most cases the video recording of the presentation starts with a certain delay with respect to the audio signal. An example of the synchronization of audiovisual streams is presented in Fig. 6.4.

At the same time, the MSAL records the speech of the participant into a *wav* file if speech boundaries are defined and there is a positive response from the module that determines the presence of the participants that there is a participant in this chair (event E_2). After the *wav* file is created, a message containing the name of its location is sent to the MSVM (event E_3). Then the system reverts to the step of determining the position of the sound source.

If there are pauses in the speech of the participant, then the audio localization system will detect and record them in separate *wav* files for each statement. As a result, the MSAL can record multiple audio files containing speech from the same user. More precisely, the audio files will contain a signal coming from the chair assigned to a certain participant. The audio file name contains the number of chair, from which the speech signal was recorded.

The creation of the *avi* file begins five seconds after the statement or, more often, after the detection of a new active speaker in another chair, either at the conference table or in the presentation area. The main difficulty of *avi* file recording is to synchronize the audio and video files.

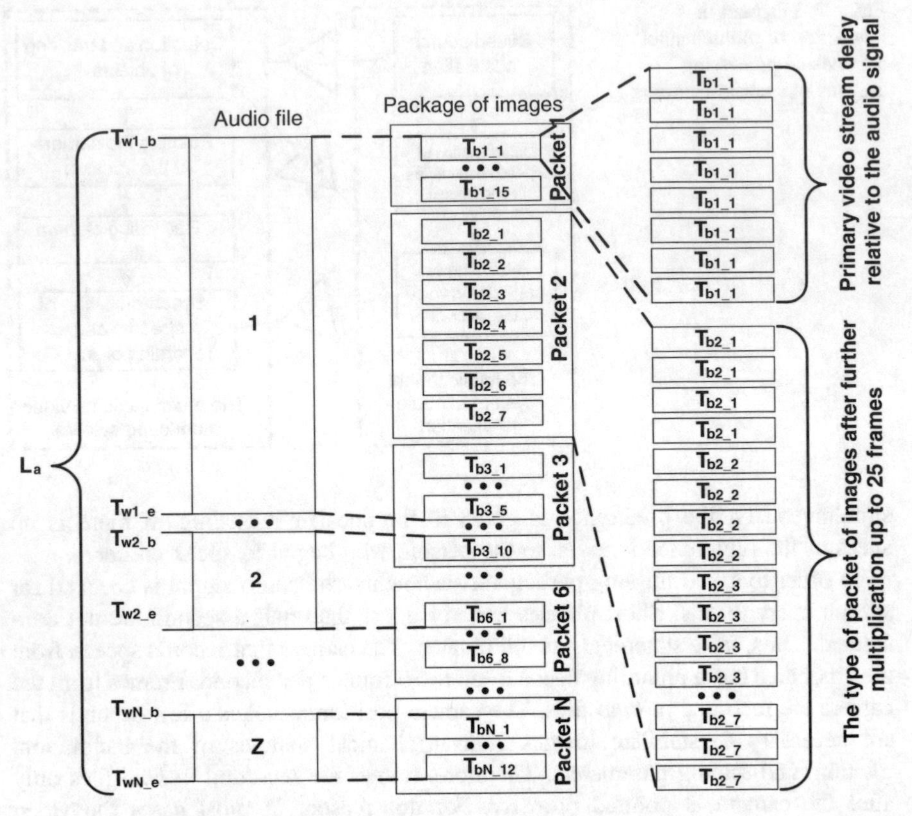

Fig. 6.4 Example of synchronization of audio and video streams in recording of the *avi* file

In order to achieve a standard frame rate of 25 frames per second, images corresponding to specific time intervals are reproduced. The *avi* file is edited during the batch processing of *bmp* files created in a time interval of one second. The data package consists of the recording time of *bmp* files, the number of the first frame, and the total number of frames in the package. The analysis of the structure is necessary in order to eliminate the asynchronous behavior that occurs when recording audio and video, since it makes it possible to calculate the number of additional frames. After the processing of all *bmp* and *wav* files selected and the addition of the multiplied images to the *avi* file, audio files are added as well.

After the event, all the presentations of the participants identified by the system are held in *avi* files. The described algorithm is used to record the presentations of the participants sitting in the chair area, which has 32 chairs and is located on the right side of the conference room. A similar algorithm is used to track the speaker in the presentation area. The method of recording the activity of the participants sitting at a conference table and the logical temporal model that merges multimedia content were developed in order to support teleconferences and to provide remote participants with multimedia content [46].

6.5 Experiments

During experiments, the assessments of the participant registration system (Sect. 6.5.1) and recording system (Sect. 6.5.2) were obtained.

6.5.1 Participant Registration System Assessment

The developed automatic participant registration system has two levels of face capture with different quality. At the first level, the rapid procedure of face recognition is used. It is based on a capture of one photo, which includes the view of all participants with low resolution and following face recognition. At this stage the image patches with participant faces have a resolution of around of 30 × 30 pixels. The faces unrecognized during the first level of processing are further separately captured by pan-tilt-zoom camera with a high resolution at the second level of registration system work. At that the captured face region has a resolution higher than 200 × 200 pixels.

There are two algorithms for image capturing and processing. In the first algorithm, a high-resolution image (1280 × 1024 pixels), which is a group photo of participants sitting in chairs, is processed for finding their faces by a face detection procedure [47] provided by Eq. 6.37, where h_t is a set of features $t = 1, 2,\ldots, T$, which are used in the Haar cascades, a_t is the features weight coefficient, I is an input image.

$$D_{v\_h}^{roi} \begin{cases} 1 & \sum_{t=1}^{T} a_t h_t(I) \geq \frac{1}{2} \sum_{t=1}^{T} a_t \\ 0 & \text{otherwice} \end{cases} \quad (6.37)$$

To reduce image processing time, (k) the region of the possible appearance of a participant's face was determined for each chair. Due to the fact that chairs have a static position, this procedure is performed once for the selected configuration of the chairs as follows.

```
{The beginning of the first algorithm}
for (k=0; k < Number_of_Chairs; k++)
    FaceRecognition[k] = FaceDetection(area[k]);
    if (FaceRegion[k])
        FaceRecognition(FaceRegion);
        SaveImage(FaceRegion);
    end if
end for
{The beginning of the second algorithm}
for (i=0; i < Number_of_Unregistered_Participants; i++)
    FaceRegion = FaceDetection(InputImage);
    if (FaceRegion)
        blurriness estimation(InputImage(FaceRegion));
{the blurriness estimation procedure}
        if (FaceRegion not blurred)
            FaceRecognition(FaceRegion);
            SaveImage(InputImage)
        end if
    end if
end for
```

Each founded region is processed by face recognition function, which allows identifying all participants with low charge of computational resource and time, because the average size of such region is around of 30 × 30 pixels. The second algorithm is aimed to identify unregistered participants, whose faces have not been recognized by the previous algorithm. The first step of this algorithm is capturing a close-up photo of a participant with a resolution of 704 × 576 pixels. Then the received photo is processed using face detection procedure. If the face region has been found, it needs to be estimated by blurriness criteria. For such estimation, the Natural Image Quality Assessment (NIQE) [28] method was implemented. The blurriness estimation procedure allows executing blurred photos from future processing. Such blurriness happens, when the participant moves in photographing moment or particularly closed by another participant, or when the camera, from which algorithm receives image, is not focused. If a face region is not blurred, then it is being processed using the face recognition procedure. If the participant has not been identified, then procedure for registration of a new participant starts, where his/her photo is used for focusing attention on audiovisual speech synthesis system.

Introduction of the blur estimation procedure as preliminary stage of photo processing allows the registration system to exclude 22% of photos with a high resolution but insufficient quality from face recognition stage, as a result the speed of the whole system is significantly increased. Implementation of blur estimation procedure at the first level of the processing of photos with a resolution of around 30 × 30 pixels has not shown positive results because such a low resolution is insufficient to make a decision about image blurriness.

For the experimental evaluation of a method of automatic registration of participants during the events in the intelligent room, the photos of participants were accumulated only at the second stage of the method. Thus, the number of accumulated photos was more than 40,000 for 36 participants. The training database contains 20 photos for each participant.

At the preparatory stage of experiments, it was decided to determine the threshold for the three face recognition methods, such as the PCA, the LDA, and the LBP. During this experiment, a threshold was calculated for each participant added to the recognition model; a maximum value of a correct recognition hypothesis for the LBP ranged from 60 to 100, for the PCA from 1656 to 3576, for the LDA from 281 to 858. As a consequence, for further experiments minimum threshold values like 60, 1656 and 281 were selected for these methods, respectively. Table 6.3 presents average values of face recognition accuracy, as well as first (False Alarm (FA) rate) and second (Miss Rate (MR)) errors type for each selected method.

The second experiment was performed to test the hypothesis about the necessity to use images with different participant's head orientations relative to the camera. The participants were photographed while sitting in chairs equidistant from the camera. As a result, photos of 16 participants have been added to the database for training face recognition models. In addition, the testing database expanded to 55,920 images. Average result values of the second experiment are shown in Table 6.4.

The high value of false positives and miss rate errors is due to the fact that the photos were stored in the course of actual of events without a prepared scenario and participants' focusing on a single object. Thus, at the time of photographing the participants can move freely, so their faces in the photos can be blurred or partially hidden.

To estimate the influence of a change of participant's face image size on recognition rate, it was decided to divide the participant's face images into several groups. Figure 6.5a shows distribution of participants' images by differences in the face sizes in ranges from 0 to 10, from 10 to 20, and so on. Most of participants have differences between minimum and maximum face sizes in a range from 30 to 40 pixels. Figure 6.5b shows a distribution of recognition rate for three methods for groups of participants.

Figure 6.5b shows that with increasing of the participant's face size in the image, the difference in the recognition accuracy gradually decreases. This is due to the

Table 6.3 Average values of face recognition accuracy, FA and MR for selected methods

Method	FA (%)	MR (%)	Accuracy (%)
LBPH	10.3	10.3	79.4
PCA	0.8	24.5	74.7
LDA	9.8	15.6	74.6

Table 6.4 Average values of face recognition accuracy, FA and MR for selected methods

Method	FA (%)	MR (%)	Accuracy (%)
LBPH	12.0	8.5	79.5
PCA	1.3	23.5	75.2
LDA	19.2	7.8	73.0

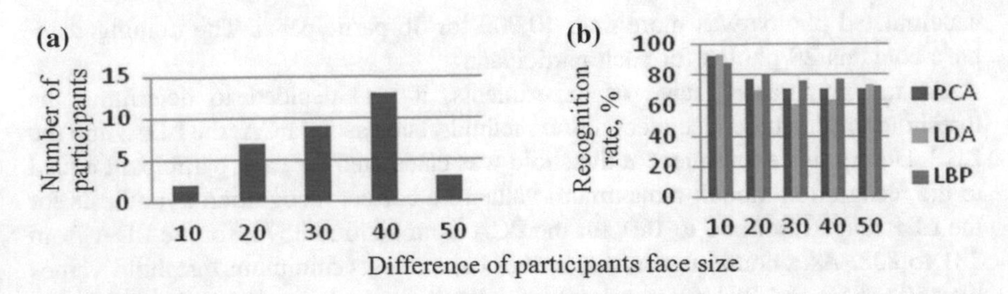

Fig. 6.5 Distribution of participants and recognition rate by differences in their face sizes

fact that at the normalization of images to a uniform size distortion in the certain facial features like the eyes, nose, mouth, and etc. may occur.

Considering experimental conditions (different distances from the camera to a participant, lighting, mobility of participants while being photographed), which influence the quality and quantity of facial features extracted from the image, and consequently the recognition accuracy and occurrence of false positives, we can conclude that the method LBP 79.5% has shown the best results.

6.5.2 Recording System Assessment

The complexity of the video monitoring system depends on the conditions and restrictions imposed by the specifics of the applied problem and geometrical characteristics of the monitored area, as well as by signal quality. Table 6.5 shows several parameters proposed in [48] and adapted to the problem of designing a video monitoring system of the intelligent meeting room. These parameters directly influence the choice of methods of video signal processing used in monitoring systems.

For the experimental assessment of the video monitoring system in the course of the activities in the intelligent meeting room, photographs of the participants taken during the registration process were collected, and experiments were conducted to assess the generation of audiovisual records containing the presentations and statements of active participants in the discussion.

During tests of the method of registration of participants of the activities in the intelligent meeting room, 21,584 photographs were made. After manual inspection, 1,749 photographs were found, in which there were no faces or faces were identified incorrectly. The main reason for the errors is the movement of the participant's head during photographing. The rest of the photographs were used to determine the speed (time) and quality of the camera pointing. Table 6.6 shows the performance evaluations of video monitoring system during the registration of participants. In order to evaluate the quality and speed of automatic photographing the relative S and absolute A face areas, as well as the time of registration (photographing) of the participant T, were calculated. In addition, two indicators were

Table 6.5 List of quantitative indicators of the system performance evaluation

Indicator	Description
A_p is the accuracy of detection of participants in the conference room	$A_p = (N_p - N_{FA\_p} - N_{MR\_p})/N_p$, where N_p is the maximum number of participants at the same time in the conference room, N_{FA_p} is the total number of mistakenly detected participants, N_{MR_p} is the number of missed participants
A_{o_ch} is the accuracy of the chair occupancy determination	$A_{o\_ch} = (N_{ch} - N_{FA\_ch} - N_{MR\_s\_p})/N_{ch}$, where N_{ch} is the number of chairs in the conference room, N_{FA_ch} is the number of wrong determinations of chair occupancy, $N_{MR\_s\_p}$ is the number of chairs with sitting participants missed by the system
A_{s_p} is the accuracy of detection of participants sitting in the chair area	$A_{s\_p} = (N_{ch} - N_{FA\_p\_f} - N_{MR\_p\_f})/N_{ch}$, where $N_{FA\_p\_f}$ is the number of mistakenly detected faces of participants sitting in the chair area, $N_{MR\_p\_f}$ is the number of faces of participants sitting in the chair area missed by the system
A_m is the accuracy of the operating mode determination	$A_m = N'_m/N_m$, where N_m is the number of mode switches detected during the activity, N'_m is the number of correctly determined modes
A_{m_s} is the accuracy of the pointing of the camera at the speaker in the presentation area	$A_{m\_s} = N'_{m\_s}/N_{m\_s}$, where N_{m_s} is the total number of frames taken while taking photographs of the speakers in the presentation area, N'_{m_s} is the number of the frames in which speaker can be detected

assessed: the accuracy of the camera towards the face and the size of the resulting photograph. At each stage of processing the displacement D of the center of the face from the center of the frame was calculated using the face detection algorithm. In order to avoid zoom influence, the displacement value was determined as a percentage of the frame size. When calculating the displacement D and size R of the faces in the frame, the formulas, described in [49], are used.

Based on obtained data, it is possible to say that automatic registration is not inferior in quality and is superior in speed than manual registration. In addition, the presence of the operator while taking pictures distracts participants from the activities and faces in such photographs are often overly formal. In the case of automatic registration, the photographs are taken discreetly to enhance the quality of the photographs in aesthetic terms and make it possible for participants to concentrate on the activities.

In evaluating the performance of the method of audiovisual recording of presentations, the following indicators were used. The initial video stream delay with respect to the audio signal L_{b_d}, which is calculated as the difference between the time of the creation of T_{w1_b} of the first *wav* file and the time of creation of *bmp* T_{b1_1} file of the corresponding time T_{w1_b}:

Table 6.6 Quality assessment results of the pointing of the camera, when photographing participants

R (%)			*D* (%)			*S* (%)			A, Mpixels			*T*, s		
Min	Max	Mean	Min	Max	Mean	Min	Max	Mean	Min	Max	Mean	Min	Max	Mean
24	73	29	0	36	9	6	54	11	0.03	0.23	0.05	0.03	49.75	1.3

$$L_{b\_d} = |T_{w1\_b} - T_{b1\_1}|. \tag{6.38}$$

The duration of the recorded *avi* file L_α is calculated by summing the lengths of all *wav* files of the presentation:

$$L_a = \sum_{i=1}^{N} (T_{wi_e} - T_{wi_b}). \tag{6.39}$$

The number of multiplied frames is equal to the sum L_{b_d} and all further multiplied images in data packets P_i provided by Eq. 6.40.

$$N_{f\_d} = \frac{L_{b\_d}}{40} + \sum_{i=1}^{N} P_i \quad P_i = P_{AF\_i} + P_{RF\_i} \quad P_{AF\_i} = \frac{(P_{FN\_i} - P_{F\_i})}{P_{F\_i}}$$
$$P_{RF\_i} = (P_{FN\_i} - P_{F\_i})\% P_{F\_i} \quad P_{NF\_i} = F_D * \frac{(T_{bN\_i} - T_{b1\_i})}{1000} \tag{6.40}$$

The average frame rate F_α in the video buffer is equal to the sum of data packet size divided by the number of packets:

$$F_a = \frac{\sum_{i=1}^{N} F_i}{N}. \tag{6.41}$$

The software system developed is regularly used during workshops and lectures. Expert reviews were made on 16 activities based on the results of the method of audiovisual recording of the participants' presentations in the intelligent meeting room, which created 212 recordings in the automatic mode, of which 93% contained statements from participants sitting in the audience. The results of the evaluation of files with presentations of participants are presented in Table 6.7.

Experimental results showed that the average recorded *avi* file consists of 137 frames, 59 of which are added optionally, and has a total length of about 5 s. The calculated average frame rate in the video buffer is 24 frames per second. This is due to the rounding of values, when calculating the required number of additional frames in the data packets. The number of multiplied frames also includes initial delay between audio and video streams. In addition, this number of additional frames is associated with varying frequency of reading of frames from cameras as a result of interference in the network equipment, as well as the limited speed of data recording on storage devices. The database of photographs of faces of participants

Table 6.7 Assessment results of performance of the algorithm for determining and recording a current speaker

L_{b_d}, ms			L_a, ms			N_{f_d}, frames		
Min	Max	Mean	Min	Max	Mean	Min	Max	Mean
80	2440	724	5312	6432	5608	32	104	59

and their presentations obtained by the system of audiovisual monitoring in the intelligent meeting room are the basis of the multimedia report of the activities.

Experimental data on the participants were collected in the course of the simulation of activities, where users held a meeting according to a given scenario, as well as during real research and education activities, when the participants were informed of audiovisual recording of their behavior but it did not affect their planned activities in an intelligent meeting room. For the simulation of activities according to a given scenario, a group of people was selected. The number of people in the group ranged from 5 to 10 people depending on the activity. The plan of artificial activities consisted of two stages. In the first phase of the experiment, the participants were asked to take any chair, and then an additional participant entered the intelligent meeting room, acting as the lecturer. The "lecturer" sequentially asked the audience ten questions without assigning a specific person to answer them. Areas of knowledge known to participants were taken into account in the selection of questions. Furthermore, the question was worded in such a way that any participant was able to answer it. Thus, a possibility of the absence of activity due to ignorance was excluded. After answering all the questions, the participants of the experiment exited the intelligent meeting room.

In the second phase of the simulated activities, the participants were asked to sit on any other chair, on which they had not yet sat. The remaining part of the experiment was similar to the first stage. The main objective of this experiment was to identify the places that users with the highest level of audio activity took during the event. Figure 6.6 shows a diagram of an intelligent meeting room with colored

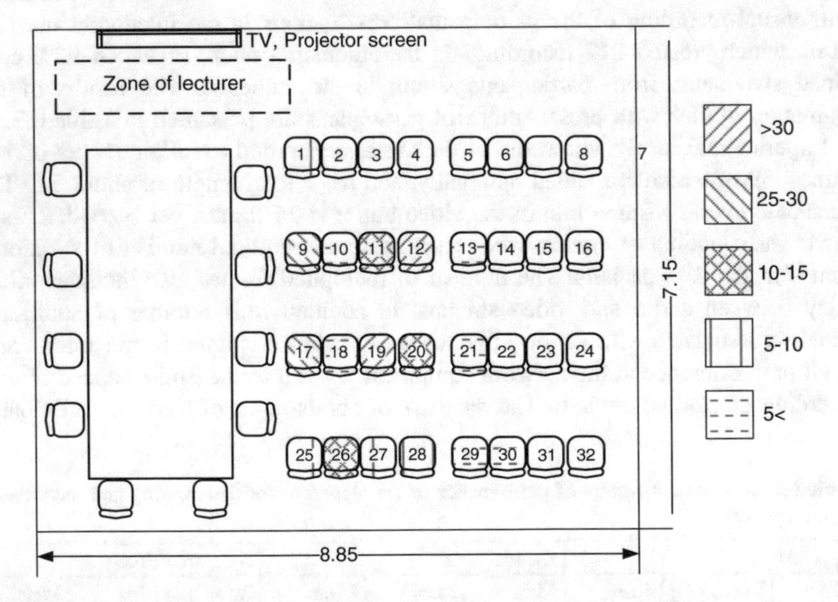

Fig. 6.6 Diagram of the intelligent meeting room with indications of audio activity of participants in the process of discussion

chairs indicating the participants, who were most active in answering questions. Chairs that were empty in the course of event were marked with white.

The second category includes activities conducted without the scenario and the list of questions for the participants. Another feature of actual activities is the high level of audio noise in the conference room, statements that do not relate to the subject, and overlapping speech in the audio signal simultaneously uttered by several active participants and/or the speaker. All these factors can affect the accuracy of the sound source positioning system and detection of false sound sources.

In automatic mode, 212 audio activities were recorded and analyzed. Expert inspection found that 15 of them contain extraneous noise or the speech of the speaker (e.g. lecturer). Thus, in order to obtain the results, only 197 speech signals with statements of participants sitting in the conference room were analyzed. As a result, it was found that almost half of the chairs with participants did not exhibit activity. A maximum density of activity was found in the center of the left half of the rows of chairs. In connection with the conducted investigation, we can plot the priorities of the initial position of the camera that records an active participant in the discussion given the current situation and the presence of the participants in the chairs.

Intelligent technologies for supporting activities and recording, archiving, and retrieving of data by created multimedia archives have been under development for over 25 years. One of the first research projects devoted to solving this problem was Nick [50], in which the requirements for intelligent technologies that are necessary to create an intelligent meeting room were first analyzed.

Based on the analysis of systems described in detail in [51], the following conclusions, useful in the design of conference automation systems, can be made. First, the recording should be unobtrusive for speakers and listeners. Second, the recording and processing should be in real time, so that information on the current situation in the audience can be transmitted to remote participants with a minimum delay. Third, the record should contain at least presentation slides and statements of all the speakers. Fourth, since the number and composition of activity participants is constantly changing, and this affects the behavior of speakers and listeners, the type of audience and people in it can help to create a convenient environment for remote participants. Fifth, during technical pauses in the speaker's presentation information about the lecture, its participants, or the general view of the audience in a multimedia report can be displayed. It is also possible to use the rules from cinematography that describe methods of camera arrangement in the room, tracking of objects, and composition of images from the cameras, as well as other rules, to make a video that meets artistic style requirements [52].

The software system of audiovisual monitoring and determination of the situation in the intelligent meeting room discussed in Sect. 6.2 meets the first five characteristics mentioned above. In addition, software automation means of recording and processing of audiovisual data are aimed at reducing the resources required for the organization, carrying out, and recording of the activities, for example, meeting, lecture, seminar, meeting of the Academic Council, and a round table in the conference room. The developed software and hardware system of the audiovisual processing can reduce the workload of secretaries and videographers and focus participants on the considered issues by automating the control of sensor and multimedia equipment. In addition, the use of the synchronization algorithm of audiovisual data streams in the developed software system eliminates the problems associated with the changing frequency of reading of frames from the cameras that arise as a result of interference in the network equipment, as well as with the limited speed of recording of information on storage devices.

At the next stages of research, methods and software should also be developed for the automation of processes of meetings in intelligent meeting rooms characterized by the use of the automatic identification of participants in activities based on the analysis of multichannel signals with an integrated background, as well as the detection and recognition of faces of participants that would reduce the cognitive load on participants and allow the use of these resources for activities.

6.6 Conclusions

The problem of realizing the support of event services on the basis of natural and unobtrusive user interaction and the intelligent space is topical for the development of the intelligent conference room. The technologies of biometrical identification on the basis of face recognition methods enables automation of the processes of registration of meeting participants. Preliminary estimation of the image blur allows one to disregard the low-quality images carrying useless information, thus relieving the burden of the automatic computer vision systems. The blur arising because of the poor photography conditions is due to incorrect focusing of the equipment or unexpected motion of the object. It can be estimated by different methods, among which one can emphasize the study of the brightness gradient of the image pixels, the ratio of brightness of pixels over a certain area, and statistical analysis of the pixel brightness coefficients.

The considered methods enable a numerical estimation of the blur, which is one of the image quality criteria. They are suitable for modeling and application in various algorithms to process visual information. Preliminary isolation of the face area in an image and estimation of its blur were used to advantage for selection of distorted frames in the problem of automatic registration of conference participants in the intelligent room.

Acknowledgements This research is supported by the Council for Grants of the President of Russia (projects № MD-254.2017.8, MK-1000.2017.8, and MK-7925.2016.9), as well as by the Russian Foundation for Basic Research (projects № 16-37-60100, 16-37-60085, 15-07-04415).

References

1. Fillinger, A., Hamchi, I., Degré, S., Diduch, L., Rose, T., Fiscus, J., Stanford, V.: Middleware and metrology for the pervasive future. IEEE Pervasive Comput. Mobile Ubiquit. Syst **8**(3), 74–83 (2009)
2. Nakashima, H., Aghajan, H.K., Augusto, J.C.: Handbook of Ambient Intelligence and Smart Environments. Springer Science + Business Media (2010)
3. Bertotti, C., Brayda, L., Cristoforetti, L., Omologo, M., Svaizer P.: The new MarkIII/IRST-Light microphone array. Research report RR-05-130 (2005)
4. Renals, S., Hain, T., Bourlard, H. Recognition and understanding of meetings the AMI and AMIDA projects. IEEE Workshop on Automatic Speech Recognition Understand, (ASRU'2007), pp. 238–247 (2007)
5. Chou H, Wang J, Fuh C, Lin S, Chen S (2010) Automated lecture recording system. Int Conf on System Science and Engineering (ICSSE'2010), Taipei, Taiwan, pp. 167–172
6. Korzun, D.G., Galov, I.V., Balandin, S.I.: Proactive personalized mobile multi blogging service on Smart M3. In: 34th International Conference on Information Technology Interfaces (ITI'2012), pp. 143–148 (2012)
7. Kadirov R, Cvetkov E, Korzun D (2012) Sensors in a smart room: preliminary study. 12th Conf. Open Innovations Framework Program FRUCT, pp. 37–42
8. Ntalampiras, S., Arsic, D., Stormer, A., Ganchev, T., Potamitis, I., Fakotakis, N.: Prometheus database: a multimodal corpus for research on modeling and interpreting human behavior. In: 16th International Conferrence on Digital Signal Processing (DSP'2009), pp. 854–861 (2009)
9. Ronzhin, A.L., Budkov, V.Y. Determination and recording of active speaker in meeting room. In: 14th International Conference on Speech and Computer (SPECOM'2011), pp. 361–366 (2011)
10. Sergeev, M.B., Solov'ev, N.V., Stadnik, A.I.: Methods to increase contrast in half-tone images for the systems of digital processing of video information. Informat.-Upravl Sist **1**(26), 2–7 (2007) (in Russian)
11. Kol'tsov, P.P.: Estimation of image blur. Komp'yut Optika **35**(1), 95–102 (2011). (in Russian)
12. Krasil'nikov, N.N.: Principles of image processing allowing for their semantic structure, Informat. -Upravl Sist. **1**(32), 2–6 (2008) (in Russian)
13. Meshcheryakov, R.V.: System for estimation of quality of speech transmission. Dokl. TUSUR **2**(1), 324–329 (2010). (in Russian)
14. Ronzhin, A.L., Budkov, VYu.: Technologies for support of the hybrid e-conferences on the basis of the methods of audiovisual processing. Vestn Komp'yut Inform Tekhnol **4**, 31–35 (2011). (in Russian)
15. Xie, X., Zheng, W.S., Lai, J., Yuen, P.C.: Face illumination normalization on large and small scale features. In: IEEE Conference on Computer Vision and Pattern Recognition (CVPR'2008), pp. 1–8 (2008)
16. Liu, J., Zheng, N., Xiong, L., Meng, G., Du, S.: Illumination transition image: parameter-based illumination estimation and re-rendering. In: 19th International Conference on Pattern Recognition (ICPR'2008), pp. 1–4 (2008)
17. Chen, T., Yin, W., Zhou, X.S., Comaniciu, D., Huang, T.S.: Total variation models for variable lighting face recognition. IEEE Trans. Pattern Anal. Mach. Intell. **28**(9), 1519–1524 (2009)
18. Xie, X., Zheng, W.S., Lai, J., Yuen, P.C., Suen, C.Y.: Normalization of face illumination based on large-and small-scale features. IEEE Trans. Image Process. **20**(7), 1807–1821 (2011)
19. Xie, X.: Illumination preprocessing for face images based on empirical mode decomposition. Sig. Process. **103**, 250–257 (2014)
20. Huang, N.E., Shen, Z., Long, S.R., Wu, M.C., Shih, H.H., Zheng, Q., Yen, N.C., Tung, C.C., Liu, H.H.: The empirical mode decomposition and the Hilbert spectrum for nonlinear and non-stationary time series analysis. In: The Royal Society of London A: Mathematical,

Physical and Engineering Sciences, vol. 454, No. 1971, pp. 903–995. The Royal Society (1998)
21. Ferzli, R., Karam, L.J.: A no-reference objective image sharpness metric based on the notion of just noticeable blur (JNB). IEEE Trans. Image Process. **18**(4), 717–728 (2009)
22. Pertuz, S., Puig, D., Garcia, M.A.: Analysis of focus measure operators for shape-from-focus. Pattern Recogn. **46**(5), 1415–1432 (2013)
23. Lorenzo, J., Deniz, O., Castrillon, M., Guerra, C.: Comparison of focus measures in face detection environments. In: 4th International Conference on Informatics in Control, Automation and Robotics (ICINCO'2007), vol. 4, pp. 418–423 (2007)
24. Krotkov, E.: Focusing. Int. J. Comput. Vision **1**(3), 223–237 (1988)
25. Gonzalez, R.C., Woods, R.E.: Digital Image Processing. Prentice Hall Press (2002)
26. Helmli, F., Scherer, S.: Adaptive shape from focus with an error estimation in light microscopy. In: 2nd International Symposium on Image and Signal Processing and Analysis (ISPA'2001), pp. 188–193 (2001)
27. Mendapara, P.: Depth map estimation using multi-focus imaging. Electronic Theses and Dissertations, University of Windsor, Ontario, Canada (2010)
28. Mittal, A., Soundarajan, R., Bovik, A.C.: Making a "completely blind" image quality analyzer. IEEE Signal Process. Lett. **20**(3), 209–212 (2013)
29. Sharifi, K., Leon-Garcia, A.: Estimation of shape parameter for generalized gaussian distributions in subband decompositions of video. IEEE Trans. Circ. Syst. Video Technol. **5**(1), 52–56 (1995)
30. Gorodnichy, D.: Video-based framework for face recognition in video. In: 2nd Canadian Conference on Computer and Robot Vision (CRV'2005), pp. 330–338 (2005)
31. Castrillon-Santana, M., Deniz-Suarez, O., Guerra-Artal, C., Hernandez-Tejera, M. Real-time detection of faces in video streams. In: 2nd Canadian Conference on Computer and Robot Vision (CRV'2005), pp. 298–305 (2005)
32. Zhang, T., Fang, B., Tang, Y.Y., He, G., Wen, J.: Topology preserving non-negative matrix factorization for face recognition. IEEE Trans. Image Process. **17**(4), 574–584 (2008)
33. Lee, D.D., Seung, H.S.: Learning the parts of objects by non-negative matrix factorization. Nature **401**, 788–791 (1999)
34. Turk, M.A., Pentland, A.P.: Face recognition using eigenfaces. In: IEEE Conference Computer Vision Pattern and Recognition (CVPR'19991), pp. 586–591 (1991)
35. Belhumeur, P.N., Hespanha, J., Kriegman, D.: Eigenfaces vs. fisherfaces: recognition using class specific linear projection. IEEE Trans. Pattern Anal. Mach. Intell. **19**(7), 711–720 (1997)
36. Ahonen, T., Hadid, A., Pietikainen, M.: Face recognition with local binary patterns. In: Pajdla, T. Matas J. (eds.) ECCV'2004, vol. 3021, pp. 469–481. LNCS (2004)
37. Georgescu, D.A.: Real-time face recognition system using eigenfaces. J Mobile Embed. Distrib. Syst. **3**(4), 193–204 (2011)
38. Yang, J., Yu, Y., Kunz, W.: An efficient LDA algorithm for face recognition. In: 6th International Conference on Control, Automation, Robotics and Vision (ICARCV'2000), pp. 1–6 (2000)
39. Fukunaga, K.: Introduction to Statistical Pattern Recognition, 2nd edn. Academic Press, New York (1990)
40. Wilks, S.S.: Mathematical Statistics. Wiley, New York (1962)
41. Ojala, T., Pietikainen, M., Harwood, D.: A comparative study of texture measures with classification based on feature distributions. Pattern Recogn. **29**(1), 51–59 (1996)
42. Ojala, T., Pietikainen, M., Maenpaa, T.: Multiresolution gray-scale and rotation invariant texture classification with local binary patterns. IEEE Trans. Pattern Anal. Mach. Intell. **24**(7), 971–987 (2002)
43. Zhao, W., Chellappa, R., Rosenfeld, A., Phillips, P.J.: Face recognition: a literature survey. Technical Report CAR-TR-948, Center for Automation Research, University of Maryland, USA (2002)
44. Gong, S., McKenna, S.J., Psarrou, A.: Dynamic Vision, from Images to Face Recognition. Imperial College Press, London, UK (2000)

45. Budkov, V., Ronzhin A.L., Glazkov, S., Ronzhin, A.: Event-driven content management system for smart meeting room. In: Balandin, S., Koucheryavy, Y., Hu H (eds.) Smart Spaces and Next Generation Wired/Wireless Networking, vol. 6869, pp. 550–560. LNCS (2011)
46. Ronzhin, A., Budkov, V., Karpov, A.: Multichannel system of audio-visual support of remote mobile participant at e-meeting. In: Balandin, S., Dunaytsev, R., Koucheryavy, Y. (eds.) Smart Spaces and Next Generation Wired/Wireless Networking, vol. 6294, pp. 62–71. LNCS (2010)
47. Viola, P., Jones, M.J., Snow, D.: Detecting pedestrians using patterns of motion and appearance. Int. J. Comput. Vis. **63**(2), 153–161 (2005)
48. Favorskaya MN (2010) Models and methods for recognizing dynamical images on the base of spatial-temporal analysis of image series. Extended Abstract of Doctoral Dissertation in Engineering Sciences, Krasnoyarsk, Russia (in Russian)
49. Yusupov, R.M., AnL, Ronzhin, Prischepa, M., AlL, Ronzhin: Models and hardware software solutions for automatic control of intelligent hall. Automat. Remote Control **72**(7), 1389–1397 (2011)
50. Cook, P., Ellis, C.S., Graf, M., Rein, G., Smith, T.: Project Nick: meetings augmentation and analysis. ACM Trans. Inf. Syst. **5**(2), 132–146 (1987)
51. Ronzhin, A.L.: Audiovisual recording system for e-learning applications. In: International Conference on Computer Graphics Theory and Applications (GRAPP'2012), pp. 515–518 (2012)
52. Rui, Y., Gupta, A., Grudin, J., He, L.: Automating lecture capture and broadcast: technology and videography. Multimedia Syst. **10**(1), 3–15 (2004)

Chapter 7
Improving Audience Analysis System Using Face Image Quality Assessment

Vladimir Khryashchev, Alexander Ganin, Ilya Nenakhov and Andrey Priorov

Abstract Video surveillance has a wide variety of applications for indoor and outdoor scene analysis. The requirements of real-time implementation with the desired degree of recognition accuracy are the main practical criteria for most vision-based systems. When a person is observed by a surveillance camera, usually it is possible to acquire multiple face images of a single person. Most of these images are useless due to the problems like the motion blur, poor illumination, small size of the face region, 3D face rotation, compression artifacts, and defocusing. Such problems are even more important in modern surveillance systems, where users may be uncooperative and the environment is uncontrolled. For most biometric applications, several of the best images are sufficient to obtain the accurate results. Therefore, there is a task for develop a low complexity algorithm, which can choose the best image from a sequence in terms of a quality. Automatic face quality assessment can be used to monitor image quality for different applications, such as video surveillance, access control, entertainment, video-based face classification, and person identification and verification. In practical situation, the normalized images at the output of face detection algorithm are post-processed and their quality is evaluated. Low quality images are discarded and only images with acceptable qualities are received for further analysis. There are several algorithms for face quality assessment that are based on estimating facial properties, such as the estimating the pose, calculating the asymmetry of the face, and non-frontal illumination to quantify the degradation of a quality. Several investigations show that application of a quality assessment component in video-based face identification system can significantly improve its performance. Another possible application of face quality assessment algorithm is to process the images with different qualities in different ways. Proposed face quality assessment method has been applied as a quality assessment component in video-based audience analysis system. Using the proposed quality measure to sort the input sequence and taking only high quality

V. Khryashchev (✉) · A. Ganin · I. Nenakhov · A. Priorov
P.G. Demidov, Yaroslavl State University, 14 Sovetskaya St.,
Yaroslavl 150000, Russian Federation
e-mail: dcslab@uniyar.ac.ru

M.N. Favorskaya and L.C. Jain (eds.), *Computer Vision in Control Systems-4*,
Intelligent Systems Reference Library 136,
https://doi.org/10.1007/978-3-319-67994-5_7

face images, we successfully demonstrated that it not only increases the recognition accuracy but also reduces the computational complexity.

Keywords Face image quality assessment · Face recognition
Gender recognition · Audience measurement system · Video surveillance

7.1 Introduction

Usually, when a person is in front of surveillance video camera, several images of his/her face are saved to storage. Most of them are useless for the biometric identification system due to several reasons: the human's movement leads to blurring, a person can be in the low-light conditions, or only a part of the face or significantly turned face may be recorded. Human identification algorithms computationally are complex enough, so recognition of the entire sequence of images can slow down the work of video surveillance systems [1, 2]. Thus, the problem of choice of the best quality images from all received images, by which an identification of a person will be performed, is important.

There are several standards that determine the face image quality—ISO/IEC 19794-5 and ICAO 9303 [3, 4]. They contain a description of the characteristics that provide a decision of the image suitability in the automatic recognition systems. All standardized characteristics are grouped into two classes: the textural features (sharpness, contrast, light intensity, compression ratio, and other distortions) and characteristics directly related to the face features (symmetry, position, rotation, eyes visibility, and the presence of glare or shadows on the face). For their automatic detection, the methods analyzed the determination of posture [3, 5], illumination, and rotation [6, 7] are described in literature.

Unfortunately in practical applications, such as video surveillance or audience analysis system, it is impossible to satisfy the requirements in the ISO/IEC and ICAO standards. The lightning conditions, face posture, compression algorithm, and choice of camera type depend on concrete application. Nevertheless, the task of selection the best quality face image form a video sequence in this case is important too and must be solved in a short period of time. To solve it, various modifications of image quality assessment algorithms are used.

The problem is commonly touched in modern scientific literature [5–16]. One of the first approaches to solve this problem is a method based on the application of clustering algorithm according to K-means approach [8]. Practical experiments have shown that it has low accuracy, when there are many low quality faces in the obtained sequence. A totally different approach searches the best quality images for face recognition by making quality evaluation of all images [2, 9]. Typical recognition system with face image quality evaluation module schema is shown in Fig. 7.1.

The quality of face images is estimated at the pre-processing stage. Depending on the algorithm used in concrete application one (Top1) or three (Top3) best

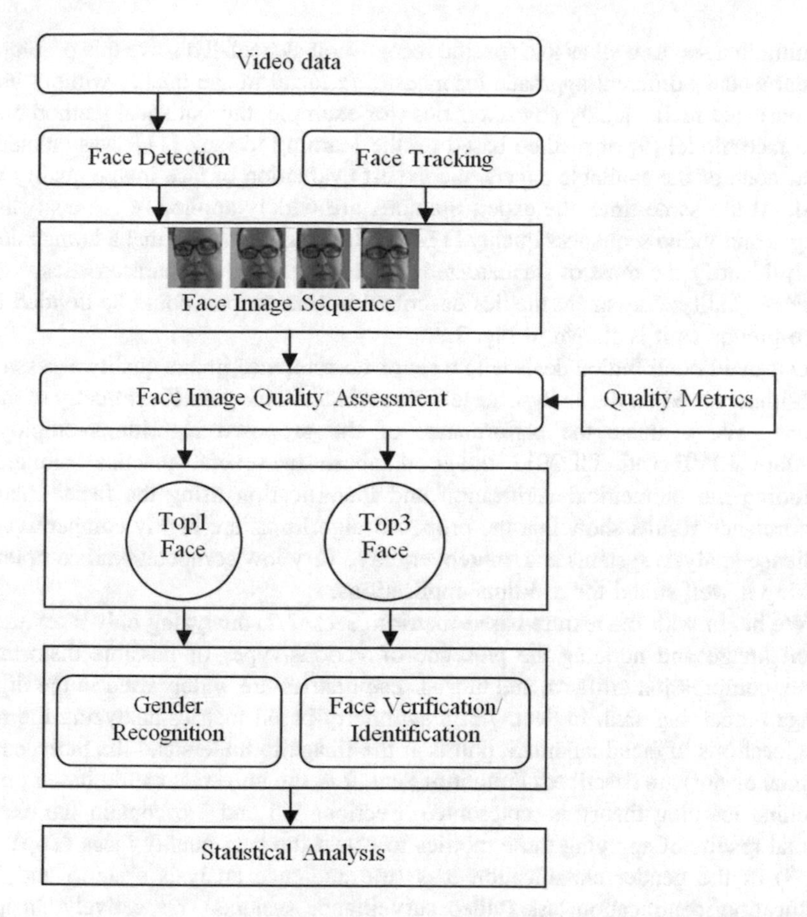

Fig. 7.1 Recognition system with face image quality evaluation module

quality images are selected. Further, they are used for gender classification in audience analysis system or for face verification/identification in video surveillance systems. Low quality images are removed or archived, then the recognition applies only to high-quality images. It is shown in [2] that the use of face image quality evaluation module increases throughput of the surveillance system.

There is a group of face image quality assessment algorithms that uses the objective methods to determine the standardized facial quality characteristics. Overall facial image quality is obtained by combining the results of these methods. This group of algorithms is called the metric fusion algorithms. The metric fusion can be done by thresholding each of characteristic values. In this case, a residual quality would be a number of characteristics above threshold. Another approach of a metric fusion assigns a weight to every measured standardized characteristic [10]. Machine learning methods are widely used to determine the metrics weights. It should be noted that the metric fusion algorithms are tied to a specific database of training images, as well as to a specific recognition system. To solve this problem, a fundamentally different approach for measuring facial image quality without using standardized facial quality characteristics (for example, the statistical method based on a face model [9] or method based on the learning to rank [11]) was created.

In none of the available papers, the expert evaluation of face image quality was used. At the same time, the expert opinions are widely applied in the analysis of images and video sequences quality [17–19]. It should be noted that a human could easily identify the most of standardized facial image quality characteristics.

Face quality assessment metrics described in this chapter could be divided into three groups as it is shown in Fig. 7.2.

Our main contribution deals with a set of no-reference image quality assessment algorithms based on the analysis of texture and landmark points' symmetry of facial images. We evaluate the performance of the proposed algorithms employing standard LIVE and TID2013 image database in several practical situations, including the biometrical verification and identification using the facial images. Performance results show that the proposed algorithms are highly competitive for audience analysis systems and, moreover, have very low computational complexity making it well suited for real-time applications.

We begin with the texture-based metrics (Sect. 7.2) analyzing only a texture of facial image and noticing the presence of various types of possible distortions: noise, compression artifacts, and blur. These metrics are widely used in the digital image processing area. In Sect. 7.3, a symmetry-based metrics analyzing the relative locations of facial landmark points at the image to understand the face posture (frontal or not) are described. Further in Sect. 7.4, the universal metric based on the machine learning theory is represented. Sections 7.5 and 7.6 contain the experimental results of applying these metrics to select the best quality faces (Top1 and Top3) in the gender classification task (the audience analysis system) and face verification/identification task (video surveillance systems), respectively. Insights and recommendations are given in the conclusions (Sect. 7.7).

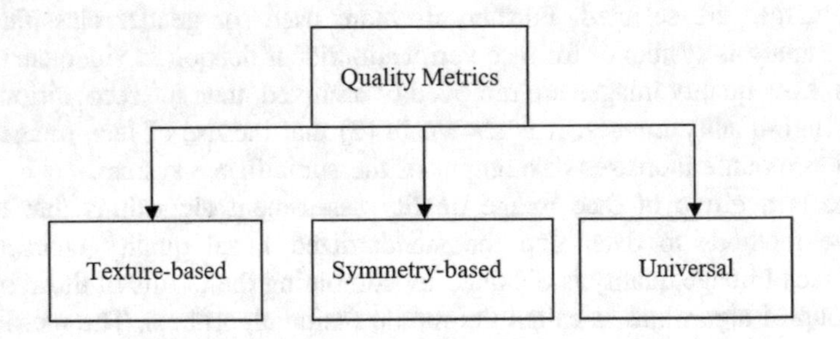

Fig. 7.2 Face quality assessment metrics classification

7.2 Texture-Based Metrics

The goal of texture-based metrics is to detect if an image is distorted or not. Typical image distortions are the compression artifacts, blur, and various types of noise. Usually no-reference image quality assessment algorithms are used as texture-based metrics [17, 19]. To use such algorithm in a face quality assessment task, we need to ensure that this algorithm is fast enough to perform in pseudo real-time mode. There are possible candidates for doing this described below: No-Reference image Quality Local Binary Pattern (NRQ LBP) (Sect. 7.2.1), Blind/Referenceless Image Spatial Quality Evaluator (BRISQUE) [19] (Sect. 7.2.2), and sharpness metric (Sect. 7.2.3).

7.2.1 No-Reference Image Quality Assessment Algorithm (NRQ LBP)

Modern no-reference quality assessment algorithms are built according to the following scheme: feature extraction → machine learning algorithm → quality score. A lot of features' types is used in no-reference algorithms, among them are Scale-Invariant Feature Transform (SIFT) descriptors [20], entropy and phase congruency [21], various natural scene statistics [19]. However, the insufficient attention is paid to local binary patterns, which are computational effective and widely used in computer vision area. A Local Binary Pattern (LBP) is a binary code, which describes the pixel's neighborhood. Parameter r is a neighborhood radius. Parameter P is a number of pixels in a neighborhood (Fig. 7.3).

Consider the image I and let g_c be a luminance value of the central pixel with (x, y) coordinates. The central pixel is a pixel, for which the LBP binary code is computed. Equations 7.1–7.3 are used to select the pixels from a neighborhood, where g_p is a luminance value of the neighborhood pixel, x_p and y_p are the coordinates of a neighborhood pixel.

$$g_p = I(x_p, y_p) \quad p = 0, \ldots, P - 1 \tag{7.1}$$

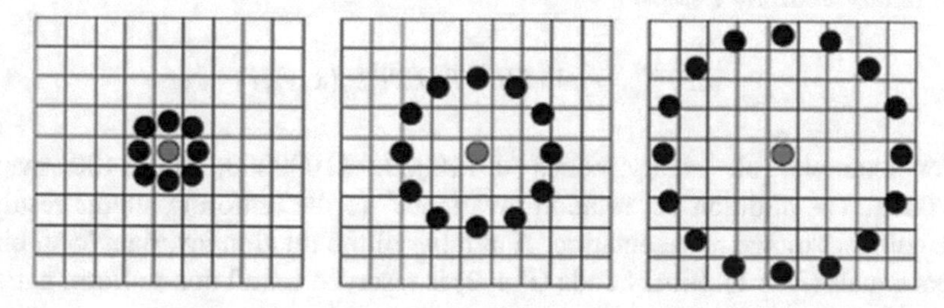

Fig. 7.3 The circular neighborhoods with appropriate (P, r) values (8, 1), (12, 2), (16, 4)

$$x_p = x + R\cos\left(\frac{2\pi p}{P}\right) \quad (7.2)$$

$$y_p = y - R\sin\left(\frac{2\pi p}{P}\right) \quad (7.3)$$

A bilinear interpolation is used, when a sampling point is not in the center of a pixel. A luminance value of the central pixel is the threshold. If a luminance value of a pixel from a neighborhood is more than the threshold value, then 1 is set in the appropriate place of the binary code; otherwise, 0 is set. Formally, the LBP can be described by Eqs. 7.4–7.5, where t_c is a luminance value of the central pixel, t_i is a luminance value of the neighborhood pixel, P is a number of the neighborhood pixels.

$$LBP_{P,r} = \sum_{p=0}^{P-1} S(t_i - t_c)2^p \quad (7.4)$$

$$S(t) = \begin{cases} 1 & t \geq 0 \\ 0 & t < 0 \end{cases} \quad (7.5)$$

The output of the LBP function is a binary code, which describes a pixel's neighborhood.

Local binary pattern is called "uniform" if it contains a number of bitwise transitions from 0 to 1 and, vise versa, less than 2, for example patterns 00000011, 00111000 are uniform and pattern 01010010 is not uniform. In average 90%, all calculated patterns are uniform [22]. In the case of uniform LBP, there is a separate output label for each uniform pattern and all non-uniform patterns are assigned to a single label. The number of different uniform patterns for a code with length p is $p\ (p + 1) + 3$, while the number of non-uniform patterns is 2^p, thus we can save a recourses working only with uniform patterns.

The rotation of image leads to the changes in local binary patterns statistics. To eliminate this issue, the special algorithm of assigning binary codes to patterns exists. Every binary code is circularly shifted to its minimum value in a view of Eq. 7.6, where *SHIFT(binary code, i)* denotes a circular bitwise right rotation of input binary code by i steps.

$$LBP^{uri}_{P,r} = \min_i SHIFT\left(LBP^u_{P,r}(x, y), i\right) \quad (7.6)$$

For example, all binary codes 00110000, 11000000, 00001100 map to 00000011. The addition of "rotation invariance" to the uniform patterns results in more compact image representation. A number of the rotation invariant local binary patters equals $(P + 1)$. Binary code $(P + 2)$ is reserved for all non-uniform patterns.

In research [22], it is shown that the LBP with $r = 1$ is not robust against the local changes in the texture. More robust local binary patterns statistics could be obtained by enlarging spatial support area of the LBP operator, in other words by increasing r parameter. On the other hand, the use of the larger r values (>3) causes the sharp decrease of the texture classification accuracy. Multi-scale LBP is a rescue: it was shown [23] that the highest accuracy achieved, when the LBP is simultaneously calculated for multiple parameters sets: ($r = 1$, $P = 8$), ($r = 2$, $P = 16$), ($r = 3$, $P = 24$). It is worth noting that the multi-scale LBPs can be computed in liner time.

The histogram of rotation invariant local binary patterns H is built using Eqs. 7.7–7.8, where k is a number of patterns [22].

$$H(k) = \sum_{m=1}^{M} \sum_{n=1}^{N} f\left(LBP_{P,r}^{uri}(m,n), k\right) \quad k \in [0, K] \tag{7.7}$$

$$f(x,y) = \begin{cases} 1 & x = y \\ 0 & x \neq y \end{cases} \tag{7.8}$$

The histograms of the rotation invariant local binary patterns calculated for standard test image from LIVE dataset [24] are depicted in Figs. 7.4 and 7.5. This image is affected by five distortion types one at a time: jpeg, jpeg2000, blur, bit errors, and white noise. Decimal numbers on histogram plots correspond to binary code values. It can be seen that a white noise causes the dip in the center of the histogram and a lot of local binary patterns fall into a 9th bin (a bin for all non-uniform patterns). On the contrary, the blur and jpeg2000 distortion types cause the rise in central part of the histogram. When an image is distorted by the compression artifacts of the JPEG codec, the histogram would contains a lot of patterns in the 8th bin. These binary patterns correspond to pixels inside block artifacts, which can be visually detected in Fig. 7.4b. We repeated this experiment for other images from LIVE dataset and obtained the similar results. Thus, we can conclude that the histogram of rotation invariant local binary patterns contains the information about the image quality.

We performed the experiment to choose the machine learning algorithm for mapping bins of the LBP histogram to quality score (Tables 7.1 and 7.2, where the best values are denoted by Bold). It can be seen that the most accurate no-reference image quality assessment algorithm based on the LBPs should be built with Extra trees regressor [25].

As a result, we have obtained a scheme presented in Fig. 7.6 for no-reference image quality assessment algorithm based on the LBP statistics features (NRQ LBP). The accuracy of this algorithm was tested on the LIVE dataset. We have found that the NRQ LBP is as accurate as the best reference algorithm Mean Structural SIMilarity (MSSIM) [26].

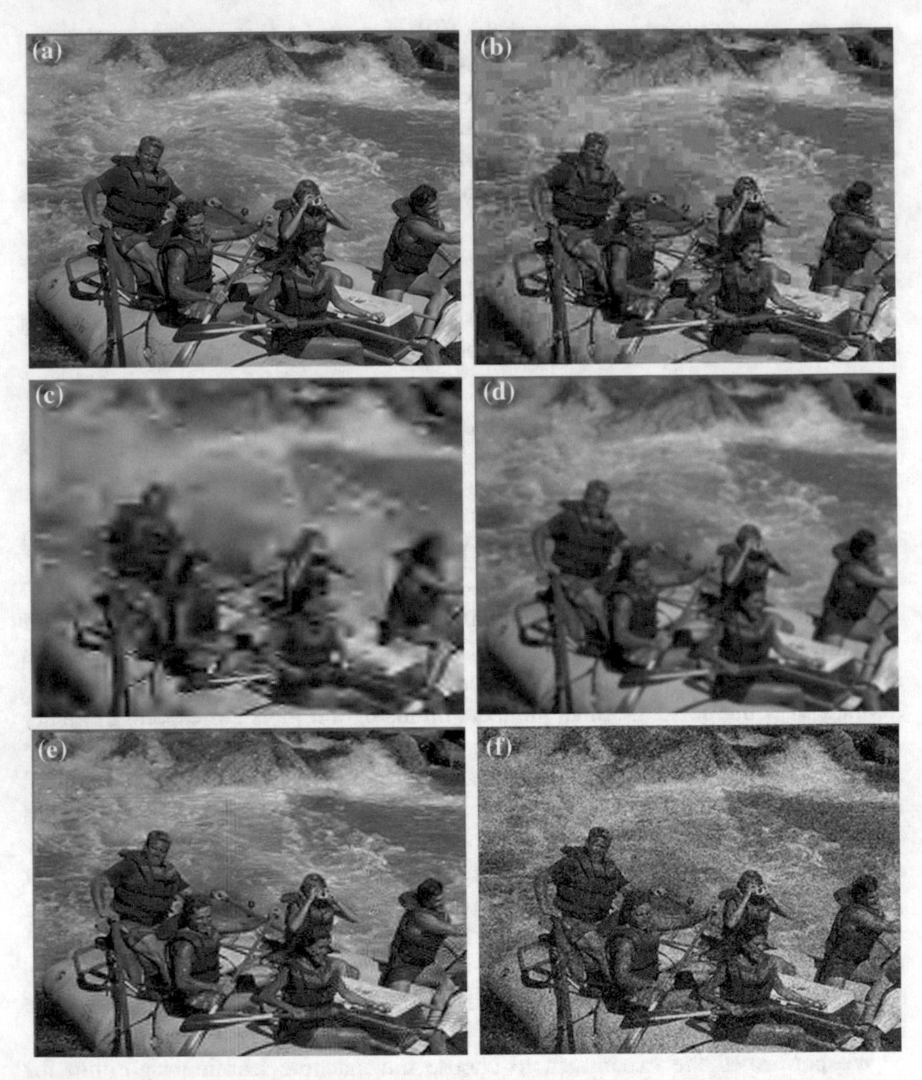

Fig. 7.4 Images with different distortions types: **a** reference image, **b** jpeg, **c** jpeg2000, **d** blur, **e** bit errors, **f** white noise

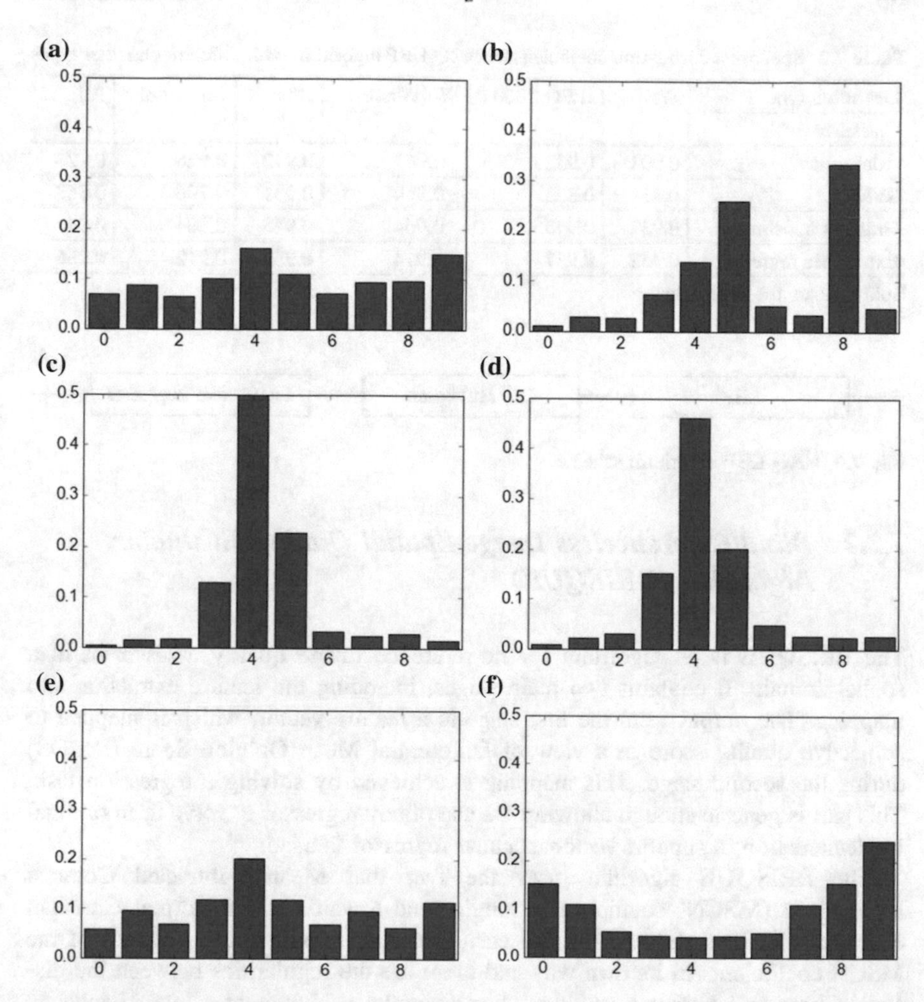

Fig. 7.5 Histogram of uniform rotation invariant local binary patterns ($r = 1$, $P = 8$): **a** reference image, **b** jpeg, **c** jpeg2000, **d** blur, **e** bit errors, **f** white noise

Table 7.1 Pearson correlation coefficient for NRQ LBP algorithm with different classifier types

Distortion type / Classifier	JPEG	JPEG 2000	White noise	Blur	Bit errors	All
AdaBoost	0.923	0.943	0.950	0.945	0.886	0.924
SVM	0.809	0.884	−0.484	0.933	0.797	0.618
Gradient boosting	0.962	0.962	0.981	0.967	0.931	0.959
Extra trees regressor	**0.970**	**0.968**	**0.987**	**0.971**	**0.936**	**0.965**

Bold denotes the best values

Table 7.2 Spearman correlation coefficient for NRQ LBP algorithm with different classifier types

Distortion type / Classifier	JPEG	JPEG 2000	White noise	Blur	Bit errors	All
AdaBoost	0.901	0.932	0.967	0.942	**0.926**	0.926
SVM	0.854	0.883	−0.364	0.931	0.796	0.567
Gradient boosting	0.937	0.945	0.964	0.945	0.904	0.947
Extra trees regressor	**0.952**	**0.957**	**0.974**	**0.955**	0.912	**0.956**

Bold denotes the best values

Fig. 7.6 NRQ LBP algorithm scheme

7.2.2 Blind/Referenceless Image Spatial Quality Evaluator Algorithm (BRISQUE)

The BRISQUE is an algorithm for no-reference image quality assessment in a spatial domain. It contains two main stages, including the feature extraction and mapping. The output from the first stage is a feature vector, which is mapped to subjective quality score in a view of Differential Mean Opinion Score (DMOS) during the second stage. This mapping is achieved by solving a regression task. This part is generic enough allowing the use of any regressor to solve it. In original implementation, a support vector machine regressor is used.

The BRISQUE algorithm uses the fact that Mean Subtracted Contrast Normalized (MSCN) coefficients strongly tend towards a unit normal Gaussian characteristic. It is shown [19] that each distortion modifies the statistics of the MSCN coefficients in its own way and there are the regularities between the distortion type and statistics variation. For example, a blur creates more Laplacian appearance, while a white-noise distortion appears to reduce the weight of the tail of the histogram.

Generalized Gaussian Distribution (GGD) can be used to effectively capture a broader spectrum of distorted image statistics provided by Eqs. 7.9–7.11, where α controls a shape, σ^2 controls a variance.

$$f\left(x;\alpha,\sigma^2\right) = \frac{\alpha}{2\beta\,\Gamma(1/\alpha)}\exp\left(-\left(\frac{|x|^{\alpha}}{\beta}\right)\right) \tag{7.9}$$

$$\beta = \sigma\sqrt{\frac{\Gamma(1/\alpha)}{\Gamma(3/\alpha)}} \tag{7.10}$$

$$\Gamma(\alpha) = \int_0^{\infty} t^{\alpha-1} e^{-t} dt \quad \alpha > 0 \tag{7.11}$$

Values of (α, σ^2) are estimated for the tested image and form the first feature set of the BRISQUE algorithm. The second feature set is generated from pairwise products of neighboring MSCN coefficients along four orientations—horizontal, vertical, main-diagonal and secondary diagonal.

Pairwise MSCN products for distorted images obey the asymmetric generalized Gaussian distribution:

$$f\left(x; \upsilon, \sigma_l^2, \sigma_r^2\right) = \begin{cases} \frac{\upsilon}{(\beta_l + \beta_r)\Gamma(1/\upsilon)} \exp\left(-\left(\frac{-x}{\beta_l}\right)^{\upsilon}\right) & x<0 \\ \frac{\upsilon}{(\beta_l + \beta_r)\Gamma(1/\upsilon)} \exp\left(-\left(\frac{-x}{\beta_r}\right)^{\upsilon}\right) & x \geq 0 \end{cases} \tag{7.12}$$

$$\beta_l = \sigma_l \sqrt{\frac{\Gamma(1/\upsilon)}{\Gamma(3/\upsilon)}} \tag{7.13}$$

$$\beta_r = \sigma_r \sqrt{\frac{\Gamma(1/\upsilon)}{\Gamma(3/\upsilon)}} \tag{7.14}$$

The BRISQUE second feature set contains values of $\left(\mu, \upsilon, \sigma_r^2, \sigma_l^2\right)$, where υ controls a shape of the distribution, σ_r^2 and σ_l^2 are the scale parameters that control the spread on each side of the mode, respectively. The first parameter from the second feature set is calculated using Eq. 7.15.

$$\eta = (\beta_r - \beta_l) \frac{\Gamma\left(\frac{2}{\upsilon}\right)}{\Gamma\left(\frac{1}{\upsilon}\right)} \tag{7.15}$$

All features are extracted in two scales, such as the original image scale and reduced scale (low pass filtered and down sampled by a factor of 2). Thus, a total 36 features are extracted for a tested image.

7.2.3 Sharpness

In real life situations, the objects in front of the surveillance camera are not static. That is why the captured images of these objects may be blurred. The task of defining of the blur degree is one of the most important challenges in quality assessment of face images. The high values of this measure should be assigned to images without blur. Discrete Laplace operator is used to calculate sharpness measure for the image:

$$L(I) = \left|\frac{\partial^2 I}{\partial x^2}\right| + \left|\frac{\partial^2 I}{\partial y^2}\right| \tag{7.16}$$

The discrete second derivatives may be computed as a convolution with the following kernels: (1, −2, 1) and $(1, -2, 1)^{\mathrm{T}}$. Sharpness metric value is the mean of the output from these convolutions. It is worth noting that the sharpness estimation is made directly inside a face area.

7.3 Symmetry-Based Metrics

Persons in front surveillance camera are not static in the most practical cases. It leads to the presence in the recorded video stream face images with different poses. Standards ISO/IEC 19794-5 and ICAO 9303 contain the requirements about the maximum acceptable head rotation angles for recognition. The goal of symmetry-based metrics is to determine if image is frontal or not. The output of this metrics is a real number showing how the current face pose differs from frontal. It worth noting that metrics from this group use facial landmarks. Thus, the accuracies depend on the landmark detector accuracy. Symmetry of landmarks points is discussed in Sect. 7.3.1, while the symmetry metric is represented in Sect. 7.3.2.

7.3.1 Symmetry of Landmarks Points

Coordinates of the landmarks points obtained by the detector described in [27] are used to determine the face symmetry in an image. Each landmark point is numbered from 1 to 68 (Fig. 7.7). Only 22 landmarks points with the numbers 9, 31, 32, 36–49, 52, 55, 58, 63, 67 are used. It allows to increase a productivity of the proposed algorithm.

Straight line through the points 9 and 31 is used to determine a degree of deviation from the front face position. Then, the distance w between points 37 and 46 is computed. The resulting measure S is calculated according to Eq. 7.17, where d_i is a distance between the point i and straight line described above.

$$S = \frac{|d_{32} - d_{36}| + |d_{40} - d_{43}| + |d_{37} - d_{46}| + |d_{49} - d_{55}| + d_{52} + d_{67}}{w} \tag{7.17}$$

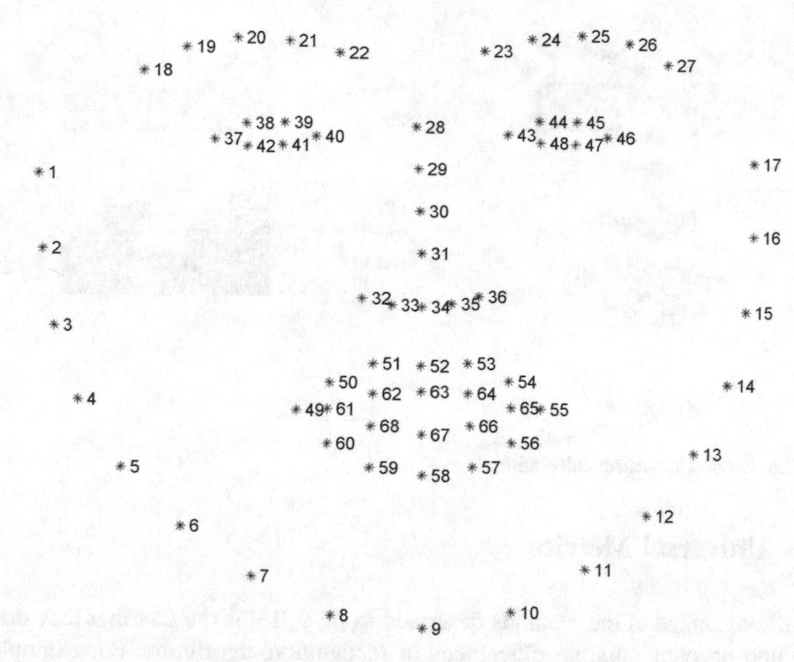

Fig. 7.7 Landmarks points' location diagram

7.3.2 *Symmetry Metric*

Symmetry metric is an assessment of how much the person's current posture is different from frontal and how much the heterogeneous lighting takes place. The symmetry metric is calculated as the intersection of directed Histograms of Oriented Gradients (HOG) descriptors in symmetric landmarks points of a face [10] provided by Eqs. 7.18–7.19, where $H_i^L(i)$ is a calculated function of the oriented gradients histogram in a landmark point with index i, *Sym* is a value of the considered symmetry metrics, N is a number of pairs of landmarks facial points.

$$d(i) = \sum_i \min\left(H_i^L(i), H_i^R(i)\right) \tag{7.18}$$

$$Sym = \frac{1}{N}\sum_{i=1}^{N} d(i) \tag{7.19}$$

Symmetrically located landmarks points have the same value i, thus $H_i^L(i)$ is calculated for the ith point of the left (Fig. 7.8).

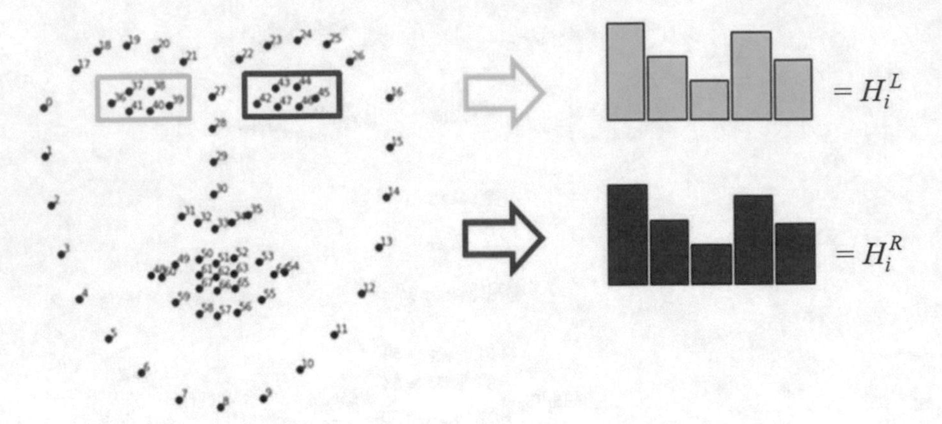

Fig. 7.8 Symmetry metric calculation

7.4 Universal Metrics

The disadvantage of the methods described in [2, 9, 13] is the fact that they do not take into account possible differences in recognition algorithms. For example, a recognition algorithm can accurately recognize faces even if part of the face covered by another object, for example, by a hand. Such algorithm faces with occlusion must not have a poor quality, whereas an algorithm, which does not work accurately for faces with occlusion, it should be the opposite. Considering the drawbacks of the existing solutions, the algorithm based on a method of learning to rank is proposed. This algorithm consists of two stages: normalization (with more or less typical) and quality control.

Assume that the face recognition algorithm is tested on the databases *A* and *B*, and algorithm based on database *A* has a higher accuracy than the one based on database *B*. In other words, the images from the database *A* are of better quality than the images from the database *B* for current recognition algorithm. Let us write it in the form $A > B$. Two images I_i and I_j are selected from *A* and *B*, respectively. The function $f(\cdot)$, which input is an image and the output is a feature vector, describes an image. Define a linear function of image quality as $S(I) = w^{\mathrm{T}} f(I)$. The goal is to find a vector *w* that would meet conditions (Eqs. 7.20–7.22) as much as possible, and we should consider that images from one database have the same quality image.

$$w^{\mathrm{T}} f(I_i) > w^{\mathrm{T}} f(I_j) \quad \forall I_i \in A \quad \forall I_j \in B \tag{7.20}$$

$$w^{\mathrm{T}} f(I_i) > w^{\mathrm{T}} f(I_j) \quad \forall I_i \in A \quad \forall I_j \in A \tag{7.21}$$

$$w^{\mathrm{T}} f(I_i) > w^{\mathrm{T}} f(I_j) \quad \forall I_i \in B \quad \forall I_j \in B \tag{7.22}$$

The description above matches with the formulas from paper [7], and it may be represented in the following terms $\xi_{i,j} \geq 0, \eta_{i,j} > 0, \gamma_{i,j} \geq 0$, respectively (Eq. 7.23).

$$\begin{aligned}
&\text{minimize}\left(\|w^{\mathrm{T}}\|_2^2 + \lambda_1 \sum \xi_{i,j}^2 + \lambda_2 \sum \eta_{i,j}^2 + \lambda_3 \sum \gamma_{i,j}^2\right) \\
&w^{\mathrm{T}}(f(I_i) - f(I_j)) \geq 1 - \xi_{i,j} \quad \forall I_i \in A \quad \forall I_j \in B \\
&w^{\mathrm{T}}(f(I_i) - f(I_j)) \leq \eta_{i,j} \quad \forall I_i \in A \quad \forall I_j \in A \\
&w^{\mathrm{T}}(f(I_i) - f(I_j)) \leq \gamma_{i,j} \quad \forall I_i \in B \quad \forall I_j \in B \\
&\xi_{i,j} \geq 0 \quad \eta_{i,j} > 0 \quad \gamma_{i,j} \geq 0
\end{aligned} \tag{7.23}$$

This approach can be extended to a larger number of databases and features. If a mixture of signs is used, then two level strategies should be used. Assume that m different feature vectors could be extracted from the image I. For ith vector, the quality will be calculated according to the formula $S_i(I) = w_i^{\mathrm{T}} f_i(I), i = 1, 2, \ldots, m$. In the first phase of learning, the vector weights w_i are calculated according to Eq. 7.23 for all of the various features. Expression $\overrightarrow{S} = [S_1(I), S_2(I) \ldots S_m(I)]^{\mathrm{T}}$ is a column vector containing various quality ratings for each attribute, respectively. In the author's implementation, m = 5 is used. Define a face image quality in second step as $S_k(I) = w_k f_\phi\left(\overrightarrow{S}\right)$, where f_ϕ is a polynomial function, which is represented by Eq. 7.24.

$$\begin{aligned}
f_\phi\left(\overrightarrow{S}\right) = \Bigg[&\left(\frac{1}{\sqrt{2}}\right) S_1 S_1^2 \left(\sqrt{2}\right) S_2 \left(\sqrt{2}\right) S_1 S_2 S_2^2 \left(\sqrt{2}\right) S_3 \left(\sqrt{2}\right) S_1 S_3 \\
&\left(\sqrt{2}\right) S_2 S_3 S_3^2 \left(\sqrt{2}\right) S_4 \left(\sqrt{2}\right) S_1 S_4 \left(\sqrt{2}\right) S_2 S_4 \left(\sqrt{2}\right) S_3 S_4 S_4^2 \\
&\left(\sqrt{2}\right) S_5 \left(\sqrt{2}\right) S_1 S_5 \left(\sqrt{2}\right) S_2 S_5 \left(\sqrt{2}\right) S_3 S_5 \left(\sqrt{2}\right) S_4 S_5 S_5^2 \Bigg]
\end{aligned} \tag{7.24}$$

7.5 Face Verification/Identification

Face recognition is the hot topic in computer science and area of active research [1]. It can be used in a variety of applications: surveillance system, human-computer interfaces, and audience analysis. Face recognition task can be divided into two separate tasks: identification and verification. During first task, we try to answer a question "Who is this person?" During a verification task we validate the claimed identity based on the facial image (one-one matching). In this section, Openface facial detector is considered in Sect. 7.5.1. Experimental results are included in Sect. 7.5.2.

7.5.1 Openface Facial Detector

There are a lot of huge companies, which are investing a lot of efforts into the face recognition systems: Facebook, Microsoft, NTechLab. However, today such systems are available for everyone with help of Openface software library [28] that contains all parts of typical facial recognition system, such as the detection, normalization, representation, and classification.

Openface facial detector is made using the histogram of oriented gradients features combined with a linear classifier. The image pyramid and sliding window detection scheme are used. This functionality is implemented in dlib library, which is used by Openface. Face detector output is the set of bounding boxes around a face. A bounding box with the biggest square is returned as a face detection result.

The variance between the detected face and faces from a database causes an accuracy decline. One way to solve this issue is to collect a larger database but this is costly and impossible, sometimes. Another way of doing this is a face normalization. Face normalization is a process of a transformation of the input image in such way that all facial landmarks are placed in predefined locations. There are multiple ways of doing this, for instance a normalized face is obtained by constructing its 3D model in a DeepFace algorithm [29]. More computational effective method based on the affine transform is used in Openface. It requires the facial landmarks as an input. Method is used for facial landmarks detection [30]. Coordinates of eyes and nose on the normalized image are calculated as a mean from the appropriate coordinates from faces in a database. In addition, an image is cropped to the size 96 × 96 pixels.

Normalized facial images are transformed to the compact vector representations with 128 dimensions by a deep neural network. The architecture of this network is presented in Table 7.3 and described in details in [31].

Table 7.3 Deep neural net architecture used in openface

Layer	Output size
Conv (7 × 7 × 3)	112 × 112 × 64
Max pooling + normalization	56 × 56 × 64
Inception (2)	56 × 56 × 192
Max pooling + normalization	28 × 28 × 192
Inception (3a)	28 × 28 × 256
Inception (3b)	28 × 28 × 320
Inception (3c)	14 × 14 × 640
Inception (4a)	14 × 14 × 640
Inception (4b)	14 × 14 × 640
Inception (4c)	14 × 14 × 640
Inception (4d)	14 × 14 × 640
Inception (4e)	7 × 7 × 1024
Inception (5a)	7 × 7 × 1024
Inception (5b)	7 × 7 × 1024
Average pooling	1 × 1 × 1024
Fully connected	1 × 1 × 128
L2 normalization	1 × 1 × 128

The neural network was trained on two large publicly available datasets FaceScrub [32] and CASIA-WebFace [33]. Stochastic gradient decent was used as an optimization algorithm. Loss function was chosen in a view of Eq. 7.25, where x_i^a is an input image of person i, x_i^p is an image of the same person from a training database, x_i^n is an image of other person. The α value was selected during training process and was equal to 0.2.

$$L = \sum_{i}^{N} f\left(x_i^a\right) - f(x_i^p)_2^2 - f\left(x_i^a\right) - f\left(x_i^n\right)_2^2 + \alpha \tag{7.25}$$

Person identity from training database is found during the classification step. Openface does not provide any classification functionality but simple classifier can be build without any external libraries by measuring Euclidian distance between facial image representations obtained in the previous step.

7.5.2 Experimental Results

Two test databases were collected for experiments: Khryashchev Face Comparison Database (KFCD) and 60 Person Face Comparison Database (60PFCD). First database contains 10 test video sequences for every person recorded at different illuminance conditions under 20 lx, 50 lx, 75 lx, 130 lx, 180 lx, and 500 lx. Not every frame from the recorded videos was placed in the database. Two methods for image extraction are used: manual (made by the experts) and automatic extraction (every 25th frame). The samples of such images are presented in Figs. 7.9 and 7.10. Standard face detector [34] was used to detect the faces. The following measures were calculated for every detected face: the image resolution, sharpness, symmetry, measure of symmetry of landmarks points S, quality measure K (based on learning to rank [11]), and two no-reference image quality metrics NRQ LBP and BRISQUE. In addition, the expert quality assessment had been conducted for each image with values ranging from 1 (the worst quality) to 10 (the best quality).

Fig. 7.9 Test images from KFCD dataset with different distortions: **a** face rotation, **b** blur, **c** low resolution of face image

Fig. 7.10 Test images KFCD dataset extracted automatically (every 25th frame is extracted)

Table 7.4 Rank correlation for different facial image quality metrics (KFCD dataset)

Illuminance, lx	NRQ LBP	BRISQUE	Sharpness	S	Sym-metry	K	Resolution
20	0.25	0.33	−0.17	−0.05	0.02	0.02	0.40
50	−0.13	−0.23	0.03	0.05	0.15	**0.36**	−0.03
75	−0.004	−0.28	−0.06	−0.15	**0.23**	−0.37	−0.09
130	0.33	**0.73**	−0.1	**0.45**	0.28	0.1	0.36
180	0.1	−0.25	−0.21	**0.06**	0.05	−0.03	−0.09
500	**0.79**	−0.06	−0.30	**0.28**	0.22	0.10	−0.15

Bold denotes the best values

The Spearman rank correlation coefficient was used as a similarity score between expert rank and rank for each measure for images with the same illuminance. The resulting values of the Spearman rank correlation coefficient for test images extracted by the experts are presented in Table 7.4, where the best values are denoted by Bold. The greater value of the correlation coefficient corresponds to the best correlation between the measure and the expert ratings.

The simulation results show that at normal and high illuminance levels (>130 lx) the proposed measure based on symmetry of landmarks points *S* outperforms other quality measures. Low values of correlation coefficient obtained on some test images (50 lx and 180 lx) can be explained by the fact that the efficiency of symmetry determining algorithms depends on accuracy rate of detector of landmarks points, which decreases with significant face rotation or blur caused by the movement of a person.

The values of correlation coefficient for *K* measure are low in most cases. The values of *K* measure often fall within a narrow range of values, for example, a range of values for the test set with the illuminance of 75 lx is equal to 15 units between the worst and the best image quality, although the entire range of measure values is 100 units.

In addition, we measure Top3 accuracy for each investigated metric. Top3 accuracy is a number of matches between 3 top quality images chosen by the experts and objective metrics. Top1 accuracy is measured similarly. Results presented in Table 7.5 show that the better choice of high-quality images is made based on learning to rank *K* metric. The second position belongs to the proposed symmetry of landmarks points metric *S*. This result can be explained by the fact that both subjective assessment and *S* metric are performed in the spatial domain. The

Table 7.5 Top3 accuracy of facial image quality metrics (KFCD dataset)

Illuminance, lx	Presence of glasses	NRQ LBP	BRISQUE	Sharpness	S	Symmetry	K	Resolution
20	–	0	0	1	1	1	1	1
20	+	0	2	2	1	2	2	0
50	–	0	0	1	0	1	1	1
50	+	0	1	0	2	2	3	0
75	–	0	1	0	2	0	1	1
75	+	2	1	1	1	0	2	1
130	–	1	0	1	1	0	2	1
180	–	1	0	1	2	1	3	0
180	+	1	1	0	1	1	0	0
500	+	1	0	0	1	1	2	0
Total		6	6	7	12	9	17	5

Table 7.6 Top1 accuracy of facial image quality metrics (KFCD dataset)

K	Resolution	Symmetry	Sharpness	S	NRQ–LBP	BRISQUE
3	0	0	0	4	2	0

Fig. 7.11 Facial images from 60PFCD dataset

accuracy of the no-reference NRQ LBP metric depends on the luminance level. It performs well, when illuminance is more than 100 lx. Results for Top1 metric are presented in Table 7.6.

The second database contains facial images of 60 persons (60 Person Face Comparison Database—60PFCD) obtained in real-life situation at low lighting conditions (<100 lx). There are 10 images for each person (Fig. 7.11). A group of 10 experts defined the best quality image (Top1) and the best 3 quality images for each 60 persons. To obtain a mean value, we weighted the expert results—every image from Top1 got a weight of 3, while other images from Top3 got a weight of 2

Table 7.7 Top1 accuracy of facial image quality metrics (60PFCD dataset)

NRQ LBP	BRISQUE	Sharpness	S	Symmetry	K	Resolution
4	7	7	7	10	20	6

Table 7.8 Top3 accuracy of facial image quality metrics (60PFCD dataset)

NRQ LBP	BRISQUE	Sharpness	S	Symmetry	K	Resolution
49	52	60	73	68	99	65

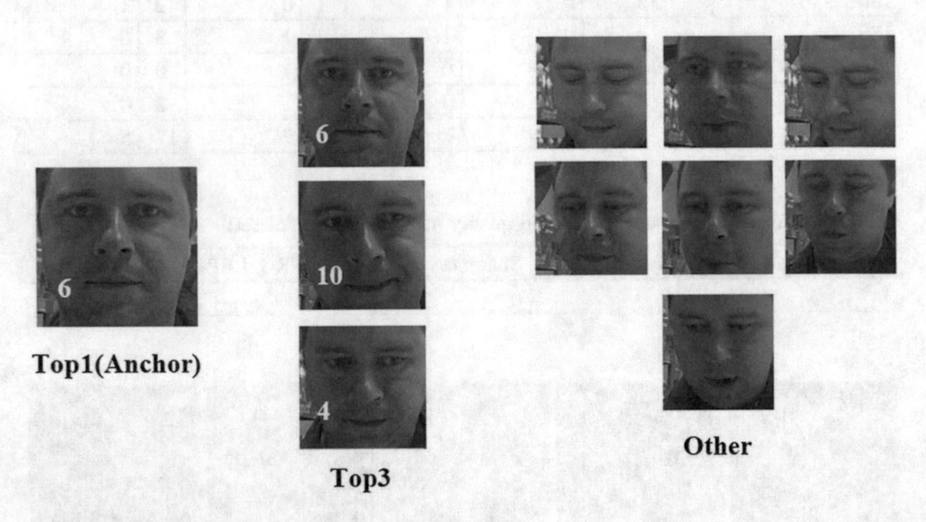

Fig. 7.12 Anchor and Top3 illustration

(no difference between second and third quality images). Like in previous experiments, we computed the investigated facial quality metrics scores for each image in the dataset. Tables 7.7 and 7.8 contain Top1 accuracy and Top3 accuracy results for objective face quality metrics. It is clear that metric *K* is more accurate than other investigated metrics.

We used 60PFCD database to measure the accuracy of face recognition system described above. By accuracy we mean a number of correctly classified images divided by a total number of the classified images. At the classification step, we used a classifier based on so-called "anchor" images. An anchor image is a single image, which was chosen from a person image sequence (Figs. 7.12 and 7.13). There were three anchor selection schemes:

- The anchor image is chosen as the highest quality image by the Mean Opinion Score (MOS) metric value.
- The anchor image is chosen as the first image in a sequence.
- The anchor image is chosen as the highest quality image by learning to rank metric value.

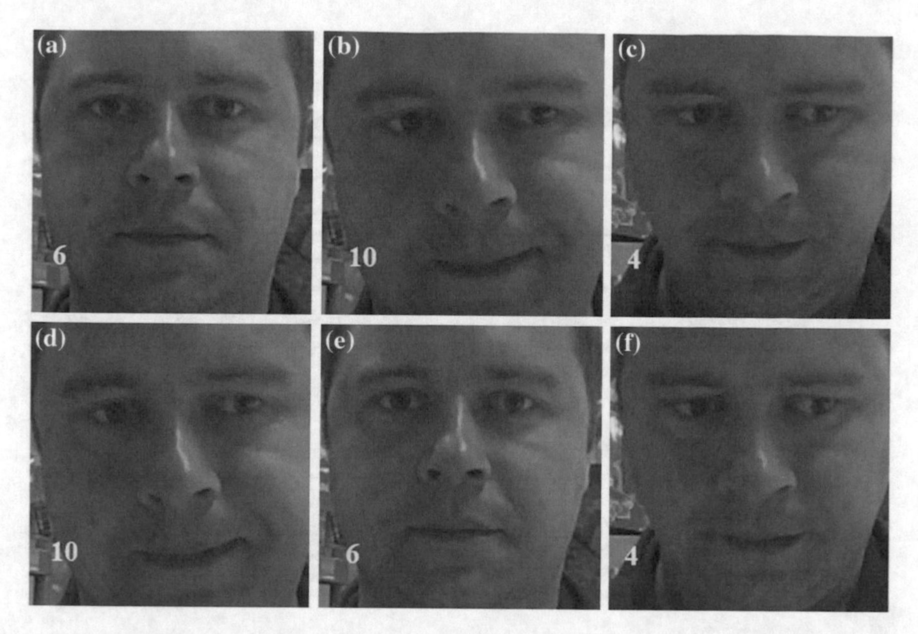

Fig. 7.13 Top3 images chosen by: **a** MOS metric, **b** K metric, **c** Symmetry of landmarks points metric, **d**–**f** Sharpness metrics

For each anchor image, we saved its vector representation to classifier memory. To classify some test image, we measured the Euclidian distance between its vector representation and the anchors vector representation. We chose the anchor representation, which had the smallest distance, and assign the test image an appropriate person id. To collect test images, we selected a pair of images from each person images (the anchor image cannot be in this pair). This pair was selected randomly or from Top3 images chosen by the MOS metric.

Figure 7.14a plots the accuracy results for two classifiers based on the "anchor" images. For the first classifier, the anchors are the highest quality images by the MOS metric. For the second classifier, the anchors are the highest quality images by the learning to rank metric. Test images were chosen from Top3 quality image by the MOS metric. It can be seen for results that classifier with anchors based on the MOS metric is a little more accurate that the classifier with anchors based on learning to rank metric. These results are confirmed by Fig. 7.14b, when test images were chosen randomly.

Figure 7.15 shows the experimental results for a classifier that uses first image as an anchor and a classifier, which uses the highest quality image based on learning to rank metric. Test images are selected randomly in this case. It can be seen from the results that the second classifier is more accurate. It means that facial image quality assessment could be used in a face recognition systems with dynamic database (new person is added during a system operation) to select the person images for storage.

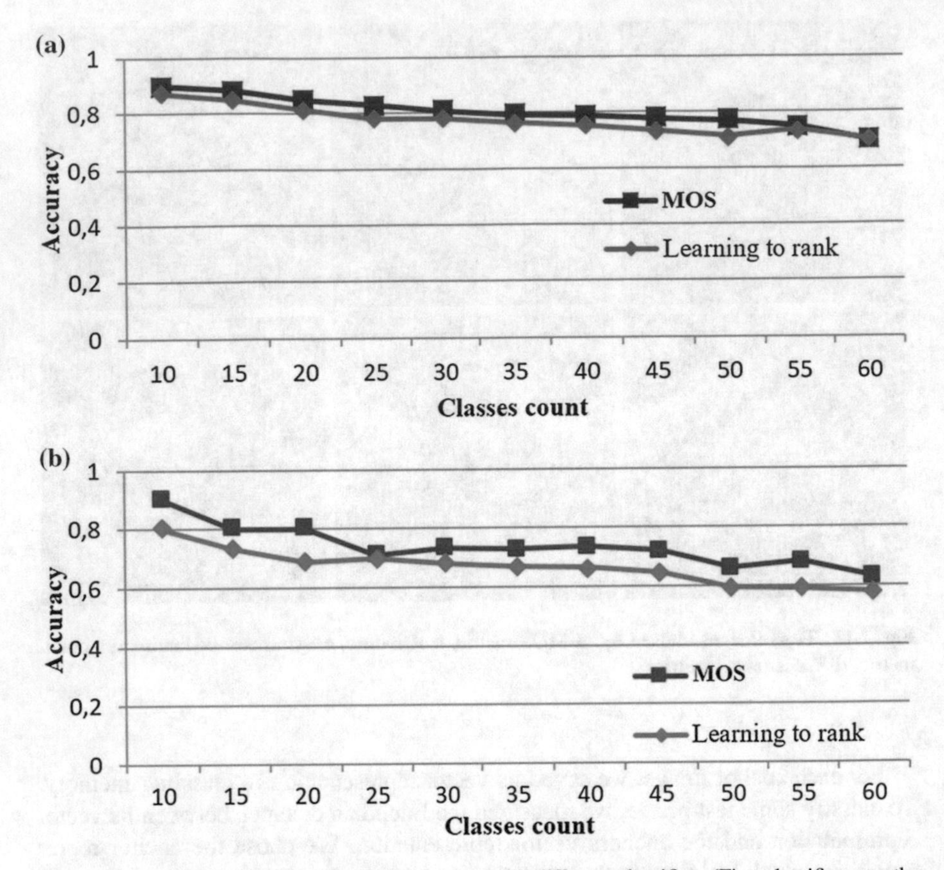

Fig. 7.14 Accuracy of a face recognition system with different classifiers (First classifier uses the highest quality image by the MOS metric as anchors. Second classifier uses the highest quality images by learning to rank metric): **a** test images are chosen from Top3 images, **b** test images are chosen randomly

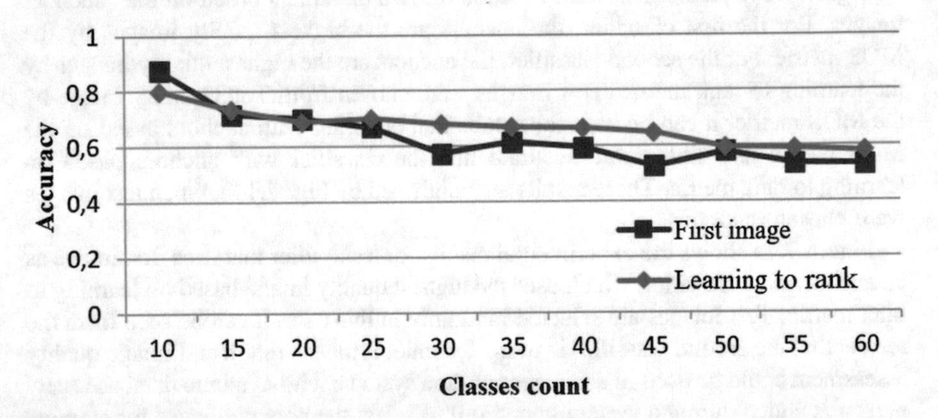

Fig. 7.15 Accuracy of a face recognition system with different classifiers. First classifier uses the first person image as anchors. Second classifier uses the highest quality images by learning to rank metric. Test images are chosen randomly

7.6 Gender Recognition

Gender recognition, for example, can be used to collect and estimate the demographic indicators [35–41]. Besides, it can be an important pre-processing step, while solving a problem of person identification. The gender recognition allows to reduce a number of candidates for analysis (in case of identical number of men and women in a database) by 2 times, and thus allows to accelerate the identification process.

Gender recognition is a fundamental task for human beings, as many social functions critically depend on the correct gender perception. Automatic gender classification has many important applications, for example, the intelligent user interfaces, visual surveillance, and demographic statistics gathering for marketing purposes. Human faces provide important visual information for gender perception. Gender classification from face images has received much research interest in the last two decades. The proposed gender classification algorithm is discussed in Sect. 7.6.1. The experimental results are represented in Sect. 7.6.2.

7.6.1 *Gender Classification Algorithm*

We propose new gender classification algorithm based on non-linear SVM classifier that has several types of features calculations shortly described below.

The Scale-Invariant Feature Transform. The feature implemented in the vlfeat library. Histogram computation size is 4 and the descriptor step is 16. For each block, the horizontal and vertical window descriptors with sizes 16 × 16 pixels are calculated with step 16.

Histogram of Oriented Gradients. The image is divided into 4 equal parts. In each part of the histogram, 16 lines and formed bins in opposite directions are calculated. Thus, it turns 8 destinations in 4 parts of the unit or 32 factors. The result is 8 destinations in 4 parts of the unit or 32 factors.

Gabor filters. This approach has been implemented to reduce the running time. The algorithm uses a filter with an aperture of 19 × 19 pixels. The convolution is performed via a Fourier transform. After receipt of the Gabor filter coefficients, the decimation occurs. The value is converted block size from 2 × 2 to 6 × 6, depending on the block size. When calculating, the Gabor filter is performed linear decimation by 2 in both directions. Further decimation occurs from 1 × 1 nearest neighbor to a 3 × 3 sizes depending on the block.

Pre-selection of blocks. Blocks are the parts of an image, where features are selected. Parts are selected randomly. In the next step, the selected blocks suitable for recognition by the AdaBoost algorithm are formed [42].

For learning the algorithm, 500 blocks was selected with each area 800 pixels. These blocks are used to calculate the characteristics of the type of HOG and GABOR. 300 blocks with each area 1200 pixels were selected for the SIFT features.

For each pair of features the feature vector f is calculated. Vector f forms a matrix F for the entire training sample, where a line number corresponds to an image number of the training sample. The vector y has 1 if the image class is "male" and −1, when the class is "female". The regression between the matrix F and vector y provides this information. To account for the displacement to the matrix F, the right column of the 1st class (F_1 = [F, 1] is added. Vector of the regression coefficients is a solution of Eq. 7.26.

$$\left(F_1' * F_1\right) * a = F_1' * y \tag{7.26}$$

Pseudo-code of proposed algorithm is represented in Fig. 7.16.

T is a number of weak classifies

Input: Training set

$$\{(x_i; y_i)\},\quad i = 1,\ldots,N;\quad x_i \in R^d,\quad y_i \in \{-1,1\}$$

1. Initialize false positive rate, classifier output, and sample weight:

$$falsePos = 0.5,\quad h = 0,\quad weights = 1/N$$

2. Repeat for $t = 1, 2, \ldots, T$:

   2.1 Update the sample weight:

$$weights = weights * e^{-y \cdot h}$$

$$weights = \frac{weights}{\sum weights}$$

   2.2 For m = 1 to M

      2.1.1 Select cyclically SIFT, HOG, GABOR + index random block.

      2.2.2 Choose the correct vector x from X.

      2.2.3 Calculate the histograms WP and WM for a vector x.

      2.2.4 For each m keep the value of the objective function

$$Z(m) = 2\sum_j \sqrt{WP_j WM_j} + CA * falsePos * C$$

   2.3 Select the feature + block:

$$h(i) = 1/2 * \ln\left(\frac{WP_j + eps}{WM_j + eps}\right)$$

   2.4 Update strong classifier:

$$F_{t+1}(x_i) = F_t(x_i) + h_t(x_i)$$

3. Output classifier:

$$F(x) = sign\left(\sum_{j=1}^{T} h_j(x)\right)$$

Fig. 7.16 Pseudo-code of proposed algorithm

Vector x can be counted for each block and features, which is nearing y and equals $x = F_1{*}a$. Each pair feature & block on the test set will match its vector, a number of test samples in a sample length. A set of vector column is an input of RealAdaBoost learning algorithm [42].

During the next step, we need to select the partition of $\{-1, 1\}$ in intervals. Each element of the training sample x will correspond to a certain interval. Further, we know that the ith test image belongs to the jth interval if $x(i)$ is included in this interval. In this embodiment, the interval is selected in 32 parts. It was observed that as a number of intervals over 16 significantly increases a quality of the model but slows the learning process.

To improve the classification, we use different approaches for pre-processing facial area. In this experiment, we used an approach Difference Of Gaussian (DOG) filter [37]. This filter allows to select the most important information from an image. The principle of the algorithm involves a subtraction of one blurred version of an original image from another image, less blurred version of the original image. In the simple case of grayscale images, the blurred images are obtained by convolving the original grayscale images with Gaussian kernels having different standard deviations. Blurring an image using a Gaussian kernel suppresses only high-frequency spatial information. The most important parameters for this algorithm are two versions of the radius. They are the easiest to specify looking at viewer. It should be remembered that the increase in short-range leads to wide borders and a decrease in long-range increases the limit, on which is determined by the border or not. In the most cases, the best results are obtained, when the first radius smaller than the second radius. The resulting image is a blurred version of the original image.

As a result, the selected 25 features are presented in pairs & block type vectors and 25 are input to the linear support vector machine. For recognition of the input image, it is necessary to extract 25 features of the image. Each feature is a scalar of each on its own vector of regression coefficients. The SVM for gender recognition works using the 25 received features. Out model works with the histogram *WP* for the elements of the vector x, which corresponding "male" in the interval $(-1..0.1)$ with a step $2/(N\,b)$. The similar histogram *WM* is a histogram only for the "women".

7.6.2 Experimental Results

We performed the gender recognition experiment at the 60PFCD database and measured how accuracy (a number of correctly recognized images divided by a total number of tested images) depends on a quality of tested images (Fig. 7.17). At first, we measured an accuracy of gender recognition for all dataset ("all" bin), after that used only the high quality images for accuracy testing (Top1, Top3 by the MOS metric). In addition, we used more practical decision making scheme, when a decision about actual gender is made using not only a single image but all Top3

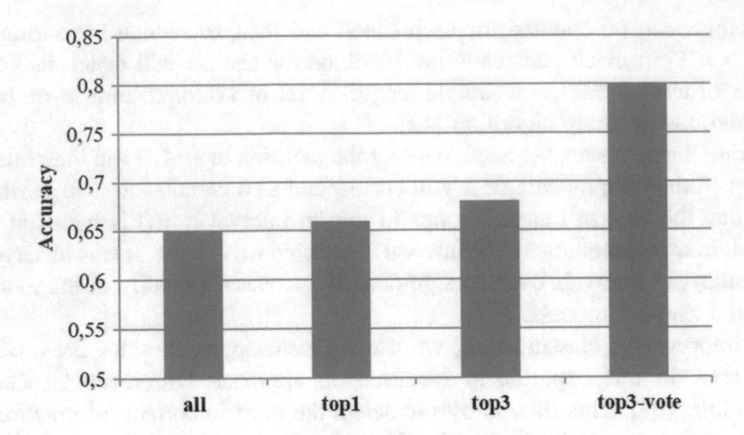

Fig. 7.17 Gender recognition accuracy on 60PFCD database

images (top3-vote). Final gender would be the gender that recognized more often in the particular sequence.

For example, assume that for sequence from three images results of gender recognition are "female", "male", "female". Thus, we see that "female" is more often and "female" would be the result. It could be seen from the results that percent of all correctly classified image is only 65%. This can be explained by the difference in train (LFW [43]) and test (60PFCD) databases for gender recognition algorithm. Images in these databases differ with resolutions, person poses, and lightning conditions. When we used Top1 and Top3 images sample overall recognition accuracy slightly increases. A drastic increase in gender recognition accuracy was obtained, when we used Top3 images sample with voting decision making scheme. In this case, the increase is equal 15% and overall accuracy is equal 80%.

7.7 Conclusions

A set of face image quality metrics was investigated in relation to the problem of selecting the best images for gender classification problem and biometric identification by facial images. The use of only the highest quality faces in a recognition process leads to accuracy improving and savings in computational resources. We obtained 15% accuracy increase in gender recognition, while using only three top quality Images, and 15–18% in face recognition. Also, the simulation results show that at normal and high illuminance levels the proposed measure based on symmetry of landmarks points outperforms other quality measures. In the experiment on a choice of three best images, the measure based on learning to rank shows the best result. The accuracy of the proposed no-reference NRQ LBP metric in face quality assessment task depends on the illuminance level. It performs well, when the illuminance is more than 100 lx. The results will be useful to engineers in building video surveillance and biometric identification using the facial images.

Acknowledgements This work was supported by Russian Foundation for Basic Research grants (№ 15-07-08674 and № 15-08-99639).

References

1. Zhao, W., Chellappa, R., Phillips, P., Rosenfeld, A.: Face recognition: a literature survey. ACM Comput. Surv. (CSUR) **35**(4), 399–458 (2003)
2. Ozay, N., Tong, Y., Frederick, W., Liu, X.: Improving face recognition with a quality-based probabilistic framework. In: International Conference on Computer Vision Pattern Recognition (CVPR'2009) Biometrics Workshop, pp. 134–141 (2009)
3. ISO/IEC 19794-5 (published version). Information technology—biometric data interchange formats (2005)
4. Machine readable travel documents. International Civil Aviation Organization. Available from: http://www.passport.go.kr/img/download/vol2.pdf (2006). Accessed 12 June 2017
5. Yang, Z., Ai, H., Wu, B., Lao, S., Cai, L.: Face pose estimation and its application in video shot selection. In: International Conference on Pattern Recognition (ICPR'2004), pp 322–325 (2004)
6. Sang, J., Lei, Z., Li, S.Z.: Face image quality evaluation for ISO/IEC standards 19794-5 and 29794-5. In: Tistarelli, M., Nixon, M.S. (eds.) Advances in Biometrics, LNCS, vol. 5558, pp. 229–238 (2009)
7. Zhang, G., Wang, Y.: Asymmetry-based quality assessment of face images. In: Bebis, G., Boyle Parvin, B., Koracin, D., Kuno, Y., Wang, J., Pajarola, R., Lindstrom, P., Hinkenjann, A., Encarnação, M.L., Silva, C.T., Coming, D. (eds.) Advances in Visual Computing, LNCS, vol. 5876, pp. 499–508 (2009)
8. Berrani, S., Garcia, C.: Enhancing face recognition from video sequences using robust statistics. In: IEEE International Conference on Video and Signal Based Surveillance (AVSS'2005), pp. 324–329 (2005)
9. Wong, Y., Chen, S., Mau, S., Sanderson, C., Lovell, B.C.: Patch-based probabilistic image quality assessment for face selection and improved video-based face recognition. In: Computer Vision and Pattern Recognition (CVPR'2011), pp. 74–81 (2011)
10. Nikitin, M., Konushin, A., Konushin, V.: Face quality assessment for face verification in video. GraphiCon'2014, pp. 111–114 (2014)
11. Chen, J., Deng, Y., Bai, G., Su, G.: Face image quality assessment based on learning to rank. IEEE Sig. Process. Lett. **22**(1), 90–94 (2015)
12. Zhu, X., Ramanan, D.: Face detection, pose estimation and landmark localization in the wild. In: IEEE Conference on Computer Vision and Pattern Recognition (CVPR'2012), pp. 2879–2886 (2012)
13. Gao, X., Li, S.Z., Liu, R., Zhang, P.: Standardization of face image sample quality. In: International Conference on Biometrics, pp. 242–251 (2007)
14. Nasrollahi, K., Moeslund, T.B.: Face quality assessment system in video sequences. In: Schouten, B., Juul, N.C., Drygajlo, A., Tistarelli, M. (eds.) Biometrics and Identity Management, LNCS, vol. 5372, pp. 10–18 (2008)
15. Chen, J., Yang, C., Deng, Y., Zhang, G., Su, G.: Exploring facial asymmetry using optical flow. IEEE Sig. Process. Lett. **21**(7), 792–795 (2014)
16. Hadid, A., Pietikainen, M.: From still image to video-based face recognition: an experimental analysis. In: Automatic Face and Gesture Recognition (AFGR'2004), pp. 813–818 (2004)

17. Nenakhov, I., Khryashchev, V., Priorov, A.: No-reference image quality assessment based on local binary patterns. In: 14th IEEE East-West Design & Test Symposium, pp. 529–532 (2016)
18. Zhang, M., Xie, J., Zhou, X., Fujita, H.: No reference image quality assessment based on local binary pattern statistics. In: Visual Communication Image Processing (VCIP'2013), pp. 1–6 (2013)
19. Mittal, A., Moorthy, A., Bovik, A.: No-reference image quality assessment in the spatial domain. IEEE Trans. Image Process. **7**(12), 4695–4708 (2012)
20. Sun, T., Ding, S., Xu, X.: No-reference image quality assessment through SIFT intensity. Appl. Math. Inf. Sci. **4**, 1925–1934 (2014)
21. Li, C., Bovik, A., Wu, X.: Blind image quality assessment using a general regression neural network. IEEE Trans. Neural Networks **22**, 793–799 (2011)
22. Pietikainen, M., Hadid, A., Zhao, G., Ahonen, T.: Computer vision using local binary patterns. Springer, London Dordrecht Heidelberg New York (2011)
23. Ojala, T., Pietikainen, M., Maenpaa, T.: Multiresolution gray-scale and rotation invariant texture classification with local binary patterns. IEEE Trans. Pattern Anal. Mach. Intell. **24**(7), 971–987 (2002)
24. Sheikh, H., Wang, Z., Cormack, L., Bovik, A.: LIVE image quality assessment database release 2. Available from: http://www.citeulike.org/user/luisette/article/11534666 (2006). Accessed 12 June 2017
25. Geurts, P., Ernst, D., Wehenkel, L.: Extremely randomized trees. Mach. Learn. **36**(1), 3–42 (2006)
26. Wang, Z., Simoncelli, E., Bovik, A.: Multi-scale structural similarity for image quality assessment. In: IEEE Asilomar Conference on Signals, Systems, and Computers, pp. 1398–1402 (2003)
27. Feng, Z., Huber, P., Kittler, J., Christmas, W., Wu, X.J.: Random cascaded-regression copse for robust facial landmark detection. IEEE Sig. Process. Lett. **22**(1), 76–80 (2015)
28. Amos, B., Ludwiczuk, B., Mahadev, S.: Openface: a general-purpose face recognition library with mobile applications. CMU-CS-16-118, CMU Sch. Comput. Sci., Tech Rep (2016)
29. Taigman, Y., Yang, M., Ranzato, M., Wolf, L.: DeepFace: closing the gap to human-level performance in face verification. In: IEEE Conference on Computer Vision Pattern Recognition (CVPR'2014), pp. 1701–1708 (2014)
30. King, D.: Dlib-ml: a machine learning toolkit. J Mach. Learn. Res. **10**, 1755–1758 (2009)
31. Schroff, F., Kalenichenko, D., Philbin, J.: FaceNet: a unified embedding for face recognition and clustering. arXiv preprint arXiv:1503.03832 (2015). Accessed 12 June 2017
32. Ng, H., Winkler, S.: A data-driven approach to cleaning large face datasets. In: IEEE International Conference on Image Processing (ICIP'2014), pp. 343–347 (2014)
33. Yi, D., Lei, Z., Liao, S., Li, S.: Learning face representation from scratch. arXiv preprint arXiv:1411.7923 (2014). Accessed 12 June 2017
34. Howse, J.: OpenCV Computer Vision with Python. Packt Publishing Ltd, Birmingem (2013)
35. Makinen, E., Raisamo, R.: An experimental comparison of gender classification methods. Pattern Recogn. Lett. **29**(10), 1544–1556 (2008)
36. Khryashchev, V., Priorov, A., Shmaglit, L., Golubev, M.: Gender Recognition via Face Area Analysis, pp. 645–649. World Congress on Engineering and Computer Science, Berkeley, USA (2012)
37. Khryashchev, V., Ganin, A., Golubev, M., Shmaglit, L.: Audience analysis system on the basis of face detection, tracking and classification techniques. In: International MultiConference of Engineers and Computer Scientists Hong Kong, LNECS, pp. 446–450 (2013)
38. Tamura, S., Kawai, H., Mitsumoto, H.: Male/female identification from 8 to 6 very low resolution face images by neural network. Pattern Recogn. Lett. **29**(2), 331–335 (1996)
39. Moghaddam, B., Yang, M.H.: Learning gender with support faces. IEEE Trans. Pattern Anal. Mach. Intell. **24**(5), 707–711 (2002)

40. Toews, M., Arbel, T.: Detection, localization, and sex classification of faces from arbitrary viewpoints and under occlusion. IEEE Trans. Pattern Anal. Mach. Intell. **31**(9), 1567–1581 (2009)
41. Gutta, S., Wechsler, H., Phillips, P.J.: Gender and ethnic classification. In: IEEE International Conference on Automatic Face and Gesture Recognition (AFGR'1998), pp. 194–199 (1998)
42. Alpaydin, E.: Introduction to machine learning. The MIT Press (2010)
43. Learned-Miller, E., Huang, G.B., RoyChowdhury, A., Li, H., Hua, G.: Labeled faces in the wild: a survey. In: Celebi, M.E., Smolka, B. (eds.) Kawulok M, pp. 189–248. Advances in Face Detection and Facial Image Analysis, Springer (2016)

Chapter 8
Real Time Eye Blink Detection Method for Android Device Controlling

Suzan Anwar, Mariofanna Milanova and Daniah Al-Nadawi

Abstract Designing systems to detect the human gestures and movements is an important area in computer vision. In this chapter, a method to detect human eye blink patterns is proposed. Our system detects the user's eye blink patterns in real time and responds with an action on a mobile device, such as the phone call, text message, and/or an alarm. In this chapter, several image processing techniques are used for detecting human eye blinks. To examine the state of the eyelid, whether it's opened or closed, the eye state value is used by computing the minimum threshold. The system is able to track the blinking of the eyes efficiently and accurately from the video using the proposed method. This system is user-friendly and easy to operate. The experiment was performed under different conditions by changing the distance from the camera and light in the room. The experimental results showed that the overall detection rate for eye blink is 98%. The proposed method takes only 8 ms as the average execution time for each frame, which makes it work more efficiently in real time applications.

Keywords Computer vision · Face detection · Eye detection · Eye tracking Android mobile control · Median blur filter · Human computer interaction

---

S. Anwar · M. Milanova (✉)
Department of Computer Science, University of Arkansas at Little Rock,
2801 S. University Avenue, Little Rock, AR 72204, USA
e-mail: mgmilanova@ualr.edu

S. Anwar
e-mail: sxanwar@ualr.edu

D. Al-Nadawi
Department of Biology, University of Arkansas at Little Rock,
2801 S. University Avenue, Little Rock, AR 72204, USA
e-mail: dtalnadawi@ualr.edu

M.N. Favorskaya and L.C. Jain (eds.), *Computer Vision in Control Systems-4*,
Intelligent Systems Reference Library 136,
https://doi.org/10.1007/978-3-319-67994-5_8

8.1 Introduction

The demand of Human Computer Interaction (HCI) or Mobile Computer Interaction (MCI) increased rapidly in the recent years. Detection of the closing and opening of the eyelid, known as eye blinking, is an important area of research across different domains. Some of these domains are human mobile interaction, computer interaction, driving safety, and health care. For instance, people with disabilities can use eye blinks to interact with computer. Another example is using eye blink detection to detect a driver's drowsiness and consciousness [1].

Also eye blink detection is important to monitor a human operator's vigilance, where a user is staring for a long time at a screen without blinking [2]. Detecting eye blink is used in face recognition system to prevent anti-spoofing [3]. There are two types of existing systems. The first type is reliable but needs an expensive hardware, such as the infrared cameras and glasses with cameras to observe the eyes. The second type is also reliable but uses standard camera only. Before starting to detect the eye blink, we have to go through some steps as shown in Fig. 8.1.

The analysis of eye blink consists of four stages, including the face detection, eye detection, eye tracking, and eye blink detection. The system starts with capturing a video from the front camera of any Android device, such as the smart phone or tablet. As a pre-processing, frames will be created from the input videos, each frame will be converted to grayscale frame using OpenCV conversion procedures. For face detection, Haar classifier is used. As for eye detection and tracking, AdaBoost and Haar feature algorithms are trained and implemented. To detect the eye blink, the minimum threshold that indicates to the binary image should be found after applying median blur filtering. Finally, the device will make an action as a response to the eye blinking. The objectives of this chapter are mentioned below:

- The overview of the recent eye tracking, face detection, eye detection, and mobile technologies.
- The implementation of designed application for an eye tracking system that detects the motion of a human eye in real time.
- The control of some mobiles device's activities and actions with human mobile interaction directly without the need of hands.

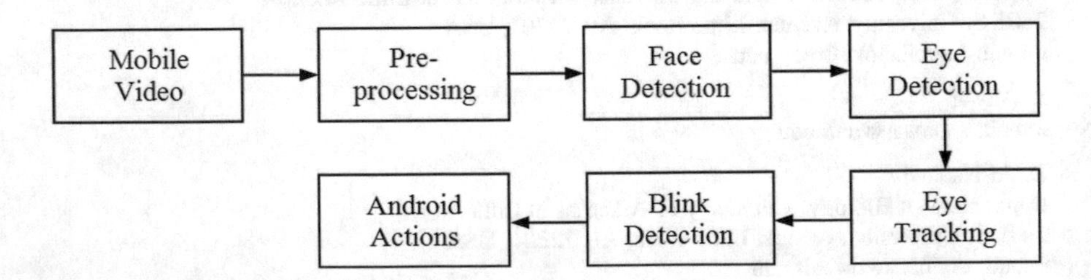

Fig. 8.1 Block diagram for eye blink detection

- The study of the effect of light and distance between the eyes and the mobile device for evaluating the accuracy detection and overall accuracy.

The interaction between a human and a mobile device can be seen in Sect. 8.2. More details about eye tracking are explained in Sect. 8.3. Some famous OpenCV algorithms and Android devices growth are described in Sect. 8.4. In Sect. 8.5, the recent literature review related to detecting face, eye, and blink is presented. Section 8.6 provides details about the proposed method for eye blink detection to control any Android device. Performance and results evaluation are discussed in Sect. 8.7. Finally, Sect. 8.8 presents the conclusions and future work.

8.2 Human Mobile Interaction

Mobile interactions played a huge role in customer's daily lives. These human mobile interactions freed customers from having use a stationary personal computer. Designing successful human mobile interactions requires a good understanding of the overlapping areas of connection or context, in which the interaction takes place. To that end, constructing a connection or context for human mobile interaction has been designed. Figure 8.2 shows how the user accompanies his/her mobile device throughout the day doing multiple tasks. Unlike home desktop, stationary work, or even laptop that cannot always be easily be taken to every places, such as appointments, meetings, schools, and airports even though they are

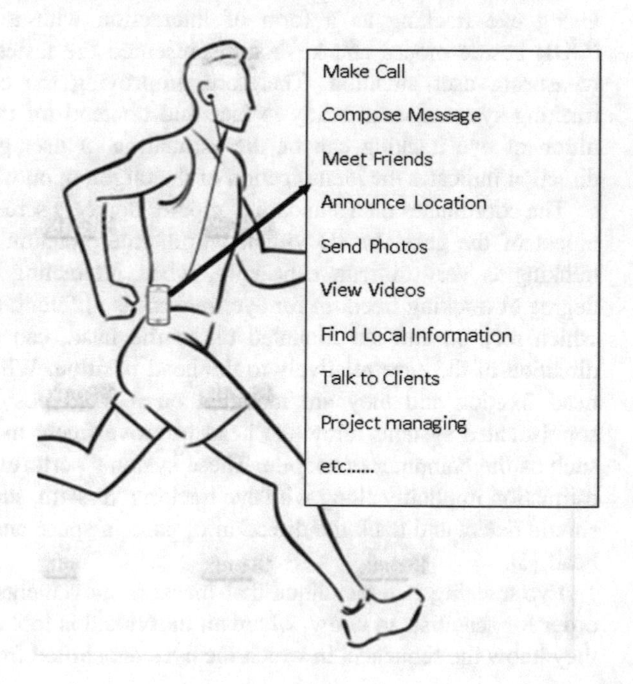

Fig. 8.2 Mobile interaction tasks

needed, the phones can travel with people in all the outdoor as well as indoor activities.

For successful human mobile interactions, the following heuristics are used. The heuristics are concluded from human computer interaction theory and practice:

- Mobile interaction is highly user-driven. According to the high and strong personal nature of mobile devices, high relevance is essential.
- The competition between new mobile experiences and old user patterns. Along with the expansion of mobile services the design of the new mobile take into account user's reliance on patterns and samples that come from prior mobile technologies.
- The main goal is ease of use. Ease of use on web can be achieved by reducing choice and guiding navigation. For mobile phone, ease of use can be accomplished by making affordances for the confusion, background noise, and hindering.
- The new technology should be calm. People valued the calm technology over discomfort or disruptions. For most of the day, the mobile device is in closer proximity to users more than the computer; the mobile device competes with other devices demands on the user's interest [4].

8.3 Eye Tracking Technology

Using eye tracking as a form of interaction with a Graphical User Interface (GUI) beside mouse and keyboard, presented the fastest non-invasive method of measuring user attention. Therefore, improving the cost and accuracy of eye tracking systems stand ready to face and contend for this role [5]. The best definition of eye tracking can be the estimation of user gaze's direction. The gaze direction indicates the identification of the target, upon which the gaze or stare falls.

The coordinates on a standard computer device's screen identify the target or the object of the gaze. In 3D virtual world, interpretation of gaze direction for eye tracking is very difficult especially, when interacting with the real world. The degree of tracking freedom for eye trackers is different. Some eye tracker systems, which rely on and are mounted on to the head, can detect and track only the direction of the gaze relatively to the head position. While other systems require a head fixation and they are mounted on a fixed position of the eyeball. More sophisticated systems allow the head to move freely in front of a specific device, such as the computer or mobile. These systems perform the head tracking or head estimation implicitly along with eye tracking. System, such as wearable eye tracker, should detect and track the direction of gaze in space and not only relatively to the head [6].

Eye tracking is a technique that measure movements of an individual's eye in order for scientists to know, where an individual is looking at and at the same time they know the sequence, in which the eyes are shifted from one location to another.

Eye tracking is a technique, where an individual's eye movements are measured so that the researcher can be able to tell, where a person is looking at any given time as well as the range, in which their eyes are strolling or shifting from one location to another. The advantage of eye tracking system is to help human computer interaction researchers understand the process of visualizing and displaying information and the factors that may impact upon the system interfaces usability.

In this way, the registers or records of eye movement can provide an impartial source of interface estimation data that can report and estimate the design of the interface. A major advantage of capturing or detecting eye tracking is to use it as control signals to help people, with special need for instance, to interact with interfaces without the need of using any input device, such as the keyboard or mouse [7].

Tracking the human eye movements and estimating the direction have already been completed. Android would use the exact implementation by dealing with the measurement, estimating, recording as well as analyzing information about the position and movements of the human eyeballs. The newest version of Android smart phones allows people to use their eyes to interact with the mobile device [8].

8.4 Face and Eye Detection Based on OpenCV Algorithms

Since the human face is considered as a dynamic object with different shapes and colors, face detection is challenging. Facial recognition is not possible if the face is not isolated from the background. To improve human computer interaction, the approaches for gaze estimation, emotion recognition, and gesture detection could be used. These approaches require the facial feature detection, tracking, and recognition. There are many existing approaches that detect the human face, with each approach having strengths and weaknesses. For face detection, some methods use the contours, templates, skin tones, or neural networks.

The problem that all these algorithms share is that they are costly in terms of computation. An image consists of a matrix of values that represent a light intensity called pixels. For face detection, these pixels should be analyzed, which analyzing pixels is a time consuming process and is even very difficult to accomplish. To do so, these pixels should be reanalyzed for both accuracy and scaling. An algorithm for object detecting, such as detecting human faces, based on the Haar classifiers had been devised by Viola and Jones. They used AdaBoost to classifier cascade rather than pixels [9]. The Haar-like features is the heart of Haar classifier. Haar-like features use the variance between neighbor groups of pixels instead of the intensity values of a pixel. The use of the variances between the pixel groups are to distinguish between the light and dark areas. Figure 8.3 shows how to form Hass-like features using two or three adjacent groups with a relative variance and finally detecting an image [10].

Scaling Haar features can easily be done by decreasing or increasing the size of examined pixel groups. This scaling process allows using features to detect objects

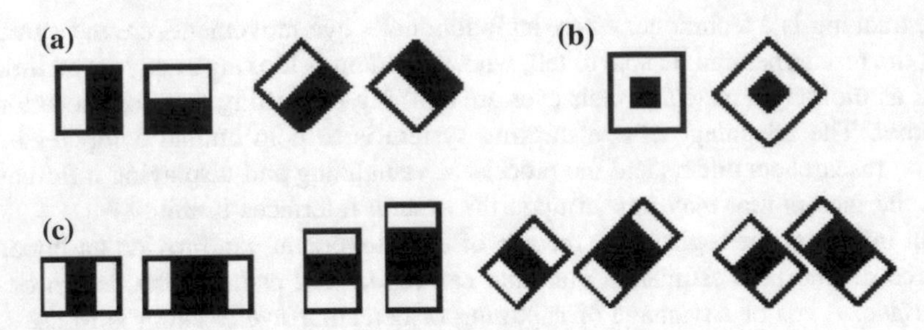

Fig. 8.3 Common Haar features [5]: **a** edge features, **b** center-surround features, **c** line features

of different sizes. Training Haar classifier cascade is required before detecting human facial features, such as the eye, mouth, and nose. In order to train the Haar classifiers cascade, the implementation of both Haar feature and AdaBoost algorithms should be taking place. Fortunately, an open source library developed by Intel is devoted to help the execution of some programs related to computer vision called Open Computer Vision Library (OpenCV). Applications that are related to areas of robotics, HCI, biometric, image processing and others employ often OpenCV library that trained and detected objects using Haar classifier cascade. The training of Haar classifier cascade is needed in negative images, which are referred to a group of images that contain scenes without the object. Also the training process is needed in another group of images that contain the visual objects, and this group is called positive images.

Image name, height, and width of the object are used to specify the position of the objects within the positive group of images. Around 5000 images from the negative group along with at least a mega-pixel resolution are involved for training facial features. The selected trained images were a set of everyday objects, such as natural scenery, paperclips, and many photographs of woods and mountains. For robust facial feature detection production, the original group of positive images must be represented of the variance among people with different race, age, and gender.

National Institute of Standards and Technology's (NIST) Facial Recognition Technology (FERET) database are the best and most used resources for these training images. This database includes more than 10,000 images of over 1000 people under different poses, lighting conditions, expressions, and angles. Around 1500 images are used for facial feature training. These training images were taken from frontal view and the angles were ranging from 0° to 45°. This angle ranging is important to provide the desire variance needed to allow object features detection, when the head is moving slightly. Classifiers for the eyes, mouth, and nose were trained separately. After training the classifier, the facial feature is detected within a different set of images from the FERET dataset using Haar classifier cascade. Table 8.1 shows the results of calculated accuracy of the classifier. As shown in Table 8.1, the classifiers have a best performance, when detecting the eye and the

Table 8.1 The accuracy of classifiers [9]

Facial feature	Positive hit rate (%)	Negative hit rate (%)
Eyes	93	23
Nose	100	29
Mouth	67	28

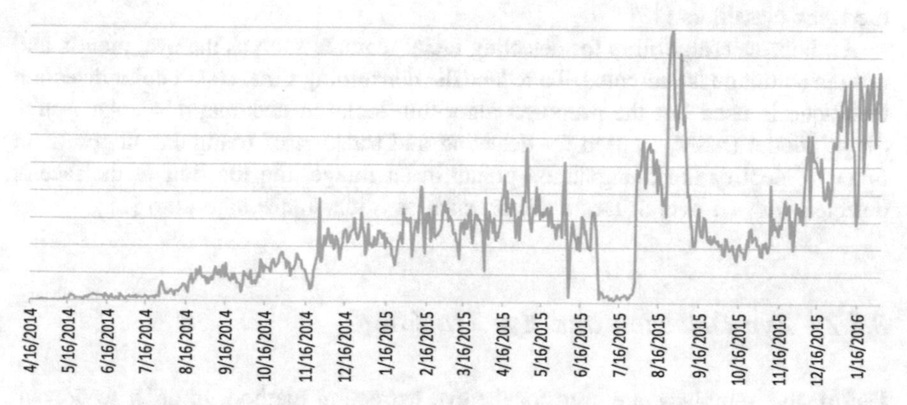

Fig. 8.4 The growth of Android device

nose features but not the mouth [9]. Some famous algorithms for face and eye detection is listed in Sect. 8.6.

More than 190 countries around the world are using Android devices, such as the phones and tablets. Android is considered as the largest installed base among all available mobile platforms in the world and it is growing very fast and everyday millions of people turn on their Android devices and look for a new apps, games, as well as services to install. Android presents a world-class platform that allows for creating games and apps for the users of Android everywhere [11]. Figure 8.4 shows the growth of Android platform.

8.5 Literature Review

There are many researchers that propose different methods for the face detection, eye detection, eye tracking, and eye blink detection, and these researches represent the background for this chapter. In Sects. 8.5.1–8.5.3, several famous and well known algorithms for the face detection, eye detection, and eye blink are reviewed, respectively.

8.5.1 Face Detection

The series list of Haar classifier and single feature filters are implemented to identify sub-region image using a fast computation of integral image technique performed very well in real time [12]. The cascade of boosted classifiers based on Haar-like features was built by two training data sets, positive samples and negative samples. A learning algorithm, Adaptive Boost, is used to construct a strong classifier from the weak classifiers [13].

An improved algorithm for detecting facial features such as the eye, mouth, and nose in an image is presented. To reduce the consuming time, a skin color detection technique is used but the proposed algorithm lacks in accuracy [14]. An Active Shape Model (ASM) is used for detecting and tracking of triangulation points. In order to identify the triangulation points in an image, the location of the face is detected with an overall face detector, such as Viola-Jones algorithm [15].

8.5.2 Eye Detection and Eye Tracking

Deformable templates are used for the eye extraction method. In order to prevent over shrinking, the eye corner detector is introduced and has improved both the speed and accuracy [16]. Composite vectors, including several pixels, are used for eye detection method with a new Biased Discriminate Analysis (BDA) [17]. A study of a new eye tracking paradigm is worked very well on continuous eye movement, analysis, as well as monitoring. The proposed method provides the context aware computing, HCI, and eye tracking techniques to discuss the classical eye tracking techniques [18].

The design and implementation of an eye tracking system on a tablet PC is discussed in [5]. This method used a neural network eye tracker for eye tracking.

In order to prevent computer vision Syndrome, an artificial system using eye movement analysis is proposed. An ElectroOculoGram (EOG) signal acquisition system is developed to record the ocular, while Daubechies order four mother wavelet is used as a signal for eye features to obtain the wavelet detail coefficients. A special hard-ware like electrodes placed on the scalp is needed to measure EOG [19].

8.5.3 Eye Blink

A method uses an image flow analysis and deterministic finite state machine is proposed for eye tracking and finding eyelid state. The blink detection, such as the blink rate, blink count, and transitional statistics, play a big role in the HCI across a wide range of disciplines [20]. Method for the face, mouth, and eye detection within the specific segments of the image was introduced to determine the drowsiness by monitoring the eye action, such as blinking [21]. The output was produced accordingly, and all activities had been considered. The proposed method considered driver's attention as well.

Doppler sensor is used for detecting eye blinking, where a unique Doppler signature can be obtained from the reflected wave of the blinking eye [22]. The classification of a close or open eye state and using Doppler sensor to set the threshold for each user was proposed. The proposed method handled the head rotation efficiently [23]. A pair of parameterized parabolic curves was used to model the shape of eyes and fit each frame to maximize the total likelihood of eye areas. The proposed system tracked the face movement and eyelid reliably in real time [24]. Use of histograms in similar way to the spirit of using pixel intensities rather than projections was introduced in [25]. Several methods yielding the template based matching to detect eye blink are presented in works [26–28].

The residual movement of the eyelid is captured by subtracting and distinguishing the movement of a whole face [29]. The idea proposed is assuming that the magnitude and orientation during head movement for all motion vectors of each pixel are similar. On the other hand, the motion vectors for eye blink are different [29]. Another idea proposed is that detecting eyelid state to be closed for two or three seconds is considered as drowsiness [30]. Detecting the eye contour using the ASM is also represented. To obtain new eye shape, the model trains the appearance surrounding each landmark and matches it in the original frame. The proposed in [31] method estimated the position of each landmark, extracted the Eye Aspect Ratio (EAR), and use Support Vector Machine (SVM) as a classifier.

Observers claim that the world just entered an era called the "post-pc", that very few people use their laptops to access the Internet, and that people are rather surfing the internet almost exclusively on their smart phones. It is hard to argue with how important those smart phones have become to our daily lives. All above methods developed a system using a PC for eye blinking. In this chapter, a new method for detecting eye blink is proposed for Android mobile and can work efficiently in real-time mobile applications. It is designed to detect eye blink in order to control Android mobile device actions, such as the making a cell phone call, sending an SMS, and setting an alarm device.

8.6 Proposed Method

The proposed method consists of five main parts including the mobile camera processing, face detection, eye detection, eye tracking, and blink detection. Figure 8.1 depicts the general steps of the proposed method. Consider the main steps of the proposed method.

Mobile camera processing. In this step, a video will be captured from the front camera of any Android device. A frame will be created from the input video; each

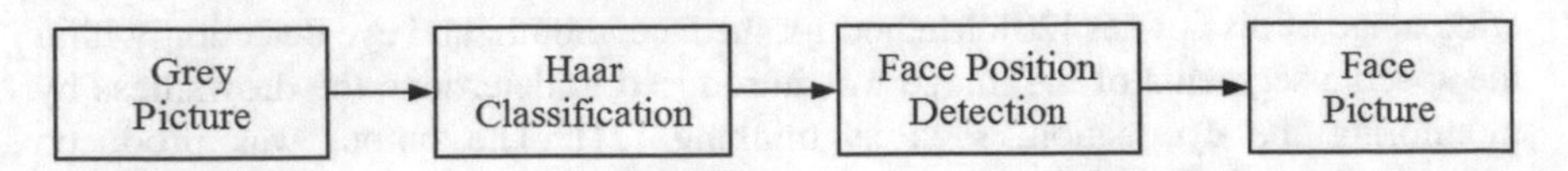

Fig. 8.5 Steps of face detection

Fig. 8.6 Detecting human's face using Haar cascade classifier

frame will be converted to grayscale frame using OpenCV conversion procedures that works on extracting only the luminance component from each frame. Samsung mobile is used to capture video for volunteers facing the front camera, which focuses on their face. The input video is a collection of color frame, which is extracted from the input video. The resulted frames will be stored in special mobile folder to be used later for face detection step.

Face detection. An image is known as a collection of color and/or light intensity values represented by pixels, and these pixels use them. This analyzing procedure is considered to be time consuming and difficult because of the wide variations of color and shape within the human face. For face detection, Haar classifier is used for some reasons as the pixels often need to be reanalyzed for precision and scaling as shown in Fig. 8.5. Based on facial detected features, Haar classifier detects faces, and then the area of the image should be regionalized to the location of the facial features with the highest probability. Finally, the detected face will be marked with colored rectangle and work in the next step (eye detection) to find an axis of the eye. Figure 8.6 shows the result of using Haar cascade classifier for face detection.

Eye detection. Before detecting the eye, the Haar cascade classifier should be trained by implementing AdaBoost and Haar feature algorithms. A set of negative scene images is used to detect the facial features. This set of images is referred to as the negative images. Another set of images is needed and called the positive images, which contain one or more instances of the object. The eyes detection is implemented by tacking the advantage of the relation between the human's facial axes and his/her eyes that the eye axes, which connect both right and left eye, is symmetric and perpendicular with human's facial axes.

Eye tracking. The most important eye features that have been used in the proposed method are the corneal reflection and pupil center. These two features helped

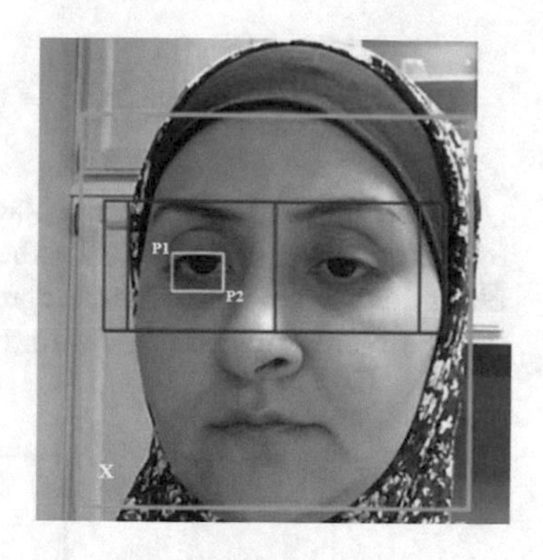

Fig. 8.7 Example result of eye detection results using the cascade of boosted classifiers based on Haar-like features

to track the movement of the eye by finding the center of the pupil and position of the cornea reflection. In order to do that, the vector between the eye's pupil and corneal is computed and measured, then the Region Of Interest (ROI) is found with further trigonometric calculations.

Figure 8.7 shows the result of using the cascade of boosted classifier based on Haar-like features to detect the eye. At the same time, Fig. 8.7 shows the success of the proposed method to make both eye's pupil (represented by P1 and P2) and human face (represented by X) move synchronously. The code below used to implement the operation of detecting X, P1, and P2, where faces are objects that will store the rectangles that bound the faces in the frame:

```
Rect[] facesArray = faces.toArray(); // Each rectangle in the faces array is a face
for (int i = 0; i < facesArray.length; i++){
P1 = facesArray[i].tl(); // Group the detections by the top-left corner
P2 = facesArray[i].br(); // Group the detections by the bottom-right corner
}
```

Eye blink detection. A few approaches and techniques are discovered to detect eye blinking and movement. These techniques used in active scene, where the camera and the human face can move in every direction independently and the eyes move freely. Since the human's eye by nature moves as combination of involuntary cognitive process, in this case care should be taken. The proposed algorithm aims to find the minimum threshold that indicates to the binary image, which has at least one black pixel after applying median blur filtering.

The eye's grayscale frames that were detected from the eye detection step and show only the eye and eyebrow picture will be used in this step to decide, whether the eye is open or closed. For the first 15 frames, the proposed algorithm will initialize them with threshold 70, which found by experiments as a best threshold value to start with. After initializing the frames a binary threshold will be applied on them, using Eq. 8.1 to decide the threshold for each frame.

$$newP(x,y) = \begin{cases} MAX & P(x,y) > threshold \\ 0 & \text{otherwise} \end{cases} \tag{8.1}$$

According to Eq. 8.1, each pixel $P(x, y)$ with threshold higher than threshold its intensity will be set to *MAX*, otherwise it should be set to zero. As a result, each black color pixel in the frame represented by 1, while each gray or white color pixel will be represented by 0. To find the eye state the length L and width W of area

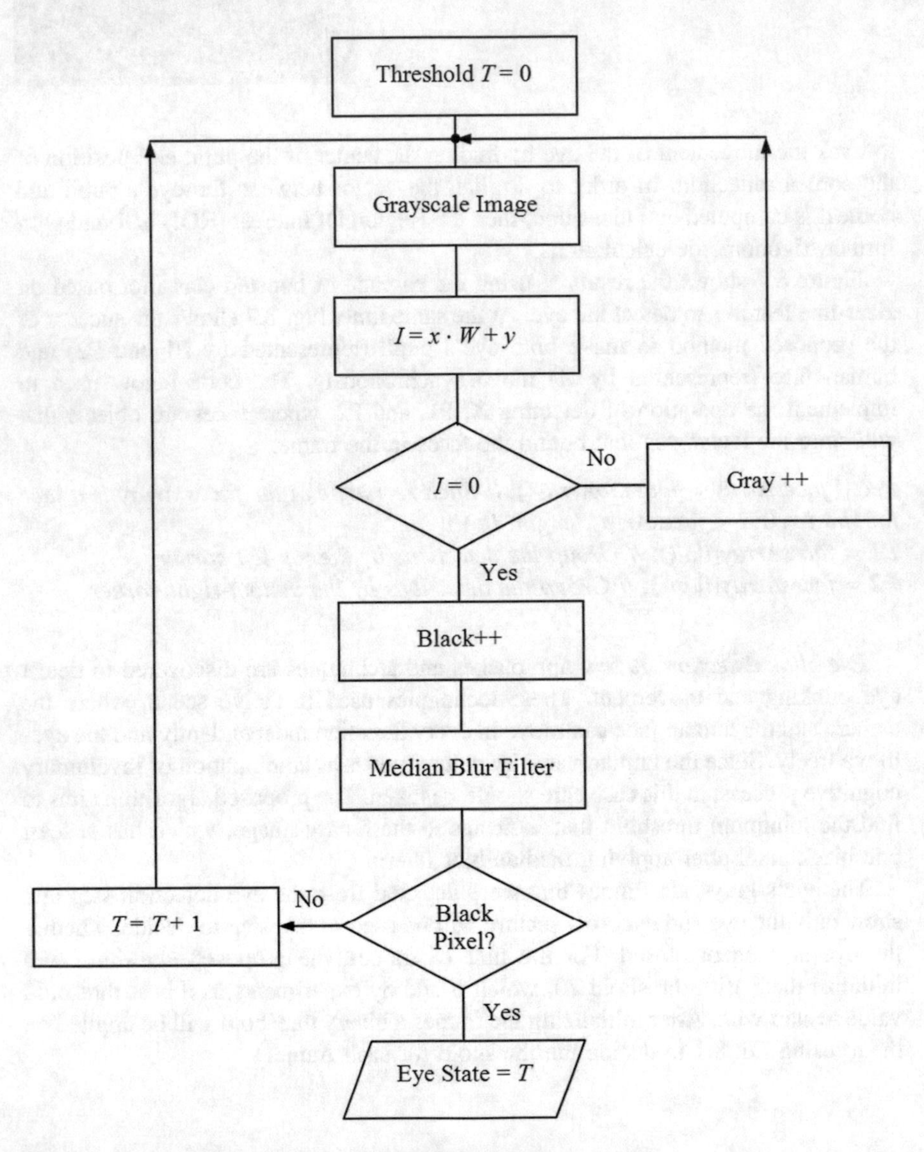

Fig. 8.8 Flow chart of proposed method to find eye state

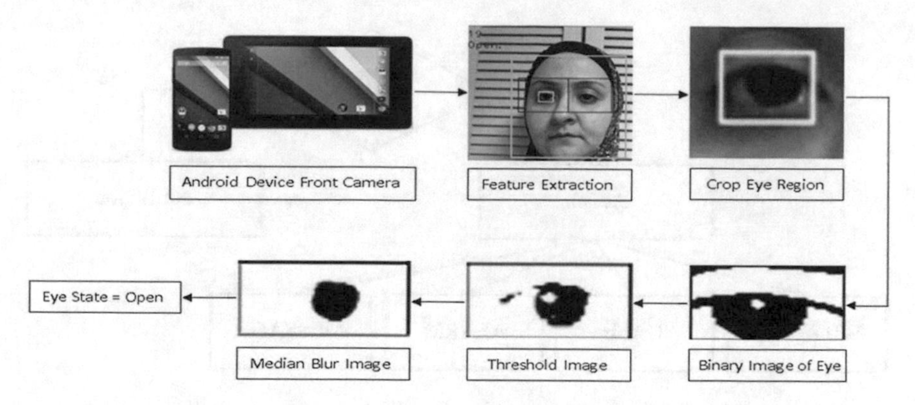

Fig. 8.9 Example of finding open eye state

under the eyebrows will be determined, and the number of gray pixel will be increased if the result of multiplying the *x* coordinate of the pixel by the *W* and adding the result to the pixel's *y* coordinate is zero. Otherwise, the number of black pixel will be increased. After that, the median blur filter is applied to replace each pixel with the median of neighboring pixel. The threshold will be increased and the same process is repeated if there is no black pixel, otherwise the eye state will get the threshold value.

Finally, by counting the number of black and white pixel we can decide whether the eye state is open or closed. Figure 8.8 shows the flow chart of finding eye state using the proposed method. An example of finding open eye state is depicted in Fig. 8.9.

Mobile Controlling with Eye Blinking. The mobile activity can be controlled in case the eye blinks by several ways, such as the sending a text message, turning on the alarm system, opening a web browser or making a phone call. The proposed method used ACTION_Message/ACTION_SET_ALARM/ACTION_CALL/ACTION_VIEW actions to control the functionality for Android device as shown in Fig. 8.10. A simple code is used to create an intent with ACTION_CALL action and to make a phone call at a given number for instance (xxxx-xxxxxxx), where the method specify tel: as URI using setData() method. Figure 8.11 shows the interface of the proposed method on Samsung note 3 phone.

The user will select the wanted action by clicking on a corresponding button (Calling, Alarm, Open Browser, or Send SMS) and if the proposed method detects an eye blink, the action will take place.

```
Intent phoneIntent = new Intent(Intent.ACTION_CALL);
CallIntent.setData(Uri.parse("tel:xxxxxxxxxxx"))
```

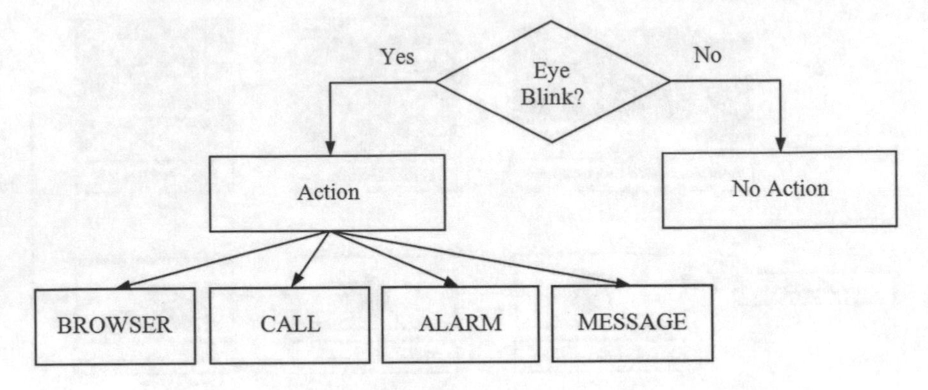

Fig. 8.10 Android mobile controlling

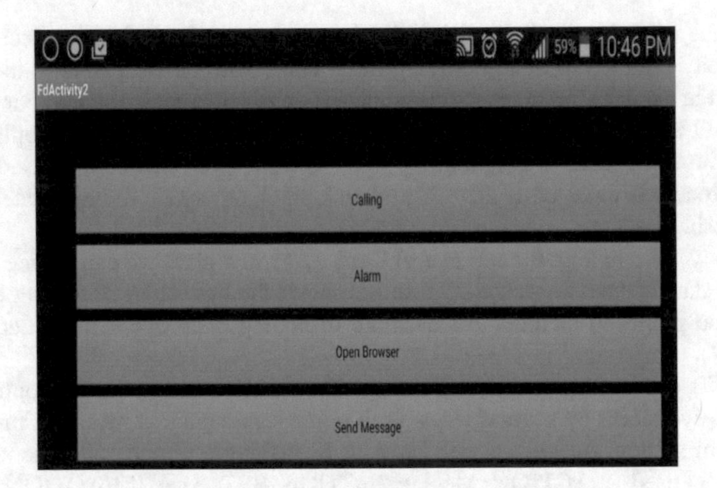

Fig. 8.11 Proposed method's interface using Samsung note 3 phone

8.7 Performance and Results Evaluation

To evaluate the performance of the proposed method, two parameters are used for this purpose. First parameter is the distance between the individual's eye and the device's screen. In the experiments, different distances are used: 15, 20, 25, 30, 35, 40, and 45 cm. The second parameter is the lightning condition of testing location, like indoor and outdoor experiments.

The result has been compared among all of the experiments according to the results of both overall accuracy detection and accuracy detection. The overall detection accuracy as well as the detection accuracy of the eye blinking detection are computed using Eqs. 8.2 and 8.3, respectively, where *TP* is a number of Blink Detected by program, *FP* is a number of Blink Detected, but they are not Blink, *FN* is a number of Blink Not Detected by program, *TN* is a number of Not Blink Detected by program.

$$Overall\,Accuracy = ((TP + TN)/(TP + FP + FN + TN)) \cdot 100\% \quad (8.2)$$

$$Detection\,Accuracy = (TP/(TP + FN)) \cdot 100\% \quad (8.3)$$

Some results of testing the performance of the proposed algorithm is shown in Fig. 8.12. The result of eye status either closed or open (eye blinks or not) is displayed in the top left part of Android device (Samsung phone) screen. Figure 8.13 shows the result of testing the algorithm, the outcomes was very satisfy, accurate, and shows that the method has high response to the eye, face, and head movements. Table 8.2 shows the best test result with specific parameters. The best results of overall and detection's accuracy was 98% when the test is hold with distance 35 cm during indoor experiment.

The test is conducted by using Intel(R) Core(TM) i7-4500U CPU with1.80 GHz, 64 Bit processer with 8.00 GB of RAM. The simulation program is compiled using Eclipse Platform Mars.2 that support both Java code, C++ code, and Windows applications, which benefits from its features to design a user interface for Android mobile application. The mobile that held the tests is Samsung Galaxy note 3 that has Android version 5 with model number SAMSUNG-SM-N900A. Finally, the

Fig. 8.12 The result of detecting eye status

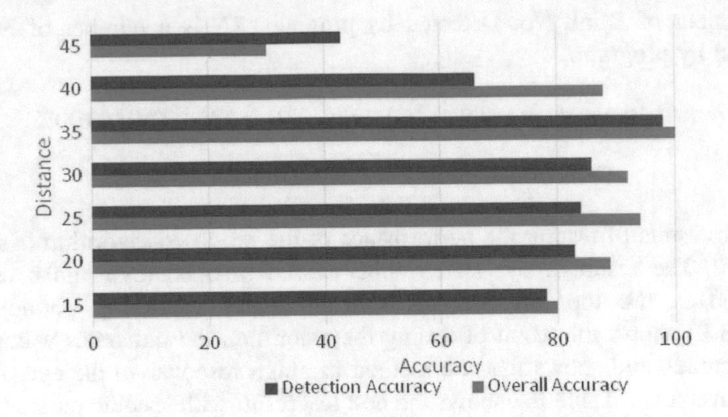

Fig. 8.13 The accuracy results of the proposed algorithm

Table 8.2 The result of detecting the eye blinks in normal room lightning and using a median filter

Distance (cm)	TP	FP	FN	TN
15	5	0	1	10
20	374	102	0	1913
25	836	77	32	1266
30	574	85	0	842
35	666	33	0	896
40	757	155	0	660
45	4	1	0	22

software executed under Windows 10 as the operating system. The test has been conducted by a number of subjects as seen in Fig. 8.12.

The subjects were asked to hold the mobile and start the application for specific amount of time. During the test, the subjects did a normal eye blinking.

8.8 Conclusions and Future Work

Detecting the eye blink to control the activity of any device, such as mobile Android device in real time, is a challenge. Two important factors in detecting eye movements are the light condition and the distance from the device's camera. In this chapter, the proposed method computes and classifies the value of the eyelid state being open or closed. To evaluate the performance of the proposed method, the experiment included a human using Samsung note 3 mobile screen. The overall and detection accuracy had reached 98%. For each frame, the average execution time was 12.30 ms, which indicates that the proposed method works efficiently for all real time applications.

Future work will include using infra-red or multispectral camera to improve the captured images. Additional techniques for the preprocessing stage, such as histogram equalization or gamma correlation, will also be included. Furthermore, using other techniques in detecting additional facial features, such as eyebrows, may enhance the accuracy of detecting eye blinking in the future.

References

1. Le, H., Dang, T., Liu, F.: Eye Blink Detection for Smart Glasses. Washington State University, Vancouver, Portland State University (2000)
2. Drutarovsky, T., Fogelton, A.: Eye blink detection using variance of motion vectors. In: Agapito, L., Bronstein, M.M., Rother. C. (eds.) Computer Vision—ECCV 2014 Workshops, LNCS, vol. 8927, Part III, pp. 436–448
3. Pan, G., Sun, L., Wu, Z., Lao, S.: Eye blink-based anti-spoofing in face recognition from a generic web camera. In: IEEE 11th International Conference on Computer Vision (ICCV'2007), pp. 1–8 (2007)
4. Savio, N., Braiterman, J.: Design sketch: the context of mobile interaction. In: Proceedings of Mobile HCI, pp. 1–3 (2004)
5. Holland, C., Komogortsev, O.: Eye tracking on unmodified common tablets: challenges and solutions. In: Symposium on Eye Tracking Research and Applications (ETRA'2012), pp. 277–280 (2012)
6. Drewes, H.: Eye gaze tracking for human computer interaction. Dissertation an der LFE Medien-Informatik der Ludwig-Maximilians-Universität, München (2010)
7. Poole, A., Ball, L.J.: Eye tracking in human-computer interaction and usability research: current status and future prospects. In: Ghaoui, C. (ed.) Encyclopedia of Human-Computer Interaction, pp. 1–13. Idea Group Inc., Pennsylvania (2005)
8. Kowalik, M.: Do-it-yourself eye tracker: impact of the viewing angle on the eye track. In: 15th Central European seminar on Computer Graphics (CESCG'2011), pp. 1–7 (2011)
9. Wilson, P.I., Fernandez, J.: Facial feature detection using Haar classifiers. J. Comput. Sci. Coll. **21**(4), 127–133 (2006)
10. Orman, Z., Abdulkadir, B., Kemer, D.: A study on face, eye detection and gaze estimation. Int. J. Comput. Sci. Eng. Surv. **2**(3), 29–46 (2011)
11. Developers: Android, the world's most popular mobile platform. http://developer.android.com/about/index.html. Accessed 29 June 2017
12. Viola, P., Jones, M.: Rapid object detection using a boosted cascade of simple features. In: International Conference on Pattern Recognition (CVPR'2001), vol. 1, pp. 511–518 (2001)
13. Adolf, F.M.: How-to build a cascade of boosted classifiers based on Haar-like features. OpenCV's Rapid Object Detection (2003)
14. Kaur, S., Singh, H.: Human eye detection using YCbCr color model, Harr-like features and template matching. Int. J. Adv. Res. Electr. Electron. Instrum. Eng. **4**(2), 825–832 (2015)
15. Anwar, S., Milanova, M., Bigazzi, A., Bocchi, L., Guazzini, A.: Real time intention recognition. In: 42nd Annual Conference of the IEEE Industrial Electronics Society (IECON'2016) (2016). doi:10.1109/IECON.2016.7794016
16. Kuo, P., Hannah, J.: An improved eye feature extraction algorithm based on de-formable templates. In: IEEE International Conference on Image Processing (ICIP'2005), pp. 1206–1209 (2005)
17. Kim, C., Turk, M.: Biased discriminant analysis using composite vectors for eye detection. In: 8th IEEE International Conference on Automatic Face and Gesture Recognition (FG'2005), pp. 17–19 (2008)

18. Bulling, A., Duchowski, A., Paiva Majaranta, P.: The first international workshop on pervasive eye tracking and mobile eye based interaction. In: 13th International Conference on Ubiquitous Computing (2014). doi:10.1145/2030112.2030248
19. Pal, M., Banerjee, A., Datta, S., Konar, A., Tibarewala, D.N., Janarthanan, R.: Electrooculography based blink detection to prevent computer vision syndrome. In: 2014 IEEE International Conference on Electronics, Computing and Communication Technologies (IEEE CONECCT'2014), pp. 1–6 (2014)
20. Heishman, R., Duric, Z.: Using image flow to detect eye blinks in color videos. In: 8th IEEE Workshop on Applications of Computer Vision (WACV'2007), pp. 52–57 (2007)
21. Raees, A., Borole, J.N.: Drowsy driver identification using eye blink detection. Int. J. Comput. Sci. Inf. Technol. **6**(1), 270–274 (2015)
22. Kim, Y.: Detection of eye blinking using Doppler sensor with principal component analysis. IEEE Antennas Wirel. Propag. Lett. **14**, 123–126 (2015)
23. Tamba, C., Tomii, S., Ohtsuki, T.: Blink detection using Doppler sensor. In: IEEE 25th Annual International Symposium on Personal, Indoor, and Mobile Radio Communication (PIMRC'2014), pp. 2119–2124 (2014)
24. Yang, F., Yu, X., Huang, J., Yang, P., Metaxas, D.: Robust eyelid tracking for fatigue detection. In: 19th IEEE International Conference on Image Processing (ICIP'2012), pp. 1–4 (2012)
25. Xu, Y., Jiang, Y., Sun, Y.: Blink detection using 3D cross model. In: 5th International Symposium on Computational Intelligence and Design (ISCID'2012), vol. 2, pp. 115–1185 (2012)
26. Awais, M., Badruddin, N., Drieberg, M.: Automated eye blink detection and tracking using template matching. In: IEEE Student Conference on Research and Development (SCOReD'2013), pp 79–83 (2013)
27. Krolak, A., Strumillo, P.: Eye-blink detection system for human–computer interaction. J. Univers. Access Inf. Soc. **11**(4), 409–419 (2012)
28. Udayashankar, A., Kowshik, A.R., Chandramouli, S., Prashanth, H.S.: Assistance for the paralyzed using eye blink detection. In: 4th International Conference on Digital Home (ICDH'2012), pp. 104–108 (2012)
29. Pauly, L., Sankar, D.: A novel method for eye tracking and blink detection in video frames. In: IEEE International Conference on Computer Graphics, Vision and Information Security (CGVIS'2015), pp. 252–257 (2015)
30. Rahman, A., Sirshar, M., Khan, A.: Real time drowsiness detection using eye blink monitoring. In: National Software Engineering Conference (NSEC'2015), pp. 1–7 (2015)
31. Soukupova, T., Cech, J.: Real-time eye blink detection using facial landmarks. In: 21st Computer Vision Winter Workshop (CVWW'2016), pp. 1–8 (2016)

Chapter 9
Techniques for Medical Images Processing Using Shearlet Transform and Color Coding

Alexander Zotin, Konstantin Simonov, Fedor Kapsargin, Tatyana Cherepanova, Alexey Kruglyakov and Luis Cadena

Abstract Image processing techniques play an important role in the diagnostics and detection of diseases and monitoring the patients having these diseases. The chapter presents the medical image processing and morphological analysis in the solution of urology and plastic surgery (hernioplasty) problems. Novel methodology for processing medical images using a color coding of contour representation obtained by Digital Shearlet Transform (DST) has been presented. The object contours in the medical urology images are obtained using the conventional filters, and then results are compared. Since medical images can contain some noise, it makes sense to suppress the noise at the preprocessing step. For this purpose, the optimized in implementation algorithms of the most frequently used filters, such as the mean filter, Gaussian filter, median filter, and 2D cleaner filter, had been developed. A comparison of the optimized and ordinary implementations of noise reduction filter shows great speed improvement of the optimized implementations

A. Zotin (✉)
Reshetnev Siberian State University of Science and Technology,
31 Krasnoyarsky Rabochy av., Krasnoyarsk 660037, Russian Federation
e-mail: zotinkrs@gmail.com

K. Simonov
Institute of Computational Modeling of the Siberian Branch of the Russian Academy of the Sciences, 50/44 Akademgorodok, Krasnoyarsk 660036, Russian Federation
e-mail: simonovkv50@gmail.com

F. Kapsargin · T. Cherepanova
V.F. Voino-Yasenetsky Krasnoyarsk State Medical University, 1 Partizana Geleznyaka St., Krasnoyarsk 660022, Russian Federation
e-mail: kapsargin@mail.ru

T. Cherepanova
e-mail: grakova@list.ru

A. Kruglyakov
Siberian Federal University, 79 Svobodny av., Krasnoyarsk 660041, Russian Federation
e-mail: piggsyy@gmail.com

L. Cadena
Universidad de las Fuerzas Armadas ESPE, Av. Gral Ruminahui s/n, Sangolqui, Ecuador
e-mail: ecuadorx@gmail.com

M.N. Favorskaya and L.C. Jain (eds.), *Computer Vision in Control Systems-4*,
Intelligent Systems Reference Library 136,
https://doi.org/10.1007/978-3-319-67994-5_9

(around 3–20 times). Additionally, the parallel implementation gives 2–3.5 times performance boost. The proposed methodology allows to improve the accuracy and decrease the error of the sought parameters and characteristics by 10–20% on average without a lack of significant details in the structural features of the examined objects. The results of the experimental study show an error decrease in data representation for the plastic surgery (hernioplasty) by 15–25%.

Keywords Medical image processing · Edge detection · Shearlet transform · Mean filter · Median filter · Gaussian filter · 2D cleaner filter · Parallel programming

9.1 Introduction

Image processing techniques play an important role in the diagnostics and detection of diseases and monitoring the patients having these diseases. Digital image processing consists of algorithmic processes that transform one image into another, in which certain information of interest is highlighted and/or information irrelevant to the application is attenuated or eliminated.

In recent years, more sophisticated means are used in the experimental and clinical medical researches that are associated with urology and plastic surgery. However, the known algorithms do not fully meet the requirements of the computational speed and quality of visual data, which are recorded with the recently designed devices [1–4]. Many papers, as it is emphasized in [5–11], describe various options for automation of medical image processing. Moreover, as shown in [6, 7] the obtained results do not always meet the requirements of the modern clinical practice. In [8], a wide range of calculation tools is given for automating the solution of the segmentation, image extraction, and noise suppression. The traditional papers on medical image processing [9–11] are also worth to mention.

Low quality image is an obstacle for the effective feature extraction, analysis, recognition, and quantitative measurements. Medical images are often contaminated by additive, impulsive, or multiplicative noise due to a number of various impacts. Therefore, there is a fundamental need of the noise reduction in medical images. Image processing techniques are used in many medical applications, such as Magnetic Resonance Imaging (MRI), ultrasound imaging, and X-ray images. The choice of filtering is often determined by the nature of physical processes, as well as the type and behavior of input data. Noise, dynamic range, color accuracy, optical artifacts, and many other details affect on the outcome of the filtering results.

In order to increase the preprocessing speed of medical images, the parallel processing at the software level is suggested. Parallel processing is a process used to accelerate the execution time of a program by dividing it into multiple fragments that are executed on the own processors or cores at the same time. The main reason for creating and using parallel computing is that parallelism is one of the best ways to bridge the bottleneck problem, which signifies the speed of a single-core processor.

One of the actual tasks in medical visual processing is the geometrical analysis of medical images. Until recently, the X-ray method of investigation has been the main one in the diagnosis of urolithiasis. However, the introduction of modern technologies of treatment cannot be only based on the insufficient possibilities of X-ray diagnostics, resulting in misleading conclusions. At present, it is necessary not only to identify the localization of the calculus but also to determine the density and configuration of the calculus and evaluate the functional state of the urinary tract above and below the obstruction. Nowadays, the Multi-Slice Computed Tomography (MSCT) with multi-planar reconstruction can accurately determine the location of a stone by the X-ray technique without using contrast agents even for those urolithiasis patients, for whom it was impossible during the initial survey.

Efficient methods for diagnostics and morphological analysis are presented in [1, 2, 12–14]. The quantitative morphological analysis is made, when urgent clinical experimental problems in the field of plastic surgery (hernioplasty) are solved [3, 4, 15]. The application of various calculation means for the solution of the above mentioned urgent problems are presented in [16–19]. However, the application of the algorithms of the DST in combination with the suggested technique of color coding (color contrast enhancement) is a novel approach in the given field.

One of the main parts of medical image analysis is the object evaluation based on the contour. The contour of an object can be described by the edge, which (in digital image processing) is defined as transitions between two regions of significantly different intensity levels. Most edge detection techniques employ the local operators based on the discrete approximations of the first and second derivatives of the grayscale images. Some of these techniques are Sobel, Prewitt, Roberts, Level of Gradients (LoG), Canny methods [20–24], as well as some transforms, for example the DST. The idea of the DST is based on the well-developed theory of wavelet analysis. Thus, the shearlet parameters are not only the bias and scale factor but the shear as well [25–31].

We believe that the use of the MSCT in combination with the DST and additional image processing allows to identify the radiological features of the changes [32]. Additionally, a use of laser in ureteritis and presence of granulation makes it possible to produce their coagulation that increases a quality of the calculus visualization. Thus, for more effective solutions in urology and plastic surgery (hernioplasty) we have suggested a modified method of geometrical analysis of visual data using the developed image color coding algorithm, which is applied after the edge detection based on the DST. This improves the accuracy of the allocation of linear structures and visual quality of the studied medical images [25, 26].

The remainder of this chapter is organized as follows. In Sect. 9.2, the related works are analyzed. Section 9.3 presents the proposed method for analysis of medical images. Section 9.4 includes a description of the experimental research and obtained results. The conclusions are given in Sect. 9.5.

9.2 Related Works

The computational schemes in urology and plastic surgery are described in several papers and reviews [33–37]. Currently, the clinical experimental material is subjected to histological study by the standard procedure [1, 2]. Usually, in these clinical studies the microscopic examination and photographing are carried out using a light microscope with an adapted digital camera. Morphometry is performed using the JMicroVision program. The investigation results are determined at the qualitative level through the morphological and morphometric parameters and indicators, such as the vessels, collagen fibers, and elastic fibers [38–41].

The initial step of the medical image processing is the noise suppression. Thus, many different studies in this field are being conducted. In [42], different medical images like the MRI, cancer, X-ray, and brain images were studied. The authors tried out various median-base filtering techniques to suppress the salt and pepper noise in the MRI and X-ray images. Sudha et al. [43] recommended a thresholding algorithm for denoising speckle noise in ultrasound images using wavelets. An improved adaptive median filtering method for denoising impulse noise was used by Mamta et al. [44]. The Hybrid Max filter, which performs significantly better than many other existing techniques for removing speckle noise, was shown by Gnanambal et al. [45]. Arin et al. [46] studied different filters, such as average (mean), Gaussian, median, etc., for a noise reduction in medical images. This paper showed the ability of restoring an image affected by different types of undesired noises. However, a universal solution has not been found yet. Based on the applied data, the authors believe that the Gaussian filter has higher ability to remove noises than other filters [46].

Many authors used different type of segmentation to detect the stone region in urology medical images [12, 14–17]. Thus, Kalannagari and Gunasundari [12] proposed segmentation in two steps for getting better results, i.e., first, to segment kidney portion and, second, to segment the stone portion. Tamilselvi and Thangaraj [14] proposed to conduct a contour processing after identifying most interested pixel (to find the calculi from the renal images) of preprocessed image by the k-means clustering. Subsequently, the pixel matching and sequence of thresholding process are performed to find the calculi. The thresholding play an important role in some other works [14, 16–19]. To solve the problem of the geometric image analysis, i.e. the allocation of linear and non-linear features (edges, borders, and contours of the objects), the methods such as Sobel, Prewitt, Roberts, Canny, LoG [20–24], and shearlet transform [25–31] can be used.

Recently, much attention has been paid to the problem of finding patterns in the visual data and image separation into morphologically different components based on the DST due to its importance for various actual applications [25–31, 47–55]. Donoho and Kutyniok [47] developed a mathematical basis, where the concept of successful separation is strictly determined and can be mathematically proved in the case of separating point and curved structures called as the geometric separation.

9.3 Proposed Method of Medical Images Analysis

Since the results of medical image analysis may be exposed to distortion due to the presence of noise, the following procedure to process the medical images are suggested (Fig. 9.1).

Step 1 Applying a noise reduction filter.
Step 2 Forming a contour representation.
Step 3 Color coding in contour representation.
Step 4 Conducting a data analysis.

In a view of the proposed method, the data analysis depends on the images and tasks. Thus, when solving the problems of analyzing urological images, the analysis includes the detection and segmentation of the texture inhomogeneity and estimation of objects' location, contours, and sizes. The analysis of the plastic surgery data (images) consists in revealing the geometric features of the texture, extracting fibers and vessels of different types and their quantifying, as well as the segmentation and contouring amorphous formations with the estimation of their sizes and shapes.

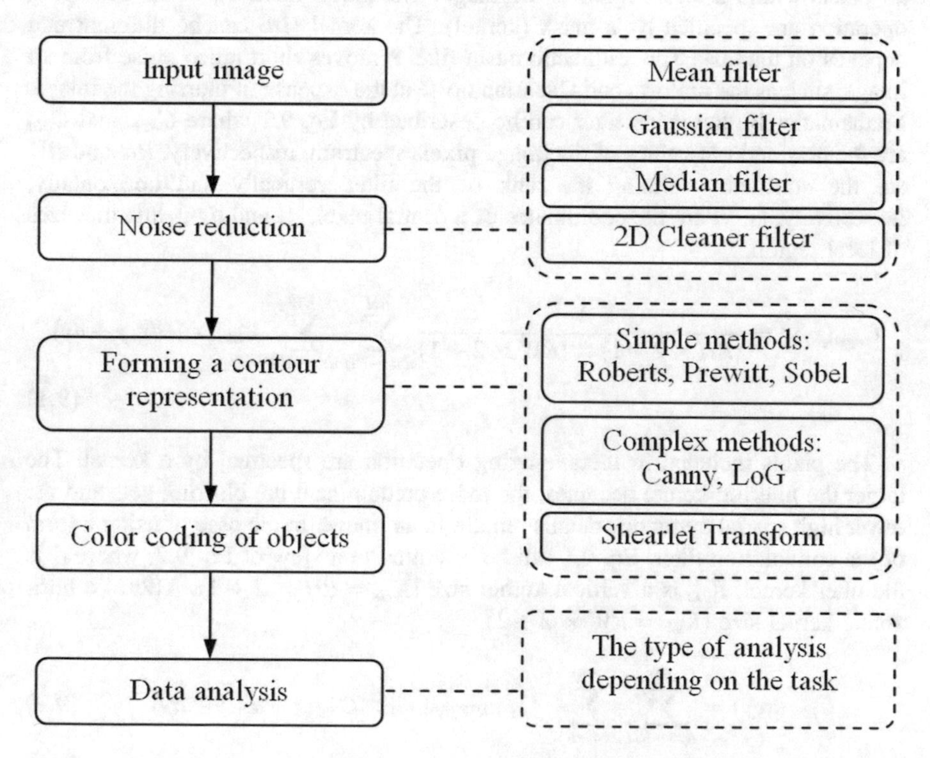

Fig. 9.1 Block diagram of the proposed method for medical images analysis

Sections 9.3.1 and 9.3.2 provide a preprocessing step of the proposed method. Section 9.3.1 briefly describes classic variant of noise reduction filter algorithms, while Sect. 9.3.2 contains information of possible optimizations. The next two sections describe steps of forming a contour representation using conventional methods (Sect. 9.3.3) and the DST (Sect. 9.3.4). A final step before data analysis (color coding of objects) is described in Sect. 9.3.5.

9.3.1 Noise Reduction Filter

The primary purpose of filters in this task is a noise reduction. However, a filter can also be used to emphasize the certain features of an image or remove other features. Almost all contemporary image processing involves a discrete or sampled signal processing. Most image processing filters can be divided into two main categories [21, 55–57]: linear and nonlinear filters. The linear filters are enough simple in their structure. The nonlinear filters include the order statistic and adaptive filters.

One of the simplest linear filters is a filter, which calculates the arithmetic mean value of a discrete spectrum. The arithmetic mean filter is defined as the average of all pixels within a local region of an image. The pixels included in the averaging operation are specified by a mask (kernel). The kernel size can be different and depends on the task. The arithmetic mean filter removes short tailed noise from an image, such as the uniform and Gaussian noise at the expense of blurring the image. Mathematically, the mean filter can be described by Eq. 9.1 where C_{new} and C_{old} are the new and old values of the image pixels spectrum, respectively, RH and RW are the constants, defining the rank of the filter vertically and horizontally, respectively, (x, y) are the coordinates of a central pixel, dx and dy define the sizes of local region.

$$C_{new}(x,y) = \frac{1}{(RH \times 2+1) \times (RW \times 2+1)} \sum_{dy=-RH}^{RH} \sum_{dx=-RW}^{RW} C_{old}(x+dx, y+dy) \tag{9.1}$$

The pixels included in the averaging operation are specified by a kernel. The larger the filtering kernel becomes, the more predominant the blurring gets and the fewer high spatial frequency details remain in an image. In the case of using a form of the convolution filter, Eq. 9.1 can be rewritten in a view of Eq. 9.2, where A is the filter kernel, K_{Hs} is a vertical kernel size ($K_{Hs} = RH \times 2 + 1$), KWs is a horizontal kernel size ($K_{Ws} = RW \times 2 + 1$).

$$C_{new}(x,y) = \sum_{dy=-RH}^{RH} \sum_{dx=-RW}^{RW} A_{dx+RW,dy+RH} \times C_{old}(x+dx, y+dy) \tag{9.2}$$

The kernel coefficients of the mean filter are calculated according to Eq. 9.3.

$$A_{i,j} = \frac{1}{K_{Hs} \times K_{Ws}} \tag{9.3}$$

The Gaussian filter (also known as the Gaussian blur) is a smoothing filter, which is used to blur the images and remove details and noise. In this sense, it is similar to the mean filter but uses a different kernel. The Gaussian filter applies the Gaussian function (which also expresses the normal distribution) for each pixel in an image. The 1D Gaussian function is expressed by Eq. 9.4, while 2D Gaussian function is a product of two Gaussians (Eq. 9.5), where x is a distance from the origin in the horizontal axis, y is a distance from the origin in the vertical axis, σ is a standard deviation of the Gaussian distribution.

$$G(x) = \frac{1}{\sqrt{2\pi} \cdot \sigma} e^{-\frac{x^2}{2\sigma^2}} \tag{9.4}$$

$$G(x, y) = \frac{1}{2\pi \cdot \sigma^2} e^{-\frac{x^2+y^2}{2\sigma^2}} \tag{9.5}$$

Since an image is represented as a collection of discrete pixels, it is necessary to produce a discrete approximation to the Gaussian function before performing the convolution. Depending on the kernel size and deviation σ, some coefficients can be out of a kernel range. Theoretically, the Gaussian distribution is a non-zero everywhere, which requires an infinitely large convolution kernel. In practice, it is effectively zero, more than about three standard deviations from the mean. Thus, it is possible to truncate the kernel size at this point. Sometimes, the kernel size is truncated even more. Thus, after the computation of the Gaussian kernel, the coefficients must be corrected so that the sum of all the coefficients is equal to 1. Once a suitable kernel has been calculated, then the Gaussian smoothing can be performed using standard convolution methods.

The best-known order-statistics filter is a median filter, which, as its title implies, replaces a value of the pixel spectrum by the median of the spectrum levels in the neighborhood of this pixel in the form of Eq. 9.6, where K_{xy} is a kernel window with the dimensions $K_{Hs} \times K_{Ws}$ centered at coordinates (x, y).

$$C_{new}(x, y) = \underset{(kx,ky) \in K_{xy}}{\text{med}} \{C_{old}(kx, ky)\} \tag{9.6}$$

The original pixel value is included in the computation of a median. Median filters are quite popular because for certain types of random noise they provide excellent noise-reduction capabilities with considerably less blurring than a linear smoothing filters of a similar size. Median filters are particularly effective in the presence of both bipolar and unipolar impulse noise. It is particularly useful in removing the speckle and salt and pepper noise. The pattern of the neighbors is defined by the kernel called a "window", which slides, entry by entry, over the entire signal. Usually the kernel size of a median filter has an odd number of entries because it is simple to define. It is just the middle value after all entries in the kernel sorted numerically.

One of the adaptive filters for noise reduction is 2D cleaner filter proposed by Casaburi [58]. It is often used in video processing. The main idea of this filter is to calculate the arithmetic mean value in each color channel $s \in \{Red, Green, Blue\}$ if a deviation from central pixel is less than a threshold. 2D cleaner filter is implemented by Eq. 9.7, where $Sv_s(i, j, TS)$ is a spectrum cut-off function of the threshold value (Eq. 9.8) and $Cc_s(i, j, TS)$ is a function indicating the suitability of the spectrum according to the threshold value (Eq. 9.9). In Eqs. 9.8 and 9.9, parameter $Sv_s(i, j)$ is a value of the spectrum considered a color channel $s \in \{Red, Green, Blue\}$, Ts is a threshold value.

$$C_{new}(s,x,y,Ts) = \frac{\sum_{dy=-RH}^{RH} \sum_{dx=-RW}^{RW} Sv_s(x+dx, y+dy, Ts)}{\sum_{dy=-RH}^{RH} \sum_{dx=-RW}^{RW} Cc_s(x+dx, y+dy, Ts)} \tag{9.7}$$

$$Sv_s(i,j,Ts) = \begin{cases} sv_s(i,j) & \text{if } |sv_s(i,j) - sv_s(0,0)| \leq Ts \\ 0 & \text{if } |sv_s(i,j) - sv_s(0,0)| > Ts \end{cases} \tag{9.8}$$

$$Cc_s(i,j,Ts) = \begin{cases} 1 & \text{if } |sv_s(i,j) - sv_s(0,0)| \leq Ts \\ 0 & \text{if } |sv_s(i,j) - sv_s(0,0)| > Ts \end{cases} \tag{9.9}$$

The pseudo code for 2D cleaner filter by Casaburi is mentioned below.

```
for (int y = 0; y < Image_Height; y++)
 for (int x = 0; x < Image_Width; x++)
 {
  GetRGB(source_pixel_InImg(y,x), r, g, b);
  tot_r = tot_g = tot_b = 0;
  count_r = count_g = count_b = 0;
  for (int dy = -RH; dy <= RH; dy++)
  for (int x = -RW; dx <= RW; dx++)
  {
   GetRGB(neighbour_pixel_InImg(y,x,dy,dx), r1, g1, b1);
   if (abs(r1-r) < Threshold) { tot_r += r1; count_r++;}
   if (abs(g1-g) < Threshold) { tot_g += g1; count_g++;}
   if (abs(b1-b) < Threshold) { tot_b += b1; count_b++;}
  }
  destination_pixel(y,x) = RGB
(tot_r/count_r, tot_g/count_g , tot_b/count_b);
 }
```

9.3.2 Optimization of Noise Reduction Filters

Ordinary implementation of filters, such as the arithmetic mean filter, Gaussian filter, and median filter, has a relatively high computational cost. The use of a speed optimized implementation to perform the preprocessing of medical images has been suggested.

Due to the arithmetic mean filter property to use the equal weights, it can be implemented applying a much simpler accumulation algorithm, which is significantly faster than a sliding window algorithm. Thus, the accumulation of the neighborhood of pixel *P*(*y*, *x*) shares a lot of pixels with the accumulation for pixel *P* (*y*, *x* + 1). This means that there is no need to compute a whole kernel for all the pixels except only for the first pixel in each row [59]. Successive pixel filter response values can be obtained with just the addition and subtraction of previous response values of filter. Thus, the filter computation can be conducted according Eq. 9.10.

$$C_{new}(x,y) = \begin{cases} \frac{1}{K_{Hs}\times K_{Ws}} \sum_{dy=-RH}^{RH} \sum_{dx=-RW}^{RW} C_{old}(x+dx, y+dy) & \text{if } x=0 \\ C_{new}(x-1,y) - \frac{1}{K_{Hs}} \sum_{dy=-RH}^{RH} C_{old}(x-RW-1, y+dy) & \text{otherwise} \\ +\frac{1}{K_{Hs}} \sum_{dy=-RH}^{RH} C_{old}(x+RW, y+dy) & \end{cases} \tag{9.10}$$

The pseudo code of the optimized version of the mean filter is mentioned below:

```
Kernel_size = (RH×2+1)×(RW×2+1)
Weight = 1/Kernel_size;
for (int y = 0; y < Image_Height; y++)
{
 x = 0; tot_r = tot_g = tot_b = 0;
 for (int dy = -RH; dy <= RH; dy++)
  for (int x = -RW; dx <= RW; dx++)
  {
  GetRGB(neighbour_pixel_InImg(y,x,dy,dx), r1, g1, b1);
  tot_r += r1×Weight;  tot_g += g1×Weight;
  tot_b += b1×Weight;
  }
  destination_pixel(y,x) = RGB (tot_r, tot_g, tot_b);
  for (int x = 1; x < Image_Width; x++)
  {
   for (int dy= -RH; dy <= RH; dy++)
   {
    dx = -RW-1
```

```
    GetRGB(neighbour_pixel_InImg(y,x,0,dx), r1, g1, b1);
    tot_r -= r1×Weight;  tot_g -= g1×Weight;
    tot_b -= b1×Weight;
    dx = RW;
    GetRGB(neighbour_pixel_InImg(y,x,0,dx), r1, g1, b1);
    tot_r += r1×Weight;  tot_g += g1×Weight;
    tot_b += b1×Weight;
   }
  destination_pixel(y,x) = RGB(tot_r, tot_g, tot_b);
  }
 }
```

We can optimize a performance of the Gaussian filter. The convolution can in fact be performed fairly quickly, since the equation for 2D isotropic Gaussian is separable into *y* and *x* components [60]. In some cases, the approximation of the Gaussian filter can be used instead of the ordinary version [61, 62]. The pseudo code of Gaussian 2D filter in double 1D interpretations is the following:

```
Cf1D_1[]=GaussianCoef1D(RW,sigma);
Cf1D_2[]=GaussianCoef1D(RH,sigma);
TMP_Image[][] is transposed version of the input image
for (int y = 0; y < Image_Height; y++)
 for (int x = 0; x < Image_Width; x++)
 {
  tot_r = tot_g = tot_b = 0;
  for (int dx = -RW; dx <= RW; dx++)
  {
 rx = dx + RW;
   GetRGB(neighbour_pixel_InImg(y,x,0,dx), r1, g1, b1);
   tot_r += r1×Cf1D_1[rx];  tot_g += g1×Cf1D_1[rx];
   tot_b += b1×Cf1D_1[rx];
  }
  Temp_IMG(x,y) = RGB(tot_r, tot_g, tot_b);
 }
for (int y = 0; y < TempIMG_Height; y++)
 for (int x = 0; x < TempIMG_Width; x++)
 {
  tot_r = tot_g = tot_b = 0;
  for(int dx = -RH; dx <= RH; dx++)
  {
```

```
rx = dx + RW;
   GetRGB(neighbour_pixel_TmpIMG(y,x,0,dx), r1, g1, b1);
   tot_r += r1×Cf1D_2[rx];  tot_g += g1×Cf1D_2[rx];
   tot_b += b1×Cf1D_2[rx];
  }
 destination_pixel(x,y)= RGB(tot_r, tot_g, tot_b);
}
```

In median filter, the computational costs are spent to calculate a median of the kernel. Since this filter processes every pixel in an image, the efficiency of this median calculation for large images is a critical factor in determining how fast the algorithm can run. The ordinary implementation involves sorting out every entry in the kernel to find the median. However, since only the middle value in the list of the numbers is required, much more efficient selection algorithms can be used [63]. Furthermore, the spectrum histogram in the median calculation can be far more efficient because it is simple to update the histogram from window to window and finding a median of the histogram is not particularly onerous [64–66]. The pseudo code of the optimized median filter based on the histogram is provided below:

```
Kernel_size = (RH×2+1)×(RW×2+1)
Median= Kernel_size/2;
HistR[0.0.255], HistG[0.0.255], HistB[0.0.255];
for (int y = 0; y < Image_Height; y++)
{
 Clear_histograms_data(HistR, HistG, HistB)
 medR = medG = medB = 0
 deltaR = deltaG = deltaB =0;
 for (int dy = -RH; dy < =RH; dy++)
  for (int x=-RW; dx<= RW; dx++)
  {
  GetRGB(neighbour_pixel_InIMG(y,x,dy,dx), r1, g1, b1);
  HistR[r1]++;  HistG[g1]++;  HistB[b1]++;
  }
medR = median(HistR);
medG = median(HistG);
medB = median(HistB);
for (int x = 1; x < Image_Width; x++)
{
 ind=0;
 for (int dy = -RH; dy <= RH; dy++)
{
 dx=-RW-1;
 GetRGB(neighbour_pixel(y,x,dy,dx), r1, g1, b1);
 HistR[r1]--;
```

```
If (r1 < medR) deltaR --;
else If (r1 > medR) then deltaR ++;
HistG[g1]--;
If (g1< medG) then deltaG--;
else If (g1 > medG) then deltaG ++;
HistB[b1]--;
If (b1< medB) then deltaB--;
else If (b1 > medB) then deltaB ++;
dx=RW;
GetRGB(neighbour_pixel(y,x,dy,dx), r1, g1, b1);
HistR[r1]++;
If (r1 < medR) then deltaR++;
else If (r1 > medR) then deltaR--;
HistG[g1] ++;
If (g1 < medG) then deltaG++;
else If (g1 > medG) then deltaG--
HistB[b1] ++;
   If (b1 < medB) then deltaB++;
   else If (b1 > medB) then deltaB--;
  }
  medR = median_Correction(HistR, deltaR);
  medG = median_Correction(HistG, deltaG);
  medB = median_Correction(HistB, deltaB);
  destination_pixel(y,x) = RGB(medR, medG, medB);
}
```

Also it is possible to use a parallel computing in order to get additional performance gains in image filtering. In image processing, the following types of code parallelization can be used [67]:

- Data parallelism. It corresponds to the task, which includes the repeated execution of the same algorithm with different initial data. Such calculations can be performed in parallel in the case of the division into data fragments and each fragment is send to an independent processing core.
- Algorithmic parallelism. This type of parallelization is performed in the algorithm by revealing the fragments, which can be executed in parallel (algorithmic decomposition). When an algorithmic decomposition is used, it should seek to partition tasks for large and rarely interacting branches. If it is possible, it should be provided with a uniform data distribution along the branches of the parallel algorithm.
- Functional parallelism. Parallelization is performed by a functional decomposition. For example, the Gaussian filter function can be decomposed into the following steps: the initial data input, kernel formation, processing, and output of the results. Parallelism is achieved with the parallel implementation of these

substeps and a creation of a "conveyor" (serial or serial-parallel) between them. Moreover, in each step data parallelism can be used.

The difference of the algorithmic parallelism from the functional one is that the functional parallelism involves pooling only functionally similar algorithm operators, while the algorithmic parallelism ignores the proximity of the functional operators.

For implementation of the parallel algorithms, the OpenMP standard for the parallelization of programs in the languages C, C++, and Fortran is widely spread [68, 69]. The parallelization according to the OpenMP standard is performed by inserting the special directives into the program, as well as calling the support functions. The OpenMP standard implements a parallel processing using multithreading. Thus, the "main" thread creates a set of subordinate threads with the task being distributed among them. A program is executed with a sequential code. At first, one process (thread) runs. At the entrance in the parallel fragment, the multiple threads are generated with the workload being distributed in the following code. If a forked thread completes its work before any other forked thread, it blocks. Once all the forked threads complete their work, the master thread resumes the execution. At the end of the parallel fragment, all threads except the "main" are terminated and following sequential fragment of a program starts. The OpenMP standard also supports the embedding of parallel fragments.

In order to improve a performance of image processing, it is expedient to carry out parallelization for external loops (the image height)—loop level parallelism. To do this, the OpenMP directive can be used:

```
#pragma omp parallel for
```

It offers a simple way to achieve a loop-level parallelism (data parallelism). The parallelization of loops is the most common application of OpenMP. Consider the following code, which represents the basis of any filter:

```
for (int y = 0; y < Image_Height; y++)
  for (int x = 0; x < Image_Width; x++)
  {
  // do some processing
  }
```

With OpenMP, this code can be parallelized according to the data parallelism in a simple way:

```
#pragma omp parallel for
for (int y = 0; y < Image_Height; y++)
  for (int x = 0; x < Image_Width; x++)
  {
    // do some processing
  }
```

Here, the directive instructs the compiler that the next loop is to be parallelized. The compiler will then distribute the work in a set of the forked threads. Thus, we can divide the image into lines, which are processed independently.

9.3.3 Forming a Contour Representation

Contour representation can be generated using different algorithms. We can use simple conventional methods, such as Roberts, Prewitt, Sobel, and more complex ones, for example LoG and Canny [1–5]. The Roberts cross operator provides a simple approximation of the gradient magnitude. It uses a 2×2 kernel. The method using the Sobel or Prewitt operators computes the gradients' approximation of intensity function along the horizontal OX and vertical OY directions (2D spatial) in each pixel and highlights the regions corresponding to the edges. It computes the approximations of the derivatives using two 3×3 kernels (masks) in order to find the localized orientation. The difference between the Prewitt and Sobel kernel is in the kernel coefficients. The LoG may be implemented in the following way: blurring the image using Gaussian, performing the Laplacian in this blurred image, finding the zero crossings of the Laplacian, and comparing the local variance in this point with the threshold. If a threshold value is exceeded, an edge is declared. Since both the Gaussian and Laplacian kernels are usually much smaller than a whole image, we can use the precalculated LoG kernel, which requires far fewer arithmetic operations. The 2D LoG function centered at zero and with the Gaussian standard deviation σ has the form of Eq. 9.11.

$$LoG(x, y) = -\frac{1}{\pi \cdot \sigma^2}\left[1 - \frac{x^2 + y^2}{2\sigma^2}\right] e^{-\frac{x^2+y^2}{2\sigma^2}} \tag{9.11}$$

Canny algorithm for the contour detection uses the following steps: the image smoothing, searching gradients, suppression of "false" peaks, double threshold filtering (potential contours defined by the thresholds), and tracing the area of ambiguity.

9.3.4 Shearlet Transform for Contour Detection

Directional multiscale representation of images to address the curved singularities has received much attention in harmonic analysis during last 25 years. Concepts related to shearlet transform are studied in the works [25–31]. Shearlets possess a uniform construction for both the continuous and discrete setting. They further stand out since they stem from a square-integrable group representation and have the corresponding useful mathematical properties. The continuous shearlet transform can proceed to the DST by constructing a discrete shearlet system. To do this, it is necessary to divide the shearlets parameters in a finite set of discrete values and build a system that will have an ability to maintain the properties of a continuous system including the ability to provide the inverse transform.

The shearlets produce an optimally sparse approximation in the class of piecewise smooth functions with C^2 singularity curves provided by Eq. 9.12, where f_N is a nonlinear shearlet approximation of function f obtained by taking the N largest shearlet coefficients in absolute values.

$$\|f - f_N\|_{L_2}^2 \leq CN^{-2}(\log N)^3 \quad N \to \infty \tag{9.12}$$

The continuous shearlet transform uses the dilation matrix A_a and shear matrix S_s (Eq. 9.13) with the parameters $d = 2$ and $\gamma = 1/2$.

$$A_a = \begin{pmatrix} a & 0 \\ 0 & \sqrt{a} \end{pmatrix} \quad a \in R^+ \quad S_s = \begin{pmatrix} 1 & s \\ 0 & 1 \end{pmatrix} \quad s \in R \tag{9.13}$$

The shearlets $\psi_{a,s,t}$ emerge by dilation, shear, and translation of a function $\psi \in L_2(R^2)$ [25].

$$\psi_{a,s,t}(x) = a^{-\frac{3}{4}}\psi(A_a^{-1}S_s^{-1}(x - t)) = a^{-\frac{3}{4}}\psi\left(\begin{pmatrix} \frac{1}{a} & -\frac{s}{a} \\ 0 & \frac{1}{\sqrt{a}} \end{pmatrix}(x - t)\right) \tag{9.14}$$

Assume that $\hat{\psi}$ can be written as Eq. 9.15.

$$\hat{\psi}(\omega_1, \omega_2) = \hat{\psi}_1(\omega_1)\hat{\psi}_2\left(\frac{\omega_2}{\omega_1}\right) \tag{9.15}$$

Consequently, Eq. 9.16 obtains for the Fourier transform.

$$\begin{aligned}\widehat{\psi}(\omega) &= a^{-\frac{3}{4}} F\left(\psi\left(\begin{pmatrix}\frac{1}{a} & -\frac{s}{a}\\ 0 & \frac{1}{\sqrt{a}}\end{pmatrix}(\cdot - t)\right)\right)(\omega)\\ &= a^{-\frac{3}{4}} e^{-2\pi i(\omega,t)} F\left(\psi\left(\begin{pmatrix}\frac{1}{a} & \frac{s}{a}\\ 0 & \frac{1}{\sqrt{a}}\end{pmatrix}\cdot\right)\right)(\omega)\\ &= a^{-\frac{3}{4}} e^{-2\pi i(\omega,t)}\left(a^{-\frac{3}{2}}\right)\widehat{\psi}\left(\begin{pmatrix}a & 0\\ s\sqrt{a} & \sqrt{a}\end{pmatrix}\omega\right)\\ &= a^{\frac{3}{4}} e^{-2\pi i(\omega,t)}\widehat{\psi}\left(a\omega_1, \sqrt{a}(s\omega_1+\omega_2)\right)\\ &= a^{\frac{3}{4}} e^{-2\pi i(\omega,t)}\widehat{\psi}_1(a\omega_1)\widehat{\psi}_2\left(a^{-\frac{1}{2}}\left(\frac{\omega_2}{\omega_1}+s\right)\right)\end{aligned} \tag{9.16}$$

The shearlet transform $SH_\psi(f)$ of $f \in L_2(R)$ is given as Eq. 9.17.

$$\begin{aligned}SH_\psi(f)(a,s,t) &= \langle f, \psi_{a,s,t}\rangle = \langle \hat{f}, \widehat{\psi}_{a,s,t}\rangle = \int_{R^2} f(\omega)\,\overline{\widehat{\psi}_{a,s,t}(\omega)}\,d\omega\\ &= a^{\frac{3}{4}}\int_{R^2} \hat{f}(\omega)\widehat{\psi}_1(a\omega_1)\,\widehat{\psi}_2\left(a^{-\frac{1}{2}}\left(\frac{\omega_2}{\omega_1}+s\right)\right)e^{2\pi i(\omega,t)}d\omega\\ &= a^{\frac{3}{4}} F^{-1}\left(\hat{f}(\omega)\,\widehat{\psi}_1(a\omega_1)\widehat{\psi}_2\left(a^{-\frac{1}{2}}\left(\frac{\omega_2}{\omega_1}+s\right)\right)\right)(t)\end{aligned} \tag{9.17}$$

The same formula is derived by interpreting the shearlet transform as a convolution with function $\psi_{a,s}(x) = \overline{\psi}\left(-A_a^{-1}S_s^{-1}x\right)$ using the convolution theorem.

The shearlet transform is invertible if function ψ fulfills the admissibility property (Eq. 9.18).

$$\int_{R^2} \frac{\left|\widehat{\psi}(v_1, v_2)\right|}{v_1^2}\,dv_1\,dv_2 < \infty \tag{9.18}$$

For the calculations, two implementations of the shearlet transform are used in the chapter. In [25, 26], an algorithm referred to as Fast Finite Shearlet Transform (FFST) is presented. It is based on the discrete fast direct and inverse Fourier transformation. The Meyer wavelet is used for the present implementation as the mother wavelet.

For finite discrete shearlets, digital images in $R^{M\times N}$ are considered as functions sampled on the grid $\left\{\left(\frac{m_1}{M}, \frac{m_2}{N}\right) : (m_1, m_2) \in G\right\}$ with $G = \{(m_1, m_2)\colon m_1 = 0,\ldots, M-1, m_2 = 0,\ldots, N-1\}$ and assumed a periodic continuation over the boundary. The DST does not only discretize the involved parameters a, s, and t but also considers only a finite number of discrete translations t. Additionally, the translation parameter t, which is independent from the dilation and shear parameters, is discretized on a rectangular grid.

In order to obtain the DST, let us denote the number of considered scales as $j_0 = \left[\frac{1}{2}\log_2 \max\{M,N\}\right]$. Then, the dilation, shear, and translation parameters are discretized according to Eqs. 9.19–9.21.

$$a_j = 2^{-2j} = \frac{1}{4^j} \quad j = 0,\ldots,j_0 - 1 \tag{9.19}$$

$$s_{j,k} = k2^{-j} \quad -2^j \leq k \leq 2^j \tag{9.20}$$

$$t_m = \left(\frac{m_1}{M}, \frac{m_2}{N}\right) \quad m \in G \tag{9.21}$$

With these notations, the shearlet becomes Eq. 9.22.

$$\psi_{j,k,m}(x) = \psi_{a_j,s_{j,k},t_m}(x) = \psi\left(A^{-1}_{a_j,\frac{1}{2}} S^{-1}_{a_{j,k}}(x - t_m)\right) \tag{9.22}$$

An alternative approach to calculating the shearlet transform referred as ShearLab is described in [53]. The given implementation is based on the discrete fast Fourier transformation. However, it interprets the frequency range in a somewhat differently. Study of the FFST algorithm shows that the object contours can be obtained as a sum of the coefficients of the DST for the fixed scale and all possible values of the shift and bias parameters. We suggested to use this peculiarity in solve the given approximation problem.

The objects' contours can be obtained as a sum of the coefficients of the DST with fixed parameter values shown in Eq. 9.23, where sh_ψ is a shearlet coefficient for a scale j^* cornering the orientation k and displacement m, $k_{\max}$ is a maximum number of turns, $m_{\max}$ is a maximum number of displacement.

$$f_{cont} = \sum_{k=0}^{k_{\max}} \sum_{m=0}^{m_{\max}} sh_\psi(f(j^*, k, m)) \tag{9.23}$$

The mean execution time of these is chosen as a quantitative parameter for evaluating the algorithm performance efficiency. For comparison, the calculations were carried out for the images of different size (Table 9.1).

One can see from Table 9.1 that the ShearLab algorithm is executed faster than the FFST algorithm for the images with bigger resolution. However, the FFST

Table 9.1 Mean time of the algorithm execution (s)

Algorithm	Image size, pixels			
	64 × 64	128 × 128	256 × 256	512 × 512
FFST	0.0781	0.2969	1.2969	6.8281
ShearLab	0.2969	0.3906	1.1875	3.5781

algorithm has a slight advantage in the execution time for the images with small resolution.

9.3.5 Color Coding of Objects in Contour Representation

Some propositions of the color coding are mentioned below:

- Transformations are performed element by element in order to extract the details of interest in the medical object under study.
- Some insignificant peculiarities are excluded from the image (i.e. background).
- The image is transformed into a type, which is convenient for the visual interpretation and further analysis.
- The sought parameters are estimated relative to the contour extraction and filtering for noise suppression.

Note that the algorithmic procedure of color coding—color selection and density distribution of the isolines in an image corresponds to the known technique of building elastic maps using the spatial data [70]. According to [70], consider the algorithm provision for the color coding of the objects under study. Let the analyzed object be a limited 2D manifold enclosed into a set of the observation data so that the shape and location of the manifold reflect the main peculiarities of the point distribution of the initial data in an image.

Consider 2D rectangular grid of nodes, where there are p nodes horizontally and q vertically. Enumerate the nodes of the grid using two indices $y^{i,j}$, $i = 1,\ldots,p, j = 1, \ldots,q$. The constructed grid is located in the set of the data points so that each point is associated with the nearest node on the grid. The data points have the form $t_{ij}(x,y, M)$, where (x, y) are their relative coordinates, M is an intensity (color) characterizing the peculiarities of the object under study.

Such method allows to partition a set of data into pq subsets of $K_{i,j}$, with the subset points being closer to $y^{i,j}$ than to any other node. Let $K_{i,j}$ be as Eq. 9.24.

$$K_{i,j} = \left\{ t \middle| t \in P_k, \left\| y^{i,j} - t \right\|^2 \leq \varepsilon \right\} \tag{9.24}$$

Generally, the grid can be deformed in two ways—stretching it lengthwise and bending crosswise. In the former case, it tends to keep its length, while in the latter it saves its flat form. Therefore, the grid possesses the following properties (the functional D): the stretching property (which provides the grid uniformity), smoothness property, and property of proximity to the data points.

In order to save these properties, it is necessary to add in the criterion of minimization such measures as the total stretching, total smoothness, and total aggregate measure of proximity. By adding together all three of these measures, the general criterion can be obtained, by which the grid will be attracted to the data points and aimed to minimize their stretching by taking maximum smooth shape.

Thus, the functional D (quality functional) is represented by Eq. 9.25, where the $|P_k|$ is a number of points in cell, λ and μ are the elasticity coefficients responsible for the tension and curvature of the grid, respectively, λ is a number of iterations made, D_1, D_2, D_3 are the terms that are responsible for the properties of the grid.

$$D = \frac{D_1}{|P_k|} + \lambda\frac{D_2}{pq} + \mu\frac{D_3}{pq} \rightarrow \min \tag{9.25}$$

Value of the square of the distance from the point to the nearest grid point was selected as a measure of a grid proximity to the data point (D_1) (Eq. 9.26).

$$D_1 = \sum_{i.j}\sum_{t_n \in K_{i,j}} \left\| t_n - y^{i,j} \right\|^2 \tag{9.26}$$

The greater the length of an edge, the more grid is "stretched". Thus, the functional D includes the difference between the positions of neighboring nodes, i.e. measure of stretching, which is defined by Eq. 9.27. Note that the summation boundaries are chosen in such way that in the functional D_2 one edge is not included twice.

$$D_2 = \sum_{i=1}^{p}\sum_{j=1}^{q-1} \left\| y^{i,j} - y^{i,j+1} \right\|^2 + \sum_{i=1}^{p-1}\sum_{j=1}^{q} \left\| y^{i,j} - y^{i+1,j} \right\|^2 \tag{9.27}$$

The measure of a grid smoothness (D_3) is determined by estimating the magnitude of the second derivative (Eq. 9.28).

$$D_3 = \sum_{i=1}^{p}\sum_{j=2}^{q-1} \left\| 2y^{i,j} - y^{i,j-1} - y^{i,j+1} \right\|^2 + \sum_{i=2}^{p-1}\sum_{j=1}^{q} \left\| 2y^{i,j} - y^{i-1,j} - y^{i+1,j} \right\|^2 \tag{9.28}$$

Let metric be the Euclidean one. In this case, the functional D is quadratic in the position of the nodes $y^{i,j}$. This means that with the given set of data points partitioned into taxa it is necessary to solve the system of linear equations $pq \times pq$ for its minimization. Consequently, an efficient method to minimize the functional D appears to be the following algorithm:

Step 0. The grid nodes are anyway located in the data space.
Step 1 Given the positions of the grid nodes, the data set is partitioned into taxa—subsets $K_{i,j}$.
Step 2. With the given partition of the set of data points into taxa, the functional D is minimized.
Step 3. Steps 1 and 2 are repeated until the functional D stops sensibly changing.

Method of constructing the elastic grid based on the mentioned above steps was proposed. It takes into account the input values of the physical aspects of the data for a study. The essence of the method consists in that the object of interest in a medical image is represented as the initial (central) part of the construction of the

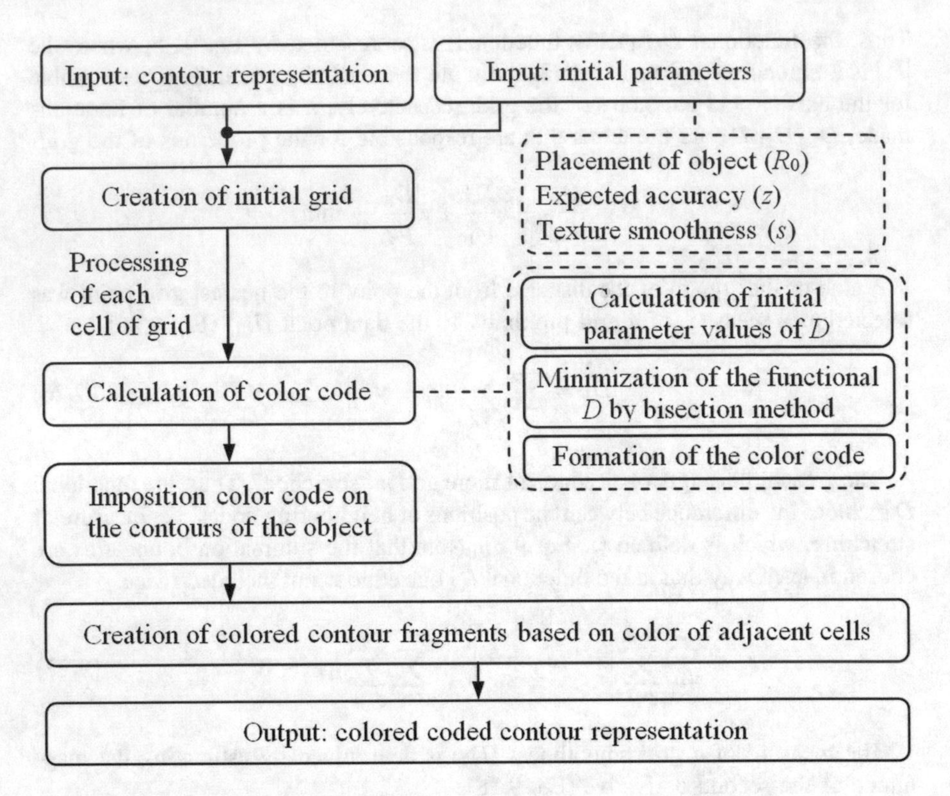

Fig. 9.2 Flow chart of the color coding method

grid. It should be noted that implementation of the computational scheme takes into account a flat image character, so the functionals D_2 and D_3 in the following calculations are not used.

Implementation of the color coding method within the grid setup includes an initial approximation, which uses a contour approximation of object $R_0(x, y)$ built by an expert in a conventional way. The base of color coding (M) is determined by the expected accuracy (z) and evaluation of texture smoothness (s). Schematically, a process of color coding is represented in Fig. 9.2.

Possible applications of the proposed method for processing of complicated medical images, in particular, in urology and plastic surgery are given in Sect. 9.4.

9.4 Experimental Research

For our experimental study, we used 120 medical images that were divided into two main categories: the urology images and plastic surgery ones (implants). The objects of interest in the urology images are the renal parenchyma and its inhomogeneity degree as well as the features of its contour, pyelocaliceal system and its features, location, sizes, shape, and density. The objects of interest in the plastic surgery (hernioplasty) images are the vessels, collagen fibers, and elastic fibers.

The experimental investigation was performed as follows. The given urology images were used to find out, in what extent the methods for forming the contour representations allow to extract the object of interest (A—contours and parameters of renal parenchyma, B—contours and parameters of calculus). For comparison, the following methods of contour representations were chosen: Roberts, Sobel, Prewitt, Canny, LoG, and the DST.

Further studies were performed on urology data to determine how a preprocessing affects the accuracy of feature extraction of the object of interest. For the step of noise reduction (preprocessing), we selected the following filters: mean, median, Gaussian and 2D cleaner filters. Before the experimental investigation, the reference images containing the urology data of interest were acquired with the help of medical experts.

In addition to the examination of the original images, the experiments were also performed with the artificial noise. In order to simulate a noise, which may occur in the equipment, the following noise characteristics were used. In the total noise map, the impulse noise part amounts to 30%, while the additive noise part is 70% with the value of the additive noise component being 15% of the dynamic range of the experimental data. The evidence concerning the extent, to which the objects of interest can be seen, i.e. their contours (as the percentage of the reference image), was used as an indicator, as well as the information about the assignment of an image fragment to the object of interest (false selection). The obtained experimental results are described in Sect. 9.4.1.

For medical data in plastic surgery (hernioplasty), the calculation accuracy was estimated for the vessels, collagen fibers, and elastic fibers in comparison with the conventional method employed in the clinical laboratory. The estimations of medical experts, as well as the experimental results based on the proposed techniques, are presented in Sect. 9.4.2. Section 9.4.3 demonstrates the experimental data of preprocessing filters optimization. The efficiency evaluation of the proposed method from a viewpoint of medical experts is presented in Sect. 9.4.4.

9.4.1 Application of Proposed Techniques in Urology

The database of medical images is a collection of urological images with a variety of urological disturbances. The computed tomography images of a kidney are represented in Fig. 9.3. Kidney parenchyma, the inhomogeneous one with clear even contour, is depicted in Fig. 9.3a. The enlarged calyces-pelvis system is given in Fig. 9.3b (pelvis is up to 4.3 cm, calices is up to 2.5 cm, inhomogeneous parenchyma is up to 0.5 cm in thickness in some parts, in the lumen of a/3 right ureter a shadow of calculus with the size of 1.5 cm and the density of 1163 HU).

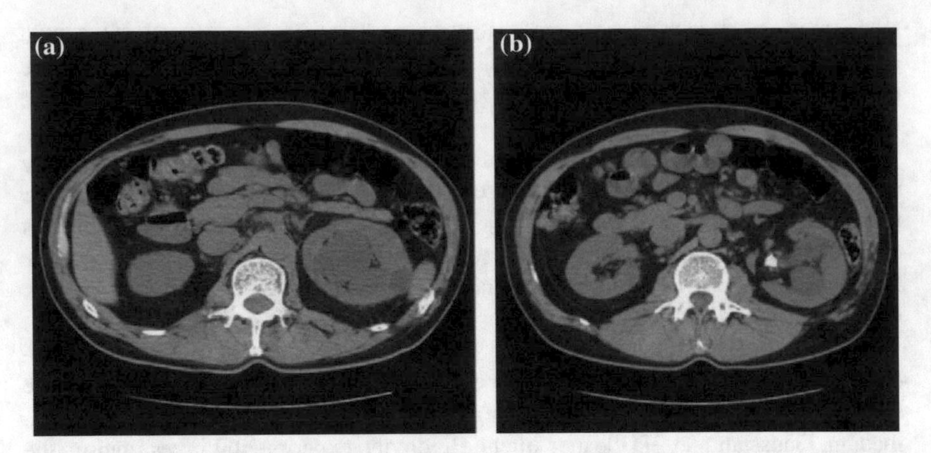

Fig. 9.3 Examples of the computed tomography images in urology: **a** inhomogeneous kidney parenchyma with clear contours, **b** inhomogeneous kidney parenchyma with enlarged calyces-pelvis system

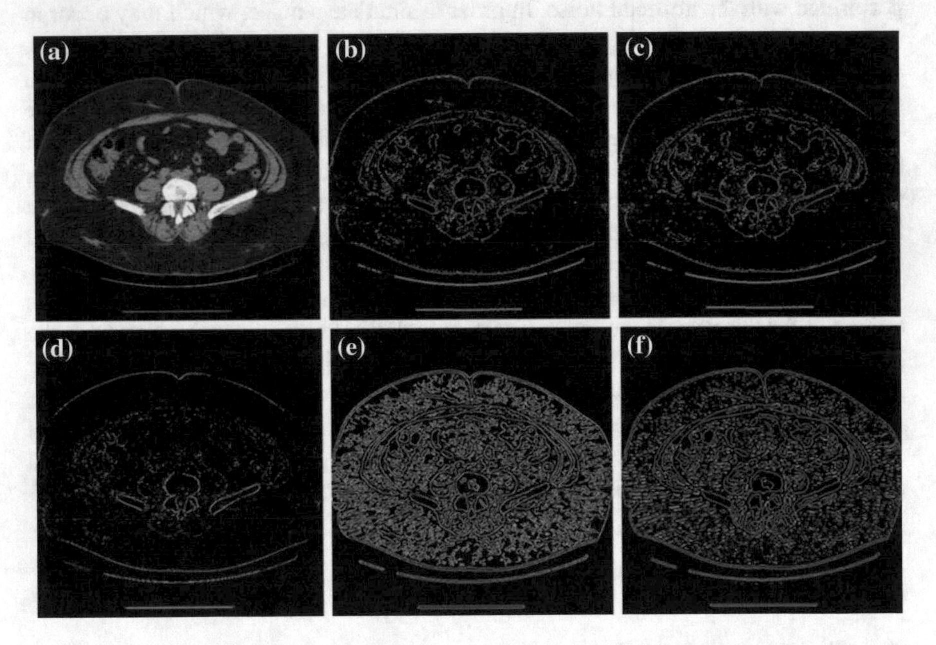

Fig. 9.4 Results of the obtained contours: **a** original image, **b** Sobel mask, **c** Prewitt mask, **d** Roberts mask, **e** Canny detector, **f** LoG detector

The examples of the image processing using conventional methods are presented below in comparison with the DST. Figure 9.4 shows the results of the obtained contour using Sobel, Prewitt, Roberts, Canny and LoG methods. In Fig. 9.5, one can see the obtained contour processed by the DST. The presented peculiarities of color detection of the studied medical objects are subject to the given task.

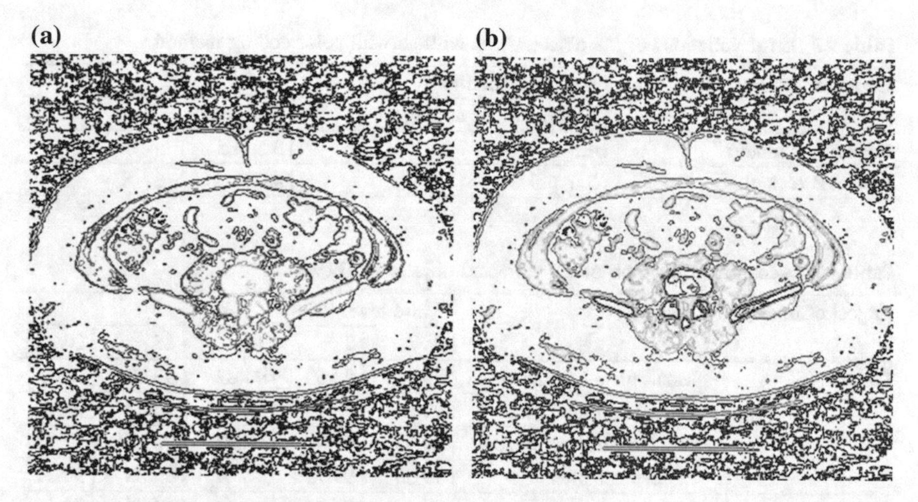

Fig. 9.5 Example of the medical image processed using the FFST algorithm: **a** with the edges in shades of *gray*, **b** edges in *color* (applying color coding)

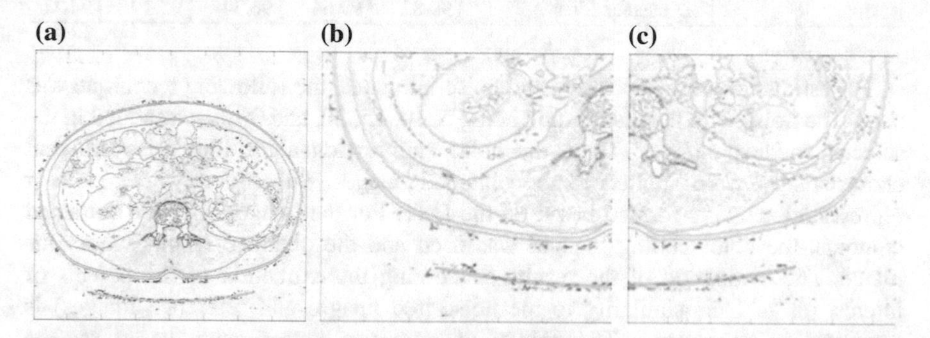

Fig. 9.6 Modified geometric analysis based on the DST combined with color coding: **a** color coding of original image from Fig. 9.6b, **b** the detailed color coding of *left part*, **c** the detailed color coding of *right part*

The analysis of the methods reveals that the Sobel and Prewitt methods provide better results in the separation characteristics of the contours in comparison to the Roberts, Canny and LoG methods. At the same time, the general approach to the image analysis using the FFST algorithm combined with color coding of the objects of interest gives better results relative to the specialized algorithms.

The homogeneity level, contour features (clear, even ones), and peculiarities of the calices-pelvis system of the kidney parenchyma, are analyzed as the objects of interest. Possible calculi are detected and their size and density are estimated. Visual results of modified geometric analysis based on the DST combined with color coding are depicted in Fig. 9.6. Table 9.2 contains the error variations' values of the objects A and B without/with color coding.

Table 9.2 Error variations of the objects A, B without/with color coding method

Object of interest	Error variation (%)	
	Without color coding	With color coding
Object A (kidney)	4.5–5.5	3.5–4.5
Object B (calculus)	5.0–6.5	4.0–5.5

Table 9.3 Accuracy estimation of the extracted objects A, B (%)

Object of interest	Filter	Noise level (%)				
		0	5	10	15	20
Object A	Mean filter optimized	99.75	98.63	94.82	93.57	92.35
	Gaussian filter 1Dx2	99.84	99.72	96.45	95.23	94.26
	Median filter optimized	99.95	99.91	98.93	98.16	97.42
	2D cleaner filter	99.86	99.75	96.47	95.35	94.31
Object B	Mean filter optimized	99.64	98.51	94.72	93.43	91.15
	Gaussian filter 1Dx2	99.83	99.63	96.17	95.17	94.08
	Median filter optimized	99.91	99.87	98.46	98.12	97.19
	2D cleaner filter	99.82	99.64	96.34	95.29	94.11

For estimation of the accuracy and noise influence, the following technique was used. The noise with the distribution being 5, 10, 15, 20, and 30%, was added in the selected medical images. Then, the noise suppression algorithms with different characteristics were applied to the obtained images. Subsequently, the contour representations were formed based on the DST. For the images with the extracted contours, the color coding method was used and the obtained results were estimated. The evaluation of the results concerning the extraction of the objects of interest (as to their similarity to the noise-free images and reference images) is presented in percentage. The results of numerical experiments based on the developed method of medical image processing are presented in Table 9.3. Example of the urology image with noise level 20% processing is shown in Fig. 9.7.

The required filter selection at the preprocessing step and application of the DST in combination with a technique of color coding of the contours eventually increase the accuracy and decrease the error of the sought parameters and characteristics in the average by 15–20% almost without loss of essential details in the structural peculiarities of the kidney under examination lacking.

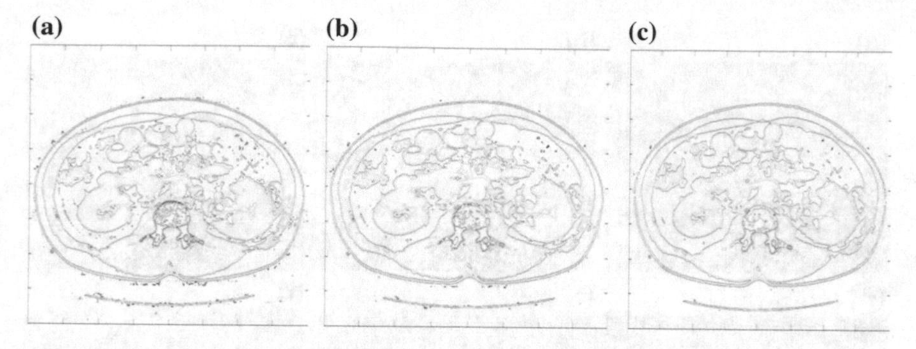

Fig. 9.7 Example of medical image processing using median filtering, the FFST algorithm, and color coding with different kernel of median filter: **a** kernel size 3 × 3, **b** kernel size 5 × 5, **c** kernel size 7 × 7

9.4.2 Processing of Plastic Surgery (Hernioplasty) Images

One of the priorities is the modern technique of anterior abdominal wall reconstruction using biocompatible hyper-elastic mesh implants from titanium nickelide [39, 40]. The new technique was developed and substantiated based on a series of experimental studies on chinchilla rabbits. As a result, a great amount of experimental material as a subject for histological examination was obtained. The results were evaluated based on the morphological and morphometric parameters [38].

In the course of the field studies in the first part of the experiment, an explant was implanted into animals, which was removed on the specified day. In order to evaluate the transplant suitability and study the morphogenesis of reparative regeneration processes in connective-tissue defects after the replacement with the "grown" material by titanium nickelide, the second part of the experiments was performed. An anterior abdominal wall defect was artificially induced in animals, then a grown implant was placed, which had been taken from the posterior hind limb of an animal at specified terms. The obtained material was subjected to the histological examination and the results were estimated regard to the morphological and morphometric parameters. Figure 9.8 demonstrates the results obtained in the first part of the experiment, after implantation of the titanium nickelide mesh in a rabbit thigh. The original images are presented in the left column, the middle column presents the results of the DST, and the results after applying color coding are given in the right column of Fig. 9.8.

The results of the morphological analysis of images using the developed technique (and the corresponding estimates) show that the amount of collagen fibers increased and by the 10th day of the implant exposition in the thigh it was almost two times as high as that as compared to the 5th day of the experiment. The results of the second part of the experiment in the transformed images are presented below. The dynamics of the morphometric parameters of the connective tissue after growing the implant in the experimental thigh and further implantation in the

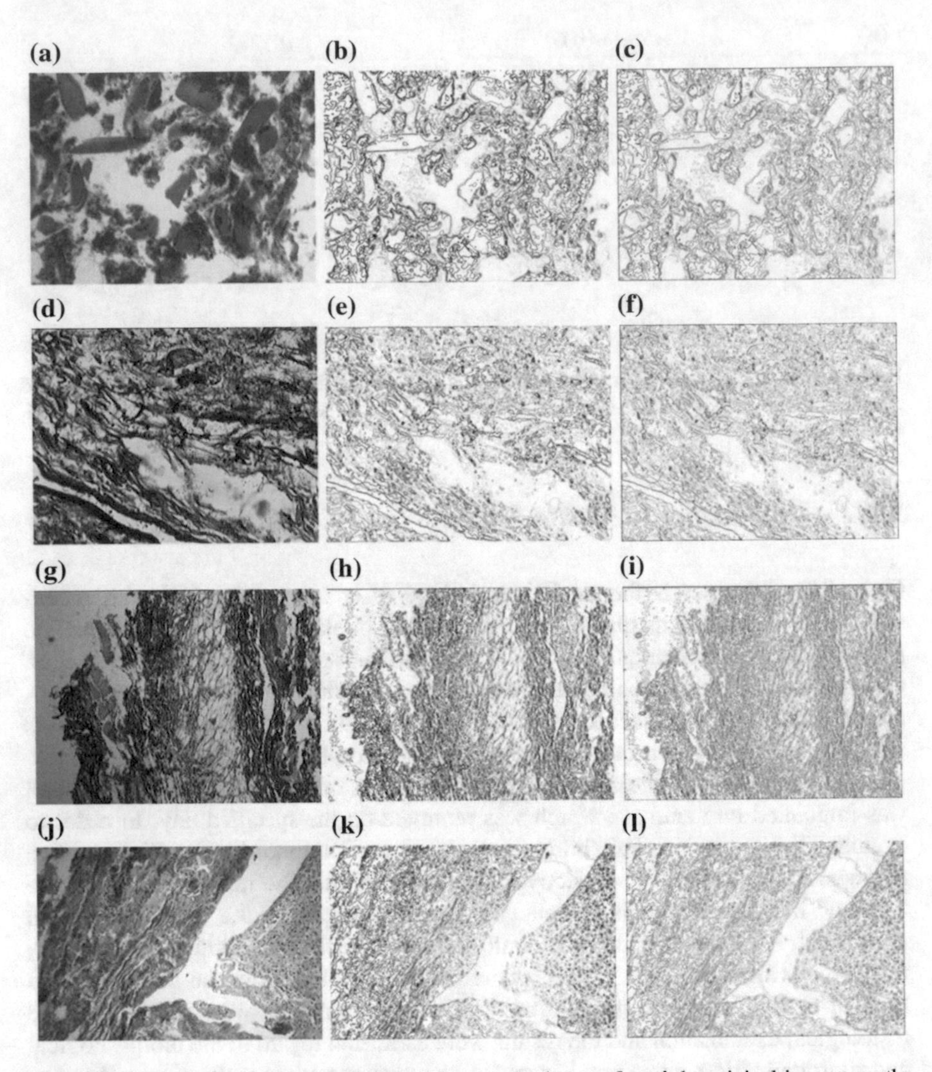

Fig. 9.8 Results of the clinical experiment data processing: **a**, **d**, **g**, **j** the original images on the 1st day, the 5th day, the 7th day, and 10th day after the implantation, respectively, **b**, **e**, **h**, **k** the results of the DST processing on the1st day, the 5th day, the 7th day, and 10th day after the implantation, respectively, **c**, **f**, **i**, **l** the results after color coding on the 1st day, the 5th day, the 7th day, and 10th day after the implantation, respectively

anterior abdominal wall defects are presented in Fig. 9.9. These images show the data of the histological pattern obtained in different periods after the material implantation into the anterior abdominal wall defects "grown during" 7 days. The left column in Fig. 9.9 shows the original images, in the middle column the results of the DST are presented, and the results after applying color coding are depicted in the right column. The characteristic variations of the corresponding parameters are given in Table 9.4.

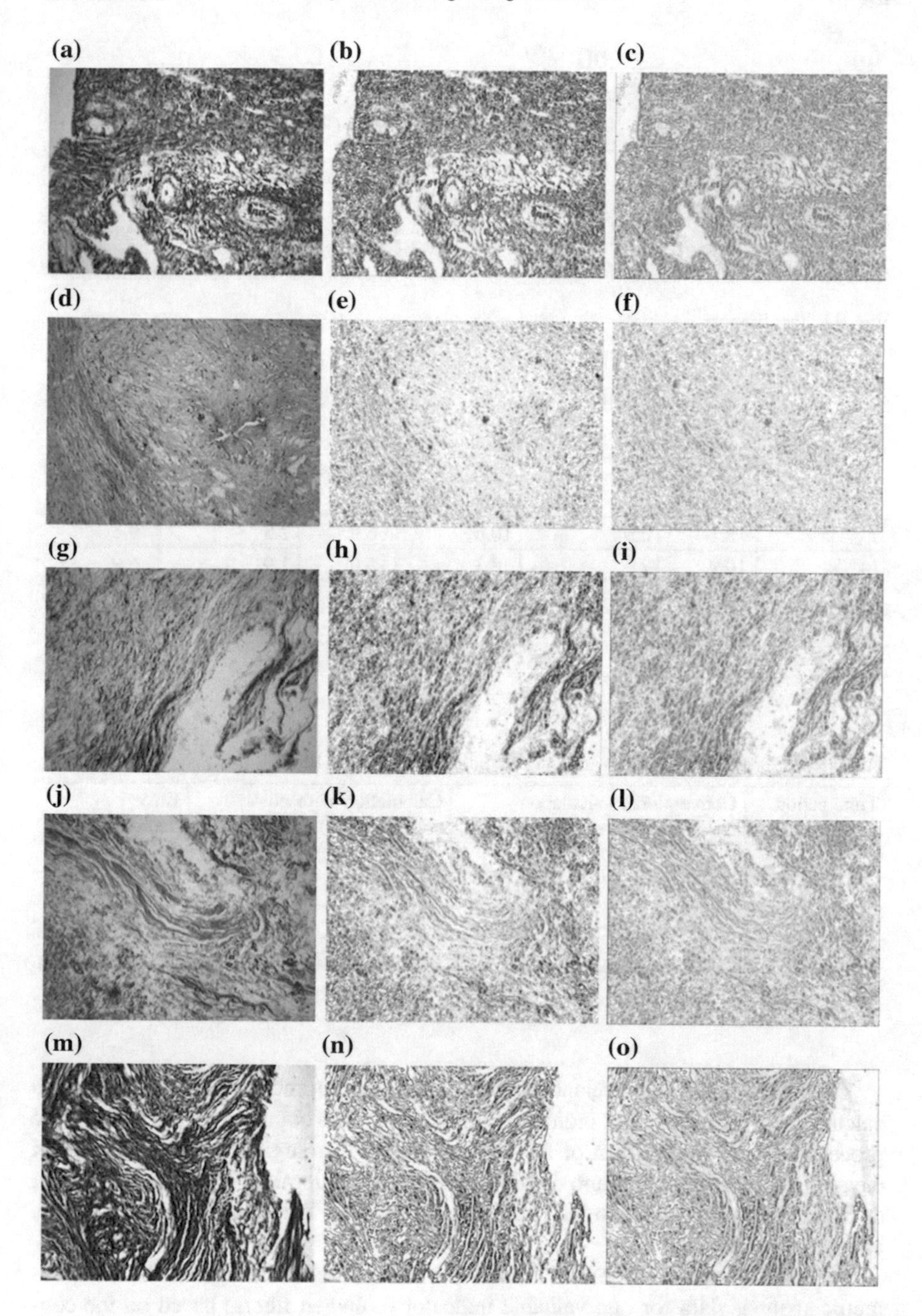

Fig. 9.9 Example of the processing results of the second part of the clinical experiment: **a**, **d**, **g**, **j**, **m**, **p** the original images on the 5th day, the 10th day, the 21th day, the 42th day, the 56th day and 208th day after the implantation, respectively, **b**, **e**, **h**, **k**, **n**, **q** the results of the DST processing on the 5th day, the 10th day, the 21th day, the 42th day, the 56th day and 208th day after the implantation, respectively, **c**, **f**, **i**, **l**, **o**, **r** the results after color coding on the 5th day, the 10th day, the 21th day, the 42th day, the 56th day and 208th day after the implantation, respectively

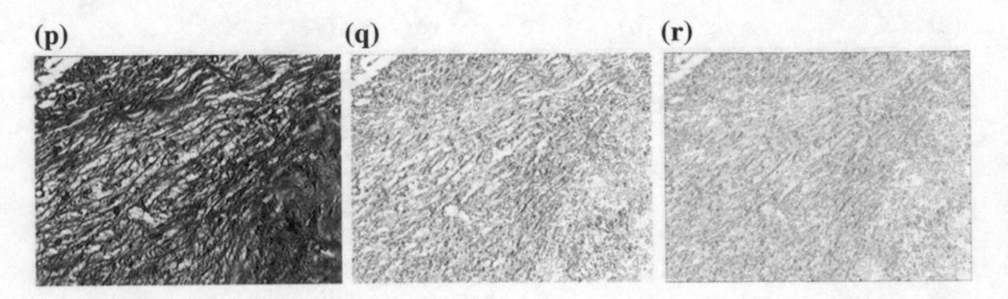

Fig. 9.9 (continued)

Table 9.4 Error estimation of the morphometric parameters in different time moments

Time period (days)	Conventional calculation technique			Calculation with color coding		
	Vessels (%)	Collagen fibers (%)	Elastic fibers (%)	Vessels (%)	Collagen fibers (%)	Elastic fibers (%)
7	8.1	2.9	10.6	6.9	2.5	9.1
14	10.9	2.3	8.3	9.0	1.9	6.9
21	13.9	3.8	14.9	11.1	3.0	11.9
28	14.0	1.9	13.6	10.6	1.4	10.3
56	7.8	1.3	8.7	5.9	0.98	6.5

Table 9.5 Estimation of error variation for indicator "collagen fibers"

Time period (days)	Conventional calculation technique (units)	Calculation with color coding (units)	Error decreasing (%)
7	0.63	0.54	14
14	0.77	0.64	17
21	1.51	1.21	20
28	0.91	0.69	24
56	0.55	0.41	25

Table 9.4 shows the evaluation data of the morphometric parameters for the calculation error for the key indicators. These data were obtained on the basis of the procedures used in the form of the analysis of the histological pattern in different days after the material implantation into the anterior abdominal wall defects "grown during" 7 days according to the following parameters: the vessels, collagen fibers, and elastic fibers.

Table 9.5 presents the estimation of the error variation range in the morphometric analysis data for one valuable indicator (collagen fibers) based on the conventional methods of analysis of the histological image in different time moments after the material implantation into the anterior abdominal wall defects "grown during" 7 days and the image processing based on the proposed method.

Table 9.6 Estimation of error variation for the indicators depending on the noise level

Noise level (%)	Error variation (units)		
	Vessels	Collagen fibers	Elastic fibers
0	1.14–1.72	1.48–2.21	0.39–0.58
5	1.04–1.27	1.09–1.63	0.29–0.43
10	1.14–1.39	1.19–1.79	0.32–0.47
15	1.19–1.45	1.26–1.87	0.33–0.49
20	1.25–1.52	1.32–1.96	0.34–0.52
30	1.28–1.55	1.35–2.01	0.35–0.53

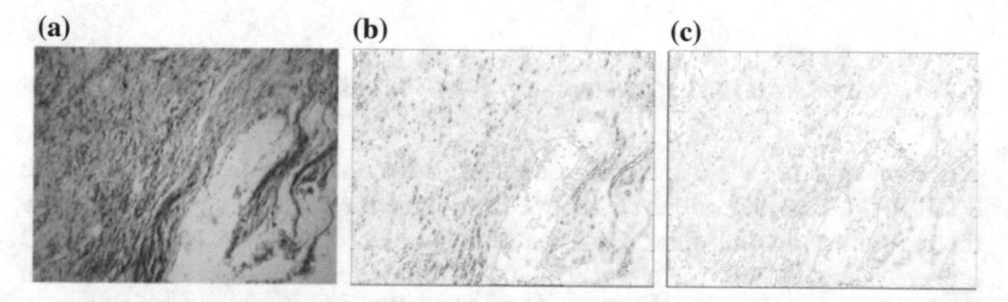

Fig. 9.10 Example of processing by the proposed method: **a** original image, **b** processed image with the 5% noise, **c** processed image with the 10% noise

In order to determine the influence of noise, we performed a study allowing to determine how a noise affects on the resultant accuracy. For these purposes, the additive noise, being the most distinctive in digital cameras, is superimposed in the original image (histological pattern on the 21th day after the material implantation into the anterior abdominal wall defects "grown during" 7 days). The obtained data of the error variation for the indicators of morphometric analysis is shown in Table 9.6.

Figure 9.10 shows an example of processing the histological pattern on the 21th day after the material implantation into the anterior abdominal wall defects "grown during" 7 days using the proposed method, when the applied noise amounts to 5 and 10%.

Table 9.7 presents the evaluation of preprocessing results using different filters (the most significant) of the indicators "collagen fibers". Table 9.7 shows the averaged data for the additive noise of 5 and 10% with the magnitude of the noise characteristics being 13% of the dynamic range. During the experimental research, 50 noisy images were generated from a single reference image.

The experimental study reveals that in most cases the best results are achieved, when preprocessing is performed using the median filter. In the case of small noise (3–7%) and low noise magnitude (5–10% of the dynamic range), the 2D cleaner filter also shows good results.

Table 9.7 Estimation of error variation for indicator "collagen fibers"

Noise level (%)	Filter	Error decrease (%)
5	Mean filter optimized	14
	Gaussian filter 1Dx2	20
	Median filter optimized	26
	2D cleaner filter	21
10	Mean filter optimized	7
	Gaussian filter 1Dx2	14
	Median filter optimized	19
	2D cleaner filter	15

9.4.3 Experimental Research of Preprocessing Filters

The comparison of the filters was made using a laptop with Intel core i5 3.1 GHz 8 GB RAM. The estimations of the processing time for the filtering algorithms like the mean filter, median filter, Gaussian filter (in both conventional and optimized implementations) and 2D cleaner filter for different kernel size are shown in Table 9.8. The data were obtained for 20 different images with the size 512 × 512 pixels. One can clearly see the advantages of the optimized algorithms. Thus, the optimized implementation of the filters is less dependent on the kernel size.

The comparison of time taken for the image processing using the selected filters with the kernel size 5 × 5 ($RH = 2$, $RW = 2$) is shown in Table 9.9. For the experimental study, the images with different sizes (from 512 × 512 pixels to 1920 × 1080 pixels) were taken. Thus, this study demonstrates that the time of image processing is proportional to the image size. Additionally, we evaluated the acceleration of processing using OpenMP for different kernel sizes with two, three and four threads (Table 9.10).

Table 9.8 Processing time of different filter implementations for images with the size 512 × 512 pixels (ms)

Filter	Kernel size, pixels				
	3 × 3	5 × 5	7 × 7	9 × 9	11 × 11
Conventional mean filter	19.26	41.11	75.29	119.36	143.10
Optimized mean filter	13.73	19.01	24.49	28.36	30.54
Conventional Gaussian filter	22.78	51.97	98.74	152.87	184.85
Gaussian filter 1D×2	17.83	24.42	31.41	37.01	41.64
Conventional median filter	147.83	427.18	917.77	1669.57	2089.21
Optimized median filter	37.74	51.72	65.36	76.56	80.30
2D cleaner filter	43.61	110.78	205.61	332.51	491.80

Table 9.9 Processing time of the images with different sizes (ms)

Filter	Image size, pixels						
	512×512 0.26 MP	640×480 0.31 MP	768×768 0.59 MP	1024×768 0.79 MP	1280×720 0.92 MP	1440×1080 1.56 MP	1920×1080 2.07 MP
Mean filter optimized	19.28	22.47	42.89	57.00	66.95	111.91	149.02
Gaussian filter 1Dx2	51.73	60.04	113.41	149.75	173.33	287.82	376.49
Median filter optimized	24.00	29.62	59.53	77.12	86.53	146.69	201.31
2D cleaner filter	111.04	128.61	244.91	324.68	379.01	637.97	846.58

Table 9.10 Evaluation of the acceleration stability (acceleration coefficient)

Filter	Threads	Kernel size, pixels				
		3 × 3	5 × 5	7 × 7	9 × 9	11 × 11
Mean filter optimized	2	1.94	1.94	1.93	1.95	1.96
	3	2.36	2.46	2.43	2.59	2.62
	4	2.36	2.52	2.55	2.84	3.13
Gaussian filter 1Dx2	2	1.87	1.86	1.88	1.89	1.86
	3	2.25	2.33	2.55	2.66	2.66
	4	2.29	2.59	2.58	3.11	3.07
Median filter optimized	2	1.94	1.94	1.93	1.96	1.96
	3	2.02	2.04	2.33	2.47	2.52
	4	2.29	2.38	2.62	2.99	3.02
2D cleaner filter	2	2.31	2.30	2.30	2.33	2.39
	3	3.00	2.98	3.03	3.12	3.17
	4	3.13	3.26	3.37	3.53	3.63

Table 9.11 Evaluation of the acceleration in processing using the optimized algorithms

Filter	Threads	Kernel size, pixels				
		3 × 3	5 × 5	7 × 7	9 × 9	11 × 11
Median filter	1	3.92	8.26	14.04	20.72	27.37
	2	7.61	16.05	27.16	40.57	53.74
	3	7.90	16.88	32.69	51.27	69.06
	4	8.95	19.62	36.80	61.94	82.72
2D Gaussian filter	1	1.28	2.13	3.14	4.02	4.96
	2	2.38	3.95	5.90	7.60	9.22
	3	2.87	4.97	8.01	10.72	12.32
	4	3.03	5.51	8.11	12.92	14.28
Mean filter	1	1.40	2.16	3.07	3.93	4.74
	2	2.72	4.19	5.87	7.68	9.29
	3	3.15	5.32	7.46	10.18	12.49
	4	3.42	5.86	8.62	11.17	14.95

The experimental results show that the increase in the processing speed for different kernel sizes is almost the same. Some stability is observed in the acceleration for two threads, as well as one can see the increase of the acceleration coefficient in the case with more than two threads having the kernel size larger than 5 × 5. The increase in the computation speed using the optimized filter algorithms utilizing the OpenMP technology in comparison to the ordinary sequential implementations is presented in Table 9.11.

The experiments revealing the advantages of the optimized parallel implementation in comparison to the conventional algorithms were made in the medical images with size 512 × 512 pixels. The highest increase in the processing speed was achieved for the median filter. Thus, for the small kernel 3 × 3 the acceleration is about 8 times and for the large kernel 11 × 11 is around 70 times. The optimized versions of mean filter and Gaussian filter provide the acceleration of about 3 times for the small kernel 3 × 3 and about 13 times for the large kernel 11 × 11.

9.4.4 *Efficiency Evaluation of the Proposed Method*

The application of the developed method for visual data processing based on the DST in combination with the color coding technique significantly improved the results of the diagnostics and treatment of the urolithic patients. This is due to the clarification of the prognosis criteria (the localization, size, structure, stone density, and functional state of the upper urinary tract) in pre-operation period. The developed calculation instrumentation allows to improve the accuracy and decrease the error of the sought parameters and characteristics by 10–20% in the average without loss of significant details in the structural features of the examined object of interest.

The proposed technique in plastic surgery (hernioplasty) allows to increase the accuracy of extracting quasi-round structures (formations) and linear structures (fibers), as well as to improve a visual quality of the examined clinical objects using the image color coding for adequate qualitative estimation of its main characteristics (the length, width, and area). Estimation of errors decreased by 10–25% on average.

9.5 Conclusions

In this chapter, it was shown that the digital filters like Sobel, Prewitt, Roberts, Canny, LoG and the DST provide good contour results in medical urology images. The color coding method based on the technique of elastic maps building was proposed. For noise reduction, a preprocessing by the mean, Gaussian, median and 2D cleaner filters was investigated. In order to increase a preprocessing speed, the optimized implementations of filters were suggested. The parallel implementation of the optimized algorithms, using OpenMP and resulting in the performance being almost two times higher for 2 threads and nearly 3 times for 3 and 4 threads, had been performed. The highest increase of the processing speed has been achieved for the median filter. Thus, for the small kernel 3 × 3 the acceleration of about 8 times and for the large kernel 11 × 11 of 70 times is observed. The optimized versions of the mean filter and Gaussian filter give the acceleration of about 3 times for the small kernel 3 × 3 and nearly 13 times for the large kernel 11 × 11.

The image accuracy of estimates in urology and plastic surgery (hernioplasty) has increased up 10–25% on averaged. In urology, the proposed color coding method increased accuracy, especially in complex cases of multiple stones in the kidney. In hernioplasty, the color coding allows to conduct more efficient analysis of the tissue regeneration by controlling the variability of texture with improved accuracy.

References

1. Dzeranov, N.K., Lopatkin, N.A.: Kidney STONES: CLINICAL GUIDELInes. Overlei, Moscow (2007). (in Russian)
2. Filimonov, G.P., Alyayev, YuG, Vasil'yev, P.V.: The treatment of nephrolithiasis: the role of spiral CT. Radiol. Pract. **4**, 34–35 (2001). (in Russian)
3. Nyhus, L.: Individualization of hernia repair: A new era. Surgery **114**(1), 1–2 (1993)
4. Gunther, V.E. (ed.): Shape memory biomaterials and implants. In: Proceedings of International Conference Tomsk Russia. STT, Northampton, MA (2001) (in Russian)
5. Brisbane, W., Bailey, M.R., Sorensen, M.D.: An overview of kidney stone imaging techniques. Nat. Rev. Urol. **16**(11), 654–662 (2016)
6. Troccaz, J., Baumann, M., Berkelman, P., Cinquin, P., Daanen, V., Leroy, A., Marchal, M., Payan, Y., Promayon, E., Voros, S., Bart, S., Bolla, M., Chartier-Kastler, E., Descotes, J.L., Dusserre, A., Giraud, J.Y., Long, J.A., Moalic, R., Mozer, P.: Medical Image Computing and Computer Aided Medical Interventions applied to soft tissues. Work in progress in Urology. Proc. IEEE **94**(9), 1665–1677 (2006)
7. Andrabi, Y., Patino, M., Das, ChJ, Eisner, B., Sahani, D.V., Kambadakone, A.: Advances in ct imaging for urolithiasis. Indian J Urol. **31**(3), 185–193 (2015)
8. Sharma, N., Aggarwal, L.M.: Automated medical image segmentation techniques. J. Med. Phys. **35**(1), 3–14 (2010)
9. Withey, D.J., Koles, Z.J.: Three generations of medical image segmentation: methods and available software. Int. J. Bioelectromagnetism **9**(2), 67–68 (2007)
10. Prince, J.L., Links, J.M.: Medical Imaging Signals and System. Pearson (Prentice Hall), Upper Saddle River, NJ, USA (2006)
11. Macovski, A.: Medical Imaging Systems. Prentice-Hall, Englewood Cliffs, New Jersey (1983)
12. Kalannagari, V., Gunasundari R.: Analysis and implementation of kidney stone detection by reaction diffusion level set segmentation using Xilinx system generator on FPGA. VLSI Design, vol. 2015, Article ID 581961 (2015)
13. Selby, M.G., Vrtiska, T.J., Krambeck, A.E., McCollough, C.H., Elsherbiny, H., Bergstralh, E. J., Lieske, J.C., Rule, A.D.: Quantification of asymptomatic kidney stone burden by computed tomography for predicting future symptomatic stone events. Urology **85**(1), 45–50 (2015)
14. Tamilselvi, P.R., Thangaraj, P.: Segmentation of calculi from ultrasound kidney images by region indicator with contour segmentation method. Glob. J. Comput. Sci. Technol. **11**(22), 1–9 (2011)
15. Rutkov, I.M., Robbins, A.W.: Hernioplasty with mesh'implantanti. Surg. Clin. Nort. Am. **73**, 413–426 (1993)
16. Dahiya, A., Dubey, R.B.: Survey of some multilevel thresholding techniques for medical imaging. Int. J. Sci. Eng. Res. **3**(7), 103–106 (2015)
17. Ravichandran, G., Palanivel, V.: Non-linear enhancement and selection of plane for segmenting the abdominal image for kidney stone identification. Int. J Adv. Res. Electron. Commun. Eng. **5**(2), 430–434 (2016)
18. Sawale, V.M., Chokhat, A.D.: An optimize mechanism for multifunction diagnosis of kidneys by using genetic algorithm. Int. J. Eng. Comput. Sci. **3**(12), 9656–9659 (2014)
19. Raja, R.A., Ranjani, J.J.: Segment based detection and quantification of kidney stones and its symmetric analysis using texture properties based on logical operators with ultra sound

scanning. In: International Conference on Computing and Information Technology (ICIT'2013), pp. 8–15 (2013)
20. Davies, E.: Machine Vision: Theory, Algorithms and Practicalities. Academic Press (2012)
21. Szeliski, R.: Computer Vision: Algorithms and Applications. Springer-Verlag London Limited, London (2011)
22. Jain, R., Kasturi, R., Schunck, B.G.: Machine Vision. McGraw-Hill Inc, New York (1995)
23. Gonzalez, R.C., Woods, R.E.: Digital Image Processing, 3rd edn. Prentice-Hall, Englewood Cliffs (2008)
24. Canny, J.: A computational approach to edge detection. IEEE Trans. Pattern Anal. Mach. Intell. **8**(6), 679–698 (1986)
25. Hauser, S.: Fast Finite Shearlet Transform: a Tutorial. University of Kaiserslautern, Kaiserslautern, Germany (2011)
26. Hauser, S.: Fast finite shearlet transform. Available from: http://www.mathematik.uni-kl.de/imagepro/software/ffst/ (2014). Accessed 10 Jul 2017
27. Guo, K., Labate, D., Lim, W.-Q.: Edge analysis and identification using the continuous shearlet transform. Appl. Comput. Harmonic Anal. **27**(1), 24–46 (2009)
28. Kutyniok, G., Sauer, T.: From wavelets to shearlets and back again. In: Neamtu, M., Schumaker, II. (eds.) Approximation Theory, vol. XII, pp. 201–209. Hashboro Press, San Antonio, TX, Nachville, TN (2007)
29. Kutyniok, G., Labate, D.: Introduction to shearlets. In: Kutyniok G, Labate D (eds.) Shearlets: Multiscale Analysis for Multivariate Data, pp. 1–38, LLC. Springer Science+Business Media (2012)
30. Labate, D., Easley, G., Lim, W.: Sparse directional image representations using the discrete shearlet transform. Appl. Comput. Harmonic Anal. **25**(1), 25–46 (2008)
31. Lim, W.Q.: The discrete shearlet transform: a new directional transform and compactly supported shearlet frames. IEEE Trans. Image Process. **19**(5), 1166–1180 (2010)
32. Cadena, L., Espinosa, N., Cadena, F., Kirillova, S., Barkova, D., Zotin, A.: Processing medical images by new several mathematics shearlet transform. In: International MultiConference of Engineers and Computer Scientists (IMECS'2016), vol. I, pp. 369–371 (2016)
33. Abirami, M.S., Sheela, T.: Kidney segmentation for finding its abnormalities in abdominal CT images. Int. J. Appl. Eng. Res. **10**(12), 32025–32034 (2015)
34. Senthil, K.N., Sathyavathy, S.: Segmentation of renal calculi from CT abdomen images by incorporating FCM and level set approaches. Int. J. Adv. Res. Comput. Commun. Eng. **5**(7), 132–138 (2016)
35. Liu, J., Wang, S., Turkbey, E.B., Linguraru, M.G., Yao, J., Ronald, M.: Computer-aided detection of renal calculi from noncontrast CT images using TV-flow and MSER features. Summers citation. Med. Phys. **42**(1), 144–153 (2015)
36. Ebrahimi, S., Mariano, V.Y.: Image quality improvement in kidney stone detection on computed tomography images. J. Image Graph. **3**(1), 40–46 (2015)
37. Tamilselvi, P.R., Thangaraj, P.: Computer aided diagnosis system for stone detection and early detection of kidney stones. J. Comput. Sci. **7**(2), 250–254 (2011)
38. Avtandilov, G.G.: Medical Morphometry. Medicine, Moscow (1990) (in Russian)
39. Zotov, V.A., Veronskiy, G.I., Vostrikov, O.V.: Clinical-morphological substantiation of a choice of implants in surgery of hernia of anterior abdominal. Morphology and Surgery Proc. Novosibirsk, pp. 88–91 (2000) (in Russian)
40. Radkevich, A.A., Gorbunov, N.A., Khodorenko, V.N., Usoltsev, D.M.: Reparative desmogenez in connective tissue defects after the replacement of NiTi implants. Implants Shape Mem. **1**, 21–25 (2008)
41. Radkevich, A.A., Vinnik, YuS, Kasparova, I.E.: Reparative desmogenez after replacement of connective tissue defects tissue implants from NiTi. Siberian Med. J **91**(8), 151–153 (2009)
42. Shinde, B., Mhaske, D., Dani, A.R.: Study of noise detection and noise removal techniques in medical images. Int. J. Image Graph. Sig. Process. **2**, 51–60 (2012)

43. Sudha, S., Suresh, G.R., Sukanesh, R.: Speckle noise reduction in ultrasound images by wavelet thresholding based on weighted variance. Int. J. Comp. Theory Eng. **1**(1), 1793–8201 (2009)
44. Mamta, J., Mohana, R.: An improved adaptive median filtering method for impulse noise detection. Int. J. Recent Trends Eng. **1**(1), 274–278 (2009)
45. Gnanambal, I., Marudhachalam, R.: New hybrid filtering techniques for removal of speckle noise from ultrasound medical images. Sci. Magna **7**(1), 38–53 (2011)
46. Arin, H.H., Hozheen, O.M., Sardar, P.Y.: Denoising of medical images by using some filters. Int. J. Biotechnol. Res. **3**(1), 10–20 (2015)
47. Donoho, D.L., Kutyniok, G.: Geometric separation using a wavelet-shearlet dictionary. In: International Conference on Sampling Theory and Applications (SAMPTA'2009), pp. 95–99 (2009)
48. Fadilli, M.J., Starck, J.L., Elad, M., Donoho, D.L.: MCALab: reproducible research in signal and image decomposition and inpainting. IEEE Comput. Sci. Eng. **12**(1), 44–63 (2010)
49. Guo, K., Kutyniok, G., Labate, D.: Sparse multidimensional representations using anisotropic dilation and shear operators. In: Chen, G., Lai, M. (eds.) Wavelets and Splines, pp. 189–201. Nashboro Press, Nashville, TN (2006)
50. Guo, K., Labate, D.: Optimally sparse 3D approximations using shearlet representations. Electron. Res. Announcements Math. Sci. **17**, 126–138 (2010)
51. Guo, K., Labate, D.: Optimally sparse multidimensional representation using shearlets. SIAM J. Math. Anal. **39**(1), 298–318 (2007)
52. Kutyniok, G., Labate, D.: Construction of regular and irregular shearlet frames. J. Wavelet Theory Appl. **1**(1), 1–10 (2007)
53. Labate, D., Lim, W.Q., Kutyniok, G., Weiss, G.: Sparse multidimensional representation using shearlets. In: Proceedings of SPIE 5914. Wavelets XI, 59140U, pp. 254–262 (2005)
54. Meyer, Y.: Oscillating patterns in image processing and nonlinear evolution equations. The 15th Dean Jacqueline B. Lewis Memorial Lectures, American Mathematical Society Boston, MA, USA (2002)
55. Starck, J.L., Elad, M., Donoho, D.: Image decomposition via the combination of sparse representation and a variation approach. IEEE Trans. Image Proc. **14**(10), 1570–1582 (2005)
56. Chandel, R., Gupta, G.: Image filtering algorithms and techniques: A review. Int. J Adv. Res. Comput. Sci. Softw. Eng. **3**(10), 198–202 (2013)
57. Gupta, B., Negi, S.S.: Image denoising with linear and non-linear filters: a review. Int. J. Comput. Science **10**(2), 149–154 (2013)
58. Jim Casaburi's free software download page. Available from: http://home.earthlink.net/~casaburi/download/. Accessed 10 Jul 2017
59. Lukin, A.: Tips & tricks: fast image filtering algorithms. In: 17th International Conference on Computer Graphics (GraphiCon'2007), pp. 186–189 (2007)
60. Pascal, G.: A survey of Gaussian convolution algorithms. Image Process. Line **3**, 286–310 (2013)
61. Young, I.T., van Vliet, L.J.: Recursive implementation of the Gaussian filter. Elsevier Sig. Process. **44**(2), 139–151 (1995)
62. Zing, A.: Extended binomial filter for fast Gaussian blur. Vienna, Austria (2010)
63. Suomela, J.: Median filtering is equivalent to sorting. Available from: http://users.ics.aalto.fi/suomela/median-filter/ (2014). Accessed 10 Jul 2017
64. Weiss, B.: Fast median and bilateral filtering. ACM Trans. Graph. **25**(3), 519–526 (2006)
65. Cline, D., White, K.B., Egbert, P.K.: Fast 8-bit median filtering based on separability. In: IEEE International Conference on Image Process (ICIP'2007), vol. 5, pp. 281–284 (2007)
66. Perreault, S., Hebert, P.: Median filtering in constant time. IEEE Trans. Image Process. **16**(9), 2389–2394 (2007)
67. Shameem, A., Jason, R.: Multi-core Programming. Intel Press, USA (2006)
68. Chandra, R., Dagum, L., Kohr, D., Maydan, D., McDonald, J., Menon, R.: Parallel Programming in OpenMP. Academic Press, USA (2001)

69. Kiessling, A.: An Introduction to Parallel Programming with OpenMP. A Pedagogical Seminar. The University of Edinburgh, UK (2009)
70. Gorban, A.N., Zinoviev, AYu., Pitenko, A.A.: Data visualization by elastic maps. Inf. Technol. **6**, 26–35 (2000). (in Russian)

Chapter 10
Image Analysis in Clinical Decision Support System

Natalia Obukhova and Alexandr Motyko

Abstract In this chapter, the methods of medical image processing and analysis in Clinical Decision Support Systems (CDSS) are discussed. The main principles of image analysis with the aim of differential diagnostics in the CDSS are determined. The implementation is given through the method of multispectral images automatic processing and analysis for TV system of cervix oncological changes diagnostics. The method provides differential diagnostics of the following changes in cervical tissues as Norm, Chronic Nonspecific Inflammation (CNI), Cervical Intraepithelial Neoplasia in various types of oncological changes (CIN I, CIN II, CIN III). In proposed method, images of different type (fluorescent images and images obtained in white light illumination) are analyzed. The decision rules in the classification task are based on data mining methods. For the border CIN/CNI sensitivity 87% and specificity 75% are achieved. The detail description of main steps is given in the chapter.

Keywords Medical images processing · Clinical decision support system
Multispectral images processing · Image color analysis · Texture analysis
Classification

10.1 Introduction

The modern trend of medical video systems development is the transition from tools, providing the possibility of examining the organ under investigation by a physician, as well as from systems, operating on the principle of a binary solution "Norm (there is no pathology)/Pathology (there is pathology)" in the CDSS.

N. Obukhova (✉) · A. Motyko
Saint Petersburg State Electrotechnical University "LETI",
Professora Popova Str. 5, Saint-Petersburg 197022, Russian Federation
e-mail: natalia172419@yandex.ru

A. Motyko
e-mail: motyko.alexandr@yandex.ru

M.N. Favorskaya and L.C. Jain (eds.), *Computer Vision in Control Systems-4*,
Intelligent Systems Reference Library 136,
https://doi.org/10.1007/978-3-319-67994-5_10

Systems of this type realize the integration of automatic image analysis results with the results obtained by the physician and also use the information of system database. This integration allows the sensitivity and specificity of the diagnosis to be higher than in the case of diagnosis by the physician alone or independently by the system.

The medical images automatic analysis in the CDSS is required to implement differential diagnostics with the degree of pathology corresponding (estimation of pathology degree must be in interval from zero up to one), which is based on multi-class analysis in conditions of fuzzy borders between classes and high variability of the original data. This determines the following principles of image analysis:

- The decision of rule construction ought to be based on the methods, which are typical for artificial intelligence systems, such as fuzzy sets theory and fuzzy logic, data mining methods, machine learning methods, and methods of pattern recognition.
- The feature vector for classification ought to include the features with different properties and different origins.

In particular, this approach has shown its effectiveness in the multispectral TV system for cervix oncological changes diagnosis. The system realizes differential diagnostics of the following changes in cervical tissues as Norm, CNI, CIN I, CIN II, CIN III (cervical intraepithelial neoplasia in various types of oncological changes). The main methods of multispectral images automatic processing and analysis for the differential diagnostics realization that were implemented in the CDSS are described and analyzed in this chapter. The main task of image analysis in multispectral TV system is to help a physician at the stage of colposcopic examination. This stage is the central stage in modern medicine scheme of cervical cancer diagnosis. The aim of a colposcopic examination is to identify and rank the severity of lesions, so that biopsies representing the highest-grade abnormality can be taken, if it is necessary. During colposcopy, 3–5% acetic acid solution is applied to the cervix, and areas with abnormal epithelium turn white. The colposcopic examination is used to direct biopsies of the abnormal white areas. Additional lesion characteristics, such as margin shape, color or opacity, blood vessel, intercapillary spacing, and distribution, are considered by the physicians to derive the clinical diagnosis.

The accuracy of colposcopy is highly dependent on the physician's individual skills. In expert hands, colposcopy has been reported to have a high sensitivity (96%) and a low specificity (48%) when differentiating abnormal tissues. Because of the low specificity, a biopsy is required to confirm disease. These avoidable biopsies cause an increased risk of infection, patient discomfort, delayed treatment, and substantially increased costs. There is a need for more objective, specific, and cost-effective screening and diagnostic techniques that could improve the accuracy of colposcopic examination, particularly in the hands of less-experienced practitioners. This leads to a requirement to develop new diagnostic systems on colposcopic stage. Image processing and computer vision technology are one of the key technologies to segment colposcopic images and detect cervical neoplasias.

In this chapter, new automatic method of multispectral colposcopic images analysis based on data mining technologies is introduced. Method demonstrated the high diagnostic characteristics during clinical investigation. This chapter is organized as follows. Related works are presented in Sect. 10.2. The main five steps of the proposed method, such as the image preprocessing in white light and fluorescence light, analysis of the fluorescent images, detection of areas corresponding to the effect of AcetoWhite (AW), as well as areas with high level of tissue heterogeneity in images obtained in white light, creation of the differential pathology map (image of the cervix, broken into areas with a definite diagnosis of Norm-CNI-CIN (I, II, III)), and creation of the biopsy map (image with markers at points, where a biopsy specimen is required for histological analysis) are discussed in Sects. 10.3–10.6, respectively. The clinical investigation description and results are represented in Sect. 10.7. Section 10.8 is concluded the chapter.

10.2 Related Works

Various image processing methods and algorithms have been developed to detect different colposcopic features. They can be divided into two large groups: implementing analysis and processing of image obtained in white light [1–13] and fluorescent images [14–16].

In the first group, Yang et al. [10] developed a segmentation algorithm to detect the AcetoWhite (AW) epithelium using a statistical optimization scheme for accurate clustering to track the boundaries of the AW regions. Xiong et al. [6] detected the AW regions on the base of chromaticity with watershed algorithm. Huang et al. [7] decided the task of the AW region detecting with color and brightness feature estimated in color spaces Lab and HSV. Gordon et al. [11] developed a segmentation algorithm for three tissue types in cervical imagery (original squamous, columnar, and the AW epithelium) based on color and texture information. The set of regions in the images was represented by a Gaussian mixture model, while an Expectation-Maximization algorithm was used to determine the maximum likelihood parameters of the statistical model in the feature space. As a result, the labeling of a pixel could be affiliated with the most probable Gaussian cluster according to Bayes rule. Ji et al. [12] presented a generalized texture analysis algorithm for classifying the vascular patterns from colposcopic images. They investigated six characteristic pathological vascular patterns including the network capillaries, hairpin capillaries, two types of punctation vessels, and two types of mosaic vessels. Furthermore, several approaches for tissue classification have been developed, for example a rule-based medical decision support system for detecting different stages of cervical cancer based on the signs and symptoms from physical examination and multivariate stochastic training algorithms [13].

The multi-spectral diagnostic systems are concerned to the second group. Methods of multi-spectral digital colposcope are developed at the University of Texas (USA) by the team of researchers led by Richards-Kortum [14–16]. Thus, the analysis of modern works in the field of colposcopic images automatic analysis shows:

- The most part of proposed method is devoted to automatic analysis of images obtained in white light in order to select image regions with the AW effect or for the purpose of analyzing vascular structures [1–13].
- The number of works aimed at developing fluorescent diagnostics is very small. The most significant are the works of Richards-Kortum [14–16].
- For classification, the supervised learning is used in all works. Most often, the following approaches are used in constructing the classifier: Mahalonobis distance [7, 16], Gaussian mixture model [2], Markov random fields [8, 9, 15], and support vector machine [1]. These are rather simple classifiers that allow performing "Norm/Pathology" analysis with a rigid boundary between classes. The need to eliminate the rigid boundary between classes, as well as the high degree of variability in the source data (medical images, in particular colposcopic ones, have significant differences for different patients due to the age, menopause, and other features of the women physical condition) causes the necessity of using more complex classification strategies.
- The authors did not find work that implements simultaneous analysis of images obtained in white light and fluorescence light, while the complexity of differential diagnostics in the CDSS task requires a multi-features analysis of different types' images.

It makes actual and relevance the proposed automatic method for the colposcopic images analysis. The main features of proposed method are:

- Simultaneous analysis of images of different types, such as the images captured in white light and fluorescent images obtained with wavelength excitation 360 and 390 nm. The use of fluorescent images increases the sensitivity and specificity of the diagnosis. Fluorescent images contain additional diagnostic information based on changes in the concentration and restructuring of endogenous fluorophores, such as Nicotinamide Adenine Dinucleotide Hydrogenase (NADH), Flavin Adenine Dinucleotide (FAD), collagen, keratin, and Protoporphyrin IX (PpIX).
- Application of the set of pathology features, such as the brightness, color, and texture changes with a quantitative evaluation calculated on the base of images of different types.
- The construction of methods and algorithms for automatic colposcopic images classification based on data mining technologies, which take into account the high variability of the initial data and the unclearness of class boundaries in the analysis.

In following sections, the main steps of the proposed methods are discussing.

10.3 Image Preprocessing for Automatic Analysis

The efficient analysis of any medical images is strongly connected with their preprocessing. The preprocessing includes the ordinary procedures to improve a quality of the images, such as noise reduction, contrast enhancement, etc. (Sect. 10.3.1) and special medical imaging procedures, such as the matching medical images taken under different lighting conditions, automatic segmentation of regions of interest, the image part of which is necessary to conduct the further analysis, and removal of highlights in the images (Sect. 10.3.2).

10.3.1 Shift Compensation Algorithm

The processing of multispectral images and their analysis in a combination with the images in white light require to make a matching between different images. The goal of matching is to eliminate the shifts (discrepancy in the coordinates) between images from different modes. Large shifts between images can noticeably confuse the classifier, because the real coordinates of a single biopsy point may vary in different modes and due to this fact there could be errors in calculation of the features vector of biopsy point. Thus, the compensation of the shifts between images in different modes is an important procedure for classification.

The task of shift compensation is more or less well studied and there are many approaches described in literature. Many of them are based on the detection of motion vectors according to the optical flow equation, Eq. 10.1, where $\mathbf{V} = \left(\frac{\delta x}{\delta t}, \frac{\delta y}{\delta t}\right)$ is an optical flow vector, $\nabla\mathbf{L} = \left(\frac{\partial L}{\partial x}, \frac{\partial L}{\partial y}\right)$ is a brightness gradient vector (spatial domain), $L_t = \frac{\partial L}{\partial t}$ is a brightness derivative with respect to time.

$$\langle \nabla\mathbf{L}, \mathbf{V}^T \rangle + L_t = 0 \tag{10.1}$$

The optical flow equation has an assumption that the brightness of a point (pixel) is constant during the motion along the trajectory provided by Eq. 10.2, where L (x, y, t) is a brightness of a pixel (x, y) at the moment t.

$$L(x, y, t) \approx L(x + \delta x, y + \delta y, t + \delta t) \tag{10.2}$$

This condition with a number of assumptions can be satisfactory, when the images that should be combined are in white light but unattainable in the case of combination of the multispectral images. Multispectral images differ from each other, both in color and brightness features, so it is impossible to use the displacement vector to match the images. An additional feature of the images is low detail presence. It leads to difficulties in vertical and horizontal brightness derivatives calculation and follows to errors in the definition of motion vectors. Thus, for combining the multispectral images with each other and with images obtained in white light, it is proposed to use an approach based on a phase correlation.

The method is based on the shifts property of the Fourier transform, which consists in the fact that a time shift of the signal $s(t)$ by t_0 changes the phase response of the spectrum $S(\omega)$ by an amount of $\pm\omega_{t0}$. In this case, the amplitude-frequency characteristic of the spectrum (i.e. the spectral density module) does not change. Therefore, in the case of 2D signals, if an image $g_b(x, y)$ with the size $[M, N]$ is the shifted copy of the image $g_a(x, y)$, then the Fourier images differ only in phase (Eq. 10.3).

$$\mathbf{G}_b(u, v) = \mathbf{G}_a(u, v)\, e^{-2\pi i \left(\frac{u\Delta x}{M} + \frac{v\Delta y}{N}\right)} \tag{10.3}$$

For the definition of the difference in phase, the cross-power spectrum is computed:

$$R(u, v) = \frac{\mathbf{G}_a\mathbf{G}_b^*}{\left|\mathbf{G}_a\mathbf{G}_b^*\right|} = \frac{\mathbf{G}_a\mathbf{G}_a^* e^{-2\pi i\left(\frac{u\Delta x}{M} + \frac{v\Delta y}{N}\right)}}{\left|\mathbf{G}_a\mathbf{G}_a^* e^{-2\pi i\left(\frac{u\Delta x}{M} + \frac{v\Delta y}{N}\right)}\right|} = \frac{\mathbf{G}_a\mathbf{G}_a^* e^{-2\pi i\left(\frac{u\Delta x}{M} + \frac{v\Delta y}{N}\right)}}{\left|\mathbf{G}_a\mathbf{G}_a^* \cdot 1\right|} \tag{10.4}$$

The phase of $\mathbf{G}_a\mathbf{G}_b^*$ is zero, then:

$$R(u, v) = e^{-2\pi i\left(\frac{u\Delta x}{M} + \frac{v\Delta y}{N}\right)} \tag{10.5}$$

The inverse Fourier transform of a complex exponential is the so called Kronecker delta (Eq. 10.6).

$$r(x, y) = F^{-1}(R(u, v)) = \delta(x + \Delta x, y + \Delta y) \tag{10.6}$$

Defining the maximum of r means, the evaluation of the shift between images $(\Delta x, \Delta y)$ can be estimated by Eq. 10.7.

$$\Delta x, \Delta y = \max_{x,y}(r) \tag{10.7}$$

The algorithm for shifts compensation includes the following steps for each mode:

- Convert the image to grayscale mode (the YCrCb color space).
- Process the image by a median filter for the high frequency noise reduction.
- Threshold the image for binarization. The adaptive threshold value for point (x, y) is calculated as the convolution of the image signal in the neighborhood of point (x, y) with the Gaussian function.
- Get the spectrums of the analyzing and reference images with the Discrete Fourier Transform (DFT).
- Compute the cross-power spectrum for the analyzing and reference images provided by Eq. 10.8, where $\mathbf{G}_a$ is a result of the DFT for analyzing image, $\mathbf{G}_b$

is a result of the DFT for the reference image, sign "∘" means the Adamar matrix multiplication.

$$R = \frac{\mathbf{G}_a \circ \mathbf{G}_b^*}{\left|\mathbf{G}_a \mathbf{G}_b^*\right|} \tag{10.8}$$

- Calculate the cross-correlation function of images using the inverse DFT and find its maximum. The maximum of cross-correlation function determines the spatial displacement.

The results of the algorithm are represented in Fig. 10.1. The first and second columns show the initial images obtained in different modes (390, 430, 360, and 360 nm with laser illumination 650 nm). The initial images have offsets from each other. To demonstrate this fact, the initial positions of the centers of each image are connected by a white line. The third and fourth columns include the image after displacement calculation and compensation.

The main advantage of the proposed approach in comparing with the modern algorithms operating in the signal domain is a possibility of matching the images, which have significantly different from each other in brightness and color, as well as resistance to noise, occlusion, and similar negative factors of medical images.

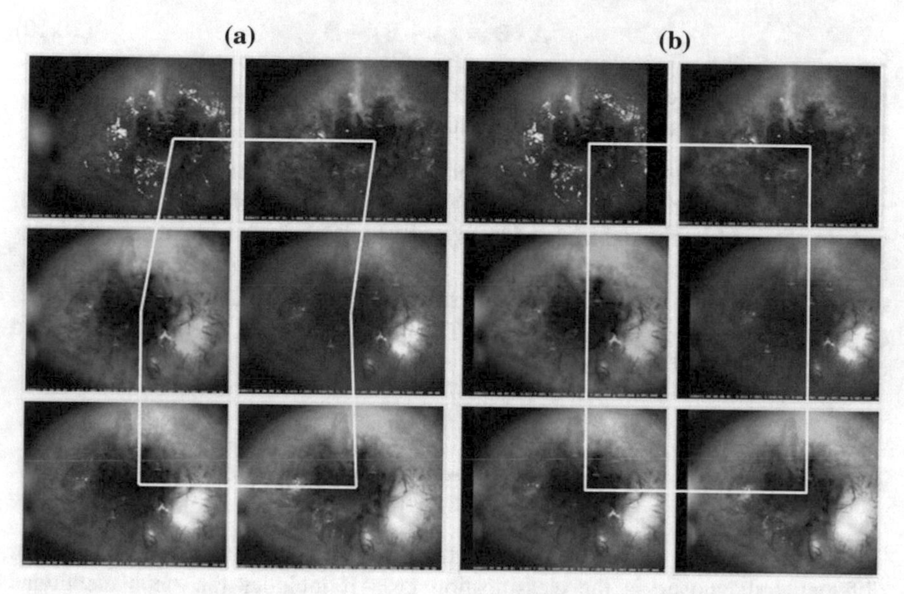

Fig. 10.1 Visual results of shift compensation algorithm: **a** initial images, **b** processed images

10.3.2 Segmentation of Regions of Interest

In real medical image, the image of cervix takes only about 70%. The rest part of image may include other tissue, medical instruments, and various artifacts. The presence of irrelevant information makes difficult the further data processing and analysis, so it is necessary to select only the image area corresponded to cervix as the region of interest. For automatic region of interest segmentation, the following algorithm is proposed:

- Image processing with Gauss filter $\mathbf{G}_\sigma$ for the high frequency noise reduction using Eq. 10.9, where sign "*" is a convolution, $L_2(x, y)$ is a brightness of pixel in the processed image, $L_1(x, y)$ is a brightness of pixel in the initial image.

$$L_2(x,y) = L_1(x,y) * \mathbf{G}_\sigma \quad \mathbf{G}_\sigma = \frac{1}{2\pi\sigma^2}e^{-\frac{(x^2+y^2)}{2\sigma^2}} \tag{10.9}$$

- Conversion to Lab color space and extraction of chromaticity component **a**.
- K-means clustering to select areas with similar features.
- Morphological filtration for edges refinement using Eq. 10.10, where **A** is a matrix of binary image, **B** is a matrix of structure element, sign "+" is a morphological operation dilatation, sign "–" is a morphological operation erosion.

$$\mathbf{A} \cdot \mathbf{B} = (\mathbf{A} + \mathbf{B}) - \mathbf{B} \tag{10.10}$$

- The largest area choosing is a Region Of Interest (ROI).
- Image cropping according to the obtained ROI mask.

The main steps results of the proposed algorithm are given in Fig. 10.2.

During experimental tests, the algorithm was implemented on the 167 series of images obtained in clinical trials of the system LuxCol. Series is the set of images taken from the same patient in different lighting conditions (modes). The experiments showed the region of interest is selected correctly for 95% of the images (some examples are given in Fig. 10.3). The compensation of image shift is solved correctly for 93% of the images.

10.4 Automatic Analysis of Fluorescent Images

The main part of image processing for fluorescent images automatic analysis for differential diagnostic is the classification task. It includes the steps mentioned below:

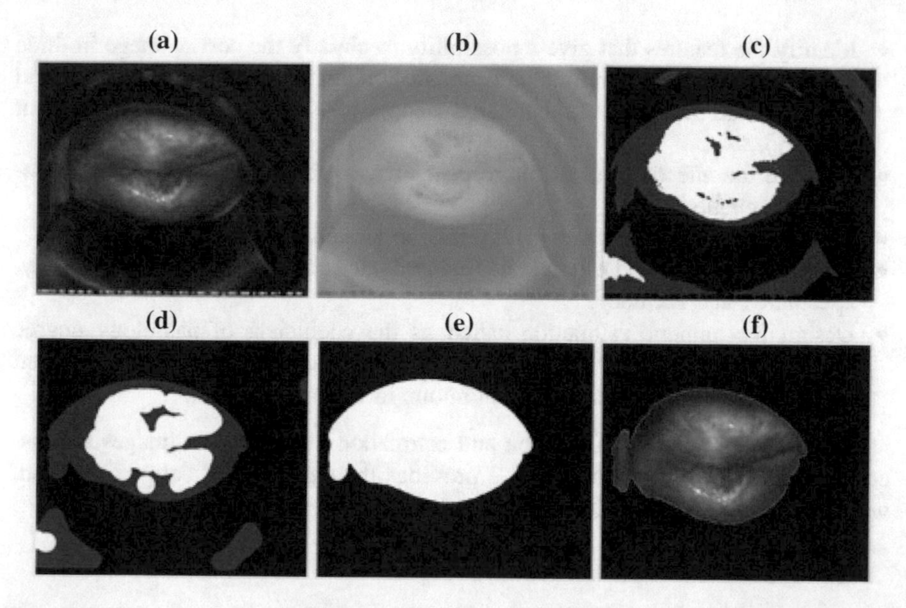

Fig. 10.2 The main steps of the ROI selection algorithm: **a** initial image, **b** image in Lab color space, **c** the result of clustering, **d** the result of edges refinement procedure, **e** the biggest area selection, **f** cropped image

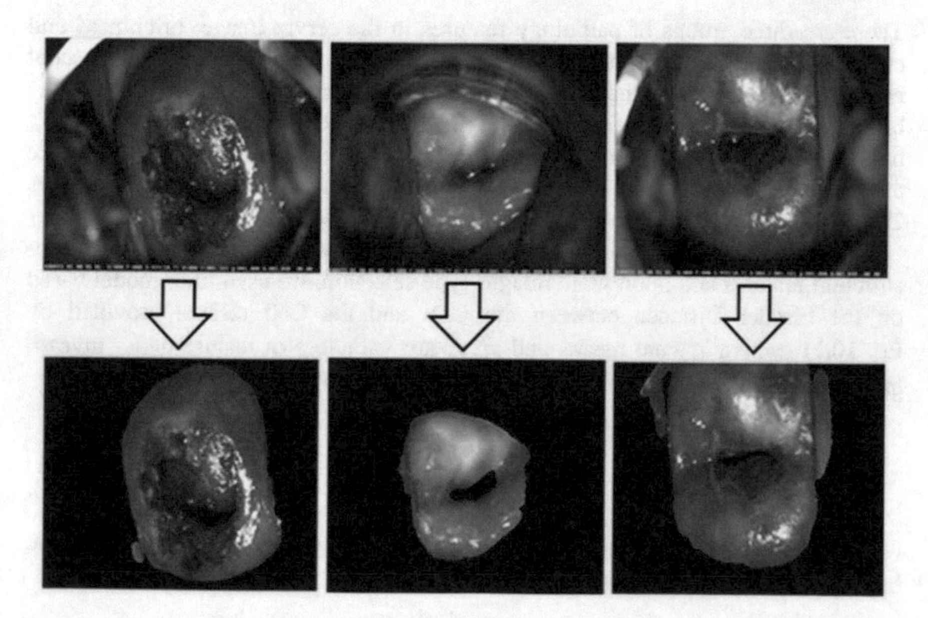

Fig. 10.3 Results of ROI selection

- Identify the features that give a possibility to classify the cervix image in three classes: norm, CNI (chronic nonspecific inflammation—inflammation), and CIN I, II, III (cervical intraepithelial neoplasia—cancer changes of different severity).
- Estimate the effectiveness of fluorescent images obtained with different wavelength excitation.
- Select the most appropriate strategy for the classification task.
- Train the classifier and estimate its main characteristics, such as the sensitivity, specificity, and accuracy.
- Design the numeric estimation called as the coefficient of pathology degree showing the belonging of particular part of image to pathology. The coefficient of pathology degree should be continuous in the range of 0–1.

The issues of features calculation and estimation of fluorescent images are discussed in Sect. 10.4.1. Section 10.4.2 provides the algorithm of color calibration, while the selection of classification rules is considered in Sect. 10.4.3.

10.4.1 Features Calculation and Estimation of Fluorescent Images Effectiveness

There are three groups of pathology features in the cervix image: brightness and color characteristics, morphological features, and texture feature. For fluorescent images analysis, we used the brightness and color features of image. To quantify the brightness and initial color features R, G, B, the ratios G/R and G/B and the (r g b), the XYZ, the YCrCb, the Lab, and the HSV color spaces have been used. To reduce a noise, all features were calculated for the local fragments with block sizes 20×20 pixels. Further, these average values were used.

For the most effective for differential diagnostic and the pathology map construction features and fluorescent images type selection, we used filter model based on the Fischer Distance between the CIN and the CNI classes provided by Eq. 10.11, where $\bar{x}, \bar{y}$ are means and s_x^2, s_y^2 are variances of feature under investigation for the CIN class and the CNI class, respectively.

$$K_D = \frac{(\bar{x} - \bar{y})^2}{s_x^2 + s_y^2} \tag{10.11}$$

According the Fischer Distance between the CIN and the CNI classes, the following results were obtained:

- The most informative is an image obtained with the exciting wavelength of 360 nm.
- The most effective features in 360 mode are the components Cr, Cb from the YCrCb color space and component b from Lab color space.

- The additional mode for improving classification is 390 mode.
- The most effective features in 390 mode are the components Cr, Cb from the YCrCb color space and component b from the Lab color space.
- The worst results were obtained for 430 mode.

In this research, the database of real colposcopic images obtained during the special clinical investigation was used. More information about database creation and properties are given in Sect. 10.7.2.

10.4.2 Color Calibration

Color is one of the main image characteristic. This feature is commonly used in different special TV systems, especially in the systems for medical applications. In classification task, a color is the feature that allows to make a decision if the image part belongs to the particular class. The use of color as a basic feature in the images analysis requires to provide the comparability of colors quantitative estimates for the images obtained by different sensors or in different conditions of observation. This requirement determines the need of special color calibration procedure.

Implementation in this case the standard correction procedure, which refers to minimizing color errors caused by the difference between the spectral sensitivity curves of real sensor and the curves of color mixing for "ideal" sensor, is not effective. Suppose, there is a sensor A and sensor B, the procedure of color correction is realized for both ones. It means that the color errors for each sensor are minimized, but it does not guarantee that the colors in image obtained by sensor A will be identical with a certain tolerance to colors in image obtained by sensor B. It is necessary to minimize error between colors in images obtained by these two sensors, i.e. to calibrate the sensor B according to the sensor A. Only then we can use the images obtained by sensors A and B all together for classification task.

The proposed calibration procedure is based on recalculation the color space of sensor B in color space of sensor A using a linear transformation matrix. This approach assumes that the image **O** generated by the sensor A is taken as a reference and the components of colors from the image **P** obtained by sensor B are recalculated, so that they differ from reference colors of the image **O** is minimal.

Let the original image **O** have N color samples, with each color sample being represented using red (R), green (G), and blue (B) intensities. Organize these original color samples into a matrix represented by Eq. 10.12.

$$\mathbf{O} = \begin{bmatrix} O\_R_1 & O\_G_1 & O\_B_1 \\ O\_R_2 & O\_G_2 & O\_B_2 \\ O\_R_N & O\_G_N & O\_B_N \end{bmatrix} \tag{10.12}$$

Let the processed image **P** also have N color samples that correspond to those in the original image:

$$\mathbf{P} = \begin{bmatrix} P\_R_1 & P\_G_1 & P\_B_1 \\ P\_R_2 & P\_G_2 & P\_B_2 \\ P\_R_N & P\_G_N & P\_B_N \end{bmatrix} \quad (10.13)$$

Then we seek the optimal linear transformation matrix $\mathbf{A_{12}}$ (4 rows $\times$ 3 columns) that maps the processed color samples **P** into the corresponding original color samples **O** in the best manner:

$$\mathbf{O} \approx \hat{\mathbf{O}} = [\mathbf{1}\,\mathbf{P}]\mathbf{A}_{12} \quad (10.14)$$

In Eq. 10.14, **1** is a column vector of N ones that provides a Direct Current (DC) offset or shift in the brightness level.

If a number of independent color components in the reference image exceeds twelve, then for the matrix $\mathbf{A}_{12}$ coefficients identification it is possible to use the method of least square using Eq. 10.15.

$$\mathbf{A}_{12} = \left([\mathbf{1}\,\mathbf{P}]^T[\mathbf{1}\,\mathbf{P}]\right)^{-1}[\mathbf{1}\,\mathbf{P}]^T\mathbf{O} \quad (10.15)$$

Equation 10.15 is the basic equation for estimating calibration matrix coefficients. The significant errors in determining of the calibration coefficients can be made by the colors having significant difference between RGB-components in the reference and calibrated images, so called “outliers”. In this case, it is advisable to use the iterative method of least squares with weights to minimize the effect of “outliers” [17].

The iterative method for estimating the matrix $\mathbf{A}_{12}$ coefficients includes the following main steps:

Step 1. Use the normal least-squares solution from Eq. 10.3 to generate an initial estimate of the color correction matrix $\mathbf{A}_{12}$.

Step 2. Generate an error vector **E**, where each element E_i is equal to the Euclidean distance between the RGB-components of color space from image **O** and the RGB-components of the fitted processed color:

$$E_i = \sqrt{\left(O\_R_i - \hat{O}\_R_i\right)^2 + \left(O\_G_i - \hat{O}\_G_i\right)^2 + \left(O\_B_i - \hat{O}\_B_i\right)^2} \quad (10.16)$$

Step 3. Generate a cost vector **C** that is the element-by-element reciprocal of the error vector **E** from Step 2 plus a small epsilon ε:

$$C_i = \frac{1}{E_i + \varepsilon} \quad (10.17)$$

The ε prevents division by zero and sets the relative weight of a point that is on the fitted line versus the weight of a point that is off the fitted line. A value $\varepsilon = 0.1$ is recommended.

Step 4. Normalize the cost vector **C** for unity norm (i.e. each element of **C** is divided by the square root of the sum of the squares of all the elements of **C**).

Step 5. Generate the cost vector $\mathbf{C}^2$ that is the element-by-element square of the cost vector **C** from Step 4.

Step 6. Generate an $N \times N$ diagonal cost matrix ($\mathbf{C}^2$) that contains the cost vector's elements ($\mathbf{C}^2$) arranged on the diagonal with zeros everywhere else.

Step 7. Using the diagonal cost matrix ($\mathbf{C}^2$) from Step 6, perform the cost-weighted least-squares fitting to determine the next estimate of the color correction matrix $\mathbf{A}_{12}$:

$$\mathbf{A}_{12} = \left([\mathbf{1}\,\mathbf{P}]^T \mathbf{C}^2 [\mathbf{1}\,\mathbf{P}]\right)^{-1} [\mathbf{1}\,\mathbf{P}]^T \mathbf{C}^2 \mathbf{O} \tag{10.18}$$

Step 8. Repeat Steps 2–7 until the elements of the correction matrix $\mathbf{A}_{12}$ converge to four decimal places.

The essential feature of the procedure is to choose a set of colors for calibration. It determines the accuracy of color calibration, in general, and accuracy the next image analysis procedures. As a base, it is advisable to use a set of colors for a typical color correction–ColorChecker. For medical TV systems, this calibration set must be changed because the medical images contain a limited number of colors but especially for them the errors of colors components should be taken into account as much as possible.

In calibration procedure realized in method of the multispectral image analysis for differential cervical cancer diagnostic, the special colors set based on the standard ColorChecker was formed allowing to minimize the errors in blue and yellow colors. These colors are present on colposcopic fluorescent images obtained by exciting with wavelength 360 nm, and these colors are the basis for the classification. Formed color set includes nine colors. The implementation of calibration procedure with special ColorChecker (Fig. 10.4) provides a possibility to use images obtained by sensors with different spectral sensitivity characteristics in the classification task.

For example, during the clinical investigation of proposed method of multispectral image analysis, the colposcopic fluorescent images were obtained by two sensors. One part was got by CCD 285 sensor, another by CMOS 236 sensor. The spectral sensitivity characteristics of these sensors have significant differences in the green channel (Fig. 10.5). This led to the fact that the images obtained by CMOS 236 sensor have not got well expressed color features, which are the basis of classification, for example, there practically have no change in the color component b for fluorescent images (Fig. 10.6).

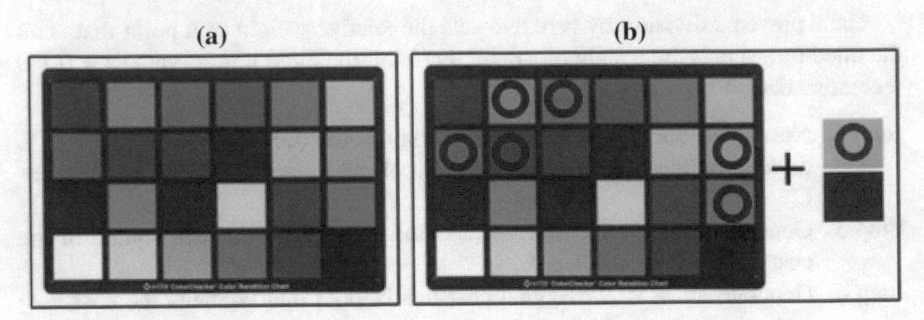

Fig. 10.4 Color palette for multispectral system calibration: **a** standard ColorChecker, **b** special ColorChecker

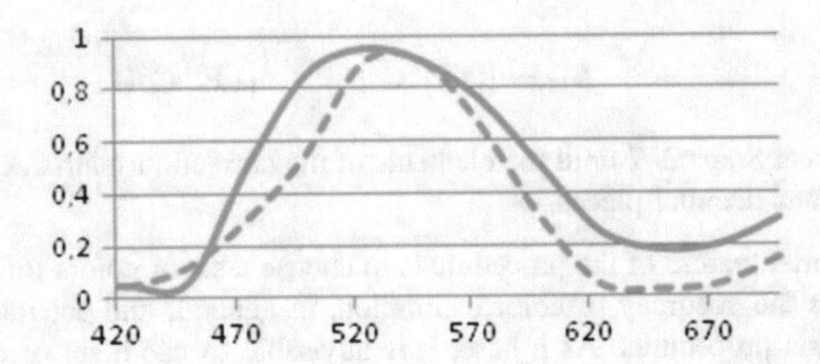

Fig. 10.5 The spectral sensitivity characteristics of reference sensor CCD 285 (*dotted line*) and calibrated sensor CMOS 236 (*solid line*)

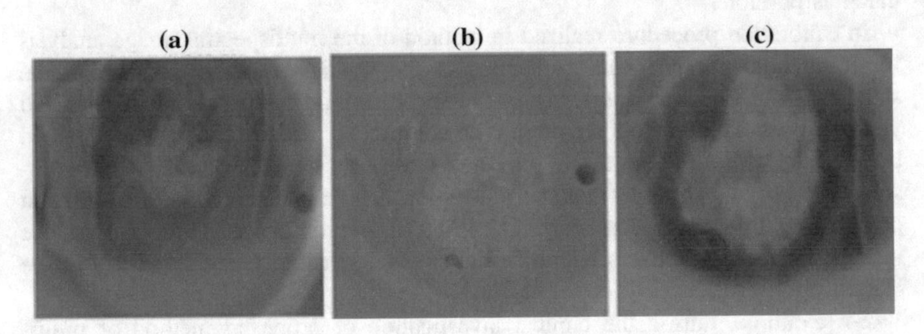

Fig. 10.6 Component b in the cervix images obtained by: **a** the reference sensor A, **b** the sensor B before calibration, **c** the sensor B after the calibration procedure. Note that images from sensor A and sensor B were obtained from different patients but both the images of sensor B from one patient

To realize a possibility to use the classifier based on the color component b and include in processing all images as from CCD 285 sensor and from CMOS 236 sensor, we applied the proposed calibration procedure. Estimation of the proposed procedure effectiveness was carried out according to the criteria: the estimation of the color component b recovery effectiveness and degree of ensuring the correct solutions in the classification task for recalculated fluorescent colposcopic images. Calibration procedure provides the effective color component b recovery (Fig. 10.6) and increases the probability of correct classification on 20% (in comparing with classification accuracy without calibration procedure).

10.4.3 Selection of Classification Rules

For selection the most appropriate strategy for the classification task, the following set of classifiers was used:

- Classification based on Mahalanobis metric (Linear Discriminate Analysis (LDA) and Quadratic Discriminate Analysis (QDA)).
- Classification based on regression model (linear multinomial regression and log regression).
- Support vector machines.
- Principle Component Analysis (PCA) with classification by Soft Independent Modeling of Class Analogy (SIMCA).
- Random Decision Forest (RDF).
- ADABoost.
- Neural networks.
- Classification based on fuzzy logic.

For classifier training and testing, we used approach called as Leave-one-out cross-validation. Leave-one-out cross-validation involves using a single data point from the original set as the validation data and the remaining points as the training set. This is repeated such that each point in the set is used once as the validation data.

For estimation the classifier validation and effectiveness, we applied basic performance measures based on the confusion matrix. The confusion matrix is a visualization tool commonly used to present performances of classifiers in classification tasks. It is used to show the relationships between real class attributes and that of predicted classes.

The level classification in the validation and effectiveness model is calculated with a number of correct and incorrect classifications in each possible value of the variables being classified in the confusion matrix. The confusion matrix is used to compute True Positives (TP), False Positives (FP), True Negatives (TN), and False Negatives (FN). There are three commonly used performance measurements including the accuracy, sensitivity, and specificity. The accuracy of classifiers is the percentage of correctness outcomes among the test sets exploited in the study provided by Eq. 10.19.

$$accuracy = \frac{TP + TN}{TP + FP + TN + FN} \tag{10.19}$$

The sensitivity is referred as the true positive rate and the specificity as the true negative rate provided by Eq. 10.20.

$$sensitivity = \frac{TP}{TP + FP} \quad specificity = \frac{TN}{TN + FN} \tag{10.20}$$

For classification strategy selection we carried out the following investigation. The special software was designed for solving the classification problems in tasks of diagnostics cervix cancer. It provides some tools for data analyzing and implements a set of classification algorithms described above. As an input data, the program uses the so called database of biopsy—a set of vectors of features (points of biopsy) with known diagnosis. Each vector of features was calculated for image block (32 × 32 elements) with known diagnosis according to the biopsy result. The quantity samples in input data was 270. In details, a process of the obtained samples is given in Sect. 7.2 and demonstrated in Fig. 10.15.

The software realizes the following functions: the data input, features selection according to Fisher distance, selection of classification strategy, setting parameters of classification strategy, classifier training, classifier testing (cross validation), and matrices of error for main classifier parameters estimation. The main windows of software realization are given in Fig. 10.7.

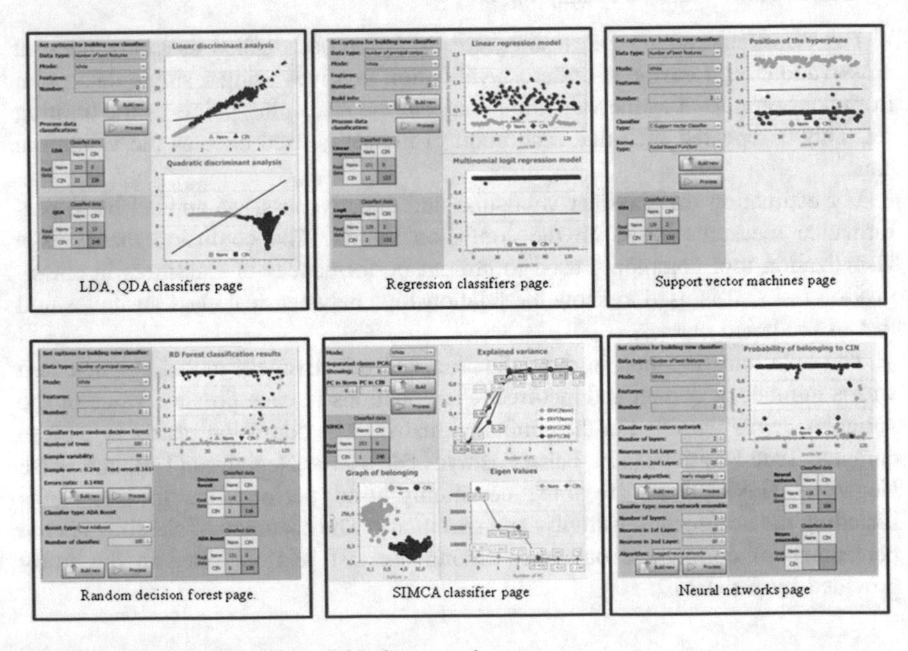

Fig. 10.7 Main windows of special software tool

Each classification strategy was realized by means the described software training, testing and estimating the sensitivity and specificity. We obtained the results, which gives us a possibility to select the RDF as a base for further investigation. For example, we obtained the sensitivity and specificity for the border NORM/CIN: LDA 094/074, SIMCA (a number of principle components is equal 6) 0.94/0.71, RDF 0.95/0.90. On the border CNI/CIN (it is more complete classification task), the RDF strategy also gave the best result.

Our investigation shows that the best specificity, sensitivity, and accuracy are obtained using the RDF strategy. This result is connected with the high degree of initial images variability (medical images, in particular, colposcopic images from different patients have significant differences not only due to pathology but also due to differences in the age, menopause, and other features of woman physical condition). Thus, the medical images demand the methods oriented on non-determined data—methods of data mining and artificial intelligence.

The classification results for the border CNI/CIN is 0.85 for the sensitivity, the specificity is 0.78. On the border Norm/CNI the sensitivity is 0.95, the specificity is 0.90. More detailed description is given in Sect. 10.8. Thus, the technique of the RDF was applied as the base for the task of differential diagnostic.

The RDF is one of the most successful ensemble learning techniques, which have been proven to be very popular and powerful techniques in the pattern recognition and machine learning for high-dimensional classification and skewed problems [18]. These studies are used the RDF to construct a collection of individual decision tree classifiers, which utilize the Classification And Regression Trees (CART) algorithms. The CART is a rule-based method that generates a binary tree through a binary recursive partitioning process that splits a node based on the "yes" and "no" answers of the predictors. The rule generated at each step maximizes the class purity within the two resulting subsets. Each subset is split further based on the independent rules. The CART algorithms use the Gini index to measure the impurity of a data partition or set of training instances. Although the aim of the CART is to maximize the difference of heterogeneity, in the real world data sets the overfitting problem that causes the classifier to have a high error of prediction in the unseen data set often encounters. Therefore, the bagging mechanism in the RDF can enable the algorithm to create the classifiers for high dimensional data very quickly. The accuracy of the classification decision is obtained by voting from the individual classifiers in the ensemble.

In our task, the three-class RDF classifier was used to separate the Norm, CNI, and CIN. We use the components Cr, Cb from the YCrCb color space and b from Lab color space for modes 360 and 390. The three-class RDF classifier means that we have three training samples with examples of all classes. For each one, the RDF response is a vector of probabilities of belonging to each class. According to our goal we use these probabilities to define for the current example as the membership degree to each class P_{NORM}, P_{CNI}, P_{CIN}. Thus, for differential diagnostics the classification was performed in mode 360 and 390 on the components Cr, Cb in the YCrCb color space and component b in Lab color space. The classification strategy is the three-class RDF. The main steps of algorithm for determining the membership degree for each image fragment includes the following ones:

Step 1. Split the source images into local fragments like blocks (20 × 20 elements).
Step 2. Define each block the features vector consisting of:

- the mean value of the components Cr, Cb (the YCrCb color space) and component b (Lab color space) calculated in the block of the fluorescent image obtained with the exciting wavelength 360 nm;
- the mean value of the components of Cr, Cb (the YCrCb color space) and component b (the Lab color space) calculated in the block of the fluorescent image obtained with the exciting wavelength 390 nm;
- the average value of the components a, b (the Lab color space) calculated in the image block obtained in white light.

Step 3. Preliminary classification of image blocks in order to select the zone of squamous epithelium for further analysis and remove the zones of the columnar epithelium, as well as the incision and the entrance to the cervical canal. The classification strategy is the RDF. The vector of features for classification is the average values of the components a and b (the Lab color space) in blocks of images obtained in white light and fluorescence light with the exciting wavelength 360 nm.
Step 4. For each block corresponding to the squamous epithelium, the classification task with three classes is solved: Class 1—tissues without pathology; Class 2—tissue with CNI; Class 3—tissue with CIN. The classification strategy is the RDF. The vector of features for classification is the mean values of components Cr, Cb (the YCrCb color space) and component b (the Lab color space) in the image blocks obtained in the fluorescence light with the excitation wavelength of 360 and 390 nm. The result of classification is the image block belonging degree to each class P_{NORM}, P_{CNI}, and P_{CIN} determined on the basis of the number of individual trees that voted for this class.

10.5 Automatic Analysis of Images Obtained in White Light Illumination

The analysis of colposcopic images obtained in white light illumination includes the AW region segmentation and texture analysis. The AW region segmentation supposes a detection part of cervix image, where changes of color for tissue with pathology after effect of acetic acid took place. Texture analysis supposes that the tissue with pathology have more inhomogeneous structure. In an image, these areas can be selected as areas with significant level of high frequency energy. Both types of analysis give us additional information, which will be used for pathology map creation together with fluorescent differential diagnostics results.

The AW region segmentation is discussed in Sect. 10.5.1. Some comments about texture analysis are involved in Sect. 10.5.2. Section 10.5.3 includes the estimation of detailed level quantity of image regions. Binary classification based on the texture feature is considered in Sect. 10.5.4.

10.5.1 The AcetoWhite Region Segmentation

The AW segmentation begins from some preprocessing procedures including the correction of non-uniform illumination, detection of the ROI that cropped the cervix area, and highlights detection with following removal. After preprocessing the AW segmentation can be realized.

The approaches for the AW segmentation might be divided in two groups. First group includes the segmentation based on the pure image processing procedures (mathematical morphology, linear filtering, etc.), while second group consists of various methods based on the supervised classification. On the base of implemented experiments, it is stated that the methods from first group provide weak results in our case. The main reasons are low contrast of the AW features and wide variability of the AW representation in our data (Fig. 10.8).

Therefore, the designed algorithm is based on classification method. As a result of the special research, the two-class RDF classifier was chosen for classification. The classes represented "Acetowhite" class and "Other" class. The features for classification are the components L, a, b in the Lab color space. In our task, we worked with image blocks 20 × 20 elements as in the task of fluorescent diagnostics. The training set included over 350 patterns of "Other" class and over 250 patterns of "Acetowhite" picked from 50 images with different cases of the AW manifestation (Fig. 10.9).

After training, the predictions $\hat{f}$ for unseen samples x' can be made by averaging the predictions $\hat{f}_b$ from all the individual regression trees on x':

$$\hat{f} = \frac{1}{B}\sum_{b=1}^{B}\hat{f}_b(x') \tag{10.21}$$

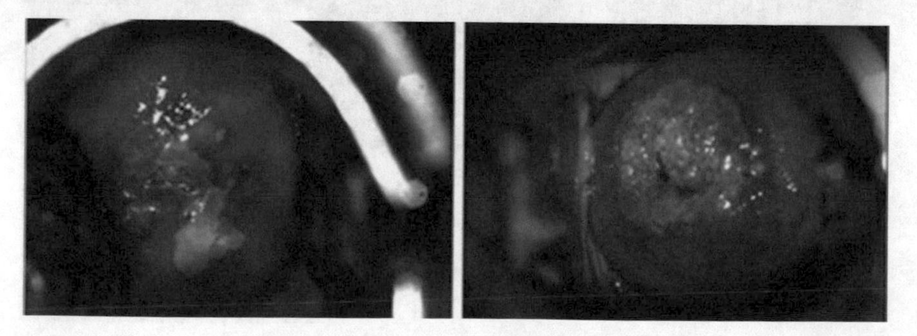

Fig. 10.8 Various representation of the AW in clinical images

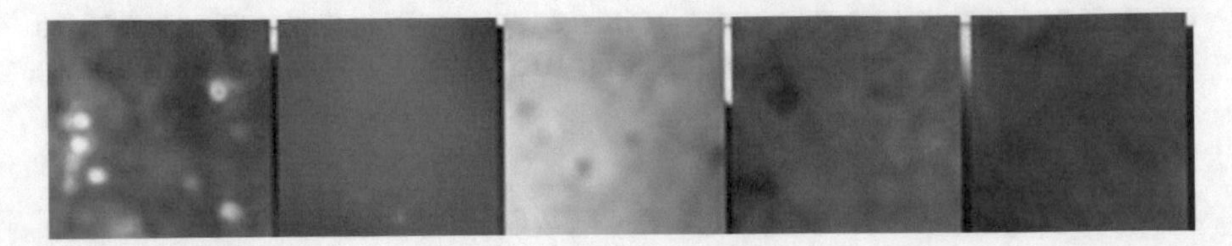

Fig. 10.9 Different patterns with of "Acetowhite" class

or by taking the majority vote in the case of decision trees.

For further investigation, the predication $\hat{f}$ as the membership degree of current image block to class "Acetowhite" P_W was found.

10.5.2 Texture Analysis

The tissue with pathology has more inhomogeneous structure. These areas can be selected in an image as areas with significant level of high frequency energy (Fig. 10.10).

The algorithm of image classification based on texture feature includes three main steps. The first step supposes a preprocessing for highlights removing from the analyzing image. It is very important step because the image regions with

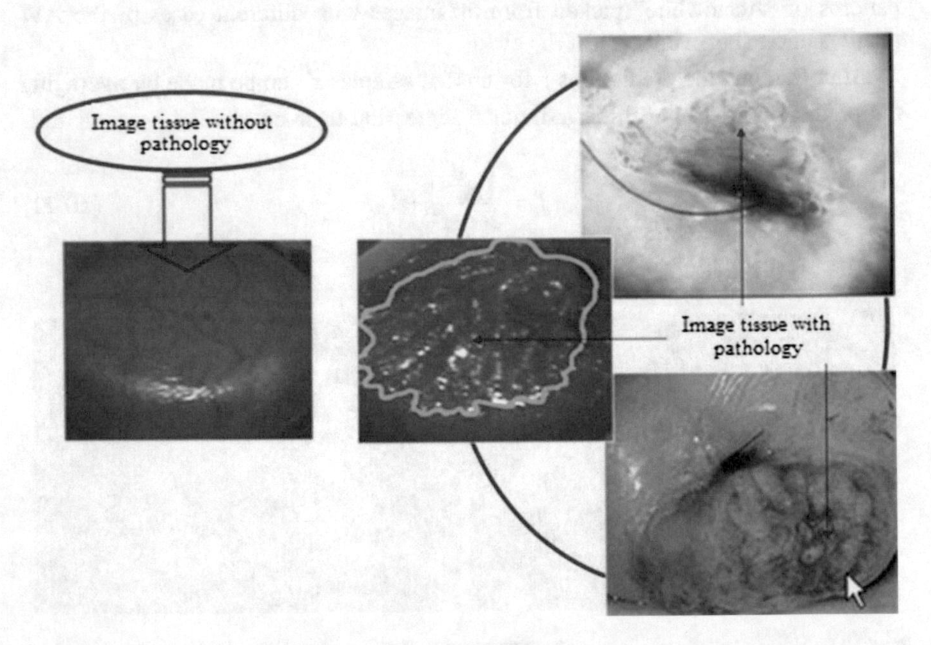

Fig. 10.10 Images of tissue with and without pathology

highlights also have significant level of high frequency energy and it can give mistakes in further image classification based on the texture feature. The second step is a creating the quantity estimation of texture feature and the last step includes a classification rule construction.

10.5.3 Estimation of Detail Level Quantity of Image Regions

To quantify the texture, the measure of Rosenfeld–Troy was used. Its definition involves the image preprocessing in order to stress high frequency energy and counting the intensity of the brightness differences in the local image fragment and expressed by Eq. 10.22, where $\Lambda(x, y)$ is a brightness of a pixel in the image after preprocessing, x_k, y_l are the coordinates of block left top corner; k, l are the block numbers in horizontal and vertical direction, N, M are the block sizes.

$$D(x_k, y_i) = (\sum_{j=1}^{M} \sum_{i=1}^{N} \Lambda(x_k + i, y_l + j))/(N * M) \tag{10.22}$$

The image preprocessing to stress high frequency energy may be realized in various methods of the following groups: linear and non-linear processing, morphological processing, and wavelet-transform. The most effective way in current task according our investigation is a morphological processing with multiscale morphological gradient using Eq. 10.23, where the signs " $\oplus$" and "Θ" are the morphological operations dilation and erosion, S_i is a group of structural square elements.

$$MG(L) = \frac{1}{3} \sum_{i=1}^{3} [((L \oplus S_i) - (L \Theta S_i)) \Theta S_{i-1}] \tag{10.23}$$

According to expression for multiscale Morphological Gradient (MG), the gradients are calculated three times using three structural elements of different dimension and then add up the results. It is important to point that before stressing high frequency in image it is useful to apply any filter for reducing noise influcnce. For this purpose, the median filter was used. The main idea of the median filter is to replace the signal by the median of neighboring entries. Thus, the step of image preprocessing for quantity texture estimation includes the median filtration and morphological multiscale gradient.

The quantity estimation of texture feature based on the measure of Rosenfield–Troy is depended from a noise level in an image. In order to reduce a noise influence, the estimation of texture in block caused by noise D_{min} is defined as moda in a distribution of the obtained texture estimations using Eq. 10.24, where

K and L are the quantity of blocks in analyzing image in the horizontal and vertical directions, respectively.

$$D_{\min} = \text{moda}\{D(x_k, y_l)\} \quad k = 1\ldots K \quad l = 1\ldots L \tag{10.24}$$

Thus, the main steps of algorithm for quantity texture estimation obtaining are the following:

- Image preprocessing with purpose to underline the high frequency energy with morphological multiscale gradient after median filter.
- Splitting image into the square units (blocks).
- Definition the detailed estimation for each block according to measure of Rosenfield–Troy (Eq. 10.22).
- Estimation D_{min} as moda in a distribution of the obtained texture estimations.
- Definition the modified estimation for each block in the analyzing image using Eq. 10.25.

$$D_M = D(x_k, y_l)/D_{\min} \tag{10.25}$$

The maps for estimation of different levels' texture are shown in Fig. 10.11. The analysis of this result gives us a possibility to make a conclusion that the obtained quantity estimation of texture feature can be used for binary classification task.

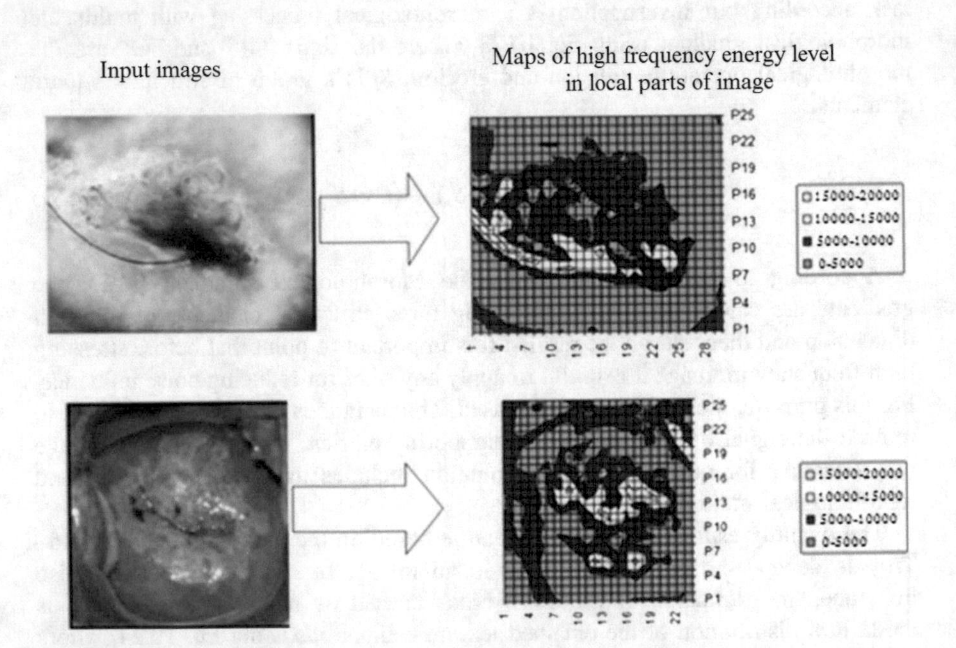

Fig. 10.11 Maps of high frequency energy level

10.5.4 Binary Classification Based on Texture Feature

The texture feature gives a possibility to select the areas with significant level of high frequency energy. However, it is not enough for differential diagnostic because it gives a possibility only to detect the suspect areas. In this case, the texture analysis is binary classification task, for which rather simple classification strategy—the Mahalanobis distance can be used. The use for classification the Mahalanobis distance assumes f finding such a linear combination of variables, which separates two classes in the best way. The classification rule of the LDA is very simple: the new object belongs to the class k (k = "Texture", "Non-Texture"), to which it is closer by the Mahalanobis metric provided by Eq. 10.26, where $\mathbf{x}$ is a vector of features, $\boldsymbol{\mu}$ is a vector of mean values, $\boldsymbol{\Sigma}$ is a covariance matrix

$$L_k = (\mathbf{x} - \boldsymbol{\mu}_\mathbf{m})\Sigma^{-1}(\mathbf{x} - \boldsymbol{\mu}_\mathbf{m})^t \tag{10.26}$$

The result of classification according to the Mahalonobis distance gives us a possibility to define the membership degrees of current image block to class "Texture" P_T according to Eq. 10.27, where L_T is the Mahalonobis distance up to class "Texture" for analyzing block, L_{NT} is the Mahalonobis distance up to class "Non-Texture" for analyzing block.

$$P_T = \frac{L_T}{L_T + L_{NT}} \tag{10.27}$$

The obtained estimations of the membership degrees of current image block to class "Texture" P_T and the membership degree of current image block to class "Acetowhite" P_W will be used as the result of image obtained in white light illumination analysis. Both estimations will be used in pathology map creation.

10.6 Creation of Differential Pathology Map

The differential pathology map is an image of the cervix divided into areas with a definite diagnosis: Norm-CNI-CIN (I, II, III). For its creation, the results of fluorescent images and images obtained in white light analysis were used. The algorithm is designed according to the rules of the fuzzy logic theory. The concept of fuzzy sets is based on the notion that the elements composing a given set, possessing a common property, can possess this property in different degrees and, consequently, belong to a given set with different degrees.

A fuzzy set $\tilde{A}$ on a universal set U is a set of pairs ($\eta_A(u)$, u), where $\eta_A(u)$ is a degree of membership of an element $u \in U$ to a fuzzy set $\tilde{A}$. The degree of membership is a number from the interval [0, 1]. The degree of membership shows for any element of the universal set the degree of it corresponding to the properties of the fuzzy set.

Let U be a universal set that includes all image blocks to be analyzed. Three fuzzy sets may be created in the universal set U. The first fuzzy set S_{flur} is a set of blocks with changes in the fluorescence. The degree of membership of any image block to this set is determined by estimation P_{CIN} obtained as the result of fluorescent images analysis. The second fuzzy set S_W includes blocks corresponding to the AW effect. The degree of membership of any image block to this set is determined by the estimation P_W. The third set S_T involves blocks corresponding to tissues with heterogeneity (texture changes). The degree of membership of any image block to this set is determined by the estimation P_T.

The blocks belonging to the pathology region must correspond to two features: the AW effect and texture changes. Thus, the analysis result of image obtained in white light can be found as intersection of the two fuzzy sets S_W and S_T. The fuzzy set obtained as a result of the intersection is S_{ADD}. The degree of membership of any image block in this set can be calculated according to the product T-norm: $P_{ADD} = P_W * P_T$.

Based on the fuzzy sets S_{flur} (analysis result of fluorescent image) and S_{ADD} (the analysis result of image obtained in white light), it is necessary to get the fuzzy set S_{CIN} of image blocks corresponding to the pathology CIN according to analysis result of all images types. The degree of membership of any image block in this set is P^M_{CIN}. The estimation P^M_{CIN} for each block is the goal of analysis, it shows corresponding degree of any block to the pathology CIN and it can be used for pathology map construction.

In the first step, for the fuzzy set S_{flur} an alpha-cut is formed at the level of 0.75. This is a set of blocks with a very high level of fluorescent changes, which allows them to be referred to the pathology area without additional information. In this case $P^M_{CIN} = P_{CIN}$. For the remaining blocks of the fuzzy set S_{flur}, the degree of membership any block in the set S_{CIN} is formed taking into account the results of the analysis in white light (the set S_{ADD}). It is necessary to increase the specificity of the fluorescent analysis. For the remaining blocks of the fuzzy set S_{flur}, the next operations are realized:

- The unification operation with the alpha-cut at the level 0.5 of fuzzy set S_{ADD} according to product S-norm $P^M_{CIN} = P_{ADD} + P_{CIN} - P_{ADD} * P_{CIN}$.
- The intersection operation with blocks that are not included in the alpha-cut of fuzzy set S_{ADD} according to product T-norm $P^M_{CIN} = P_{ADD} * P_{CIN}$.

The algorithm for determining the degree of pathology P^M_{CIN} is shown in Fig. 10.12.

The results of fluorescent analysis are three membership degrees of current block to each class: P_{NORM}, P_{CNI}, and P_{CIN}. The results of acetowhite detection are the membership degree of current block to acetowhite P_W. The result of texture analysis is the membership degree of current block to area with high texture changes P_T. Our research has shown that the analysis of fluorescent images provides very high

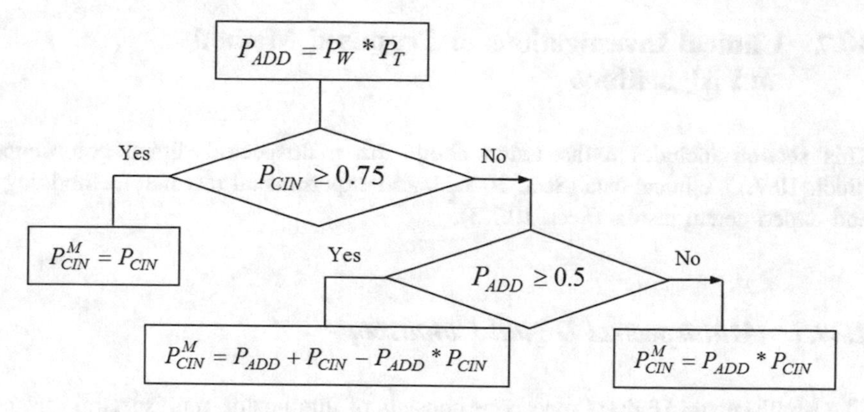

Fig. 10.12 The scheme for pathology map creation

Input data	Each block has a triplet of belonging degree to each class (P_{NORM}, P_{CNI}, P_{CIN})						
First test: find maximum in triplet Result:	P_{NORM} = maximum			P_{CNI} = maximum	P_{CIN} = maximum		
Second test: define additional value Result:	$P_{CIN} > 0.75$	$0.75 > P_{CIN} > 0.5$	$P_{CIN} < 0.5$		$P_{NORM} > 0.75$	$0.75 > P_{NORM} > 0.5$	$P_{NORM} > 0.5$
Color of the block in pathology map	Color 1	Color 2	Color 3	Color 4	Color 5	Color 6	Color 7

Fig. 10.13 The scheme for color definition for pathology map

sensitivity with moderate specificity. Therefore, the estimates of P_{NORM}, P_{CNI}, and P_{CIN} should be used as a basis for the formation of severity of pathology, and P_W and P_T as additional information to improve the characteristics of specificity. The algorithm for determining the color for displaying the analyzed image block visualization in the pathology map in accordance with the found degrees P_{NORM}, P_{CNI}, and P_{CIN}^M is shown in Fig. 10.13.

10.7 Clinical Investigations of Proposed Method and Algorithms

This section includes a discussion about the multispectral digital colposcope (Sect. 10.7.1), clinical data (Sect. 10.7.2), and experimental research methodology and experimental results (Sect. 10.7.3).

10.7.1 Multispectral Digital Colposcope

The Multispectral Digital Colposcope consists of illuminator, multispectral image–recorder unit, software and computer (Fig. 10.14). The registration unit includes the projection lens, set of detector filter, and digital video camera. Lens focal length is 100 mm and relative aperture F/2,8. Digital video camera is based on the CCD detector (format 2/3 inch) with progressive scan (ICX285AQ, SONY). Maximum frame rate of the camera is 15 Hz at resolution 1280 × 1024 elements. Its own noise is about 10e-, non-linearity characteristics of the light signal at differences of light are 100 times less than 3.5%. Signals received from the camera transmit to the computer by the serial high-speed working according USB-2.0 protocol.

Illuminator is separated into three parts: the mercury lamp, halogen lamp, and 635 nm laser. Values of excitation power density (mw/cm^2) are 26 (360 nm), 69 (390 nm), 73 (430 nm), 170 (635 nm), respectively, in the view field of 20.8 × 26 (V × H) mm and the working distance of 220 mm.

Fig. 10.14 Device LuxCol

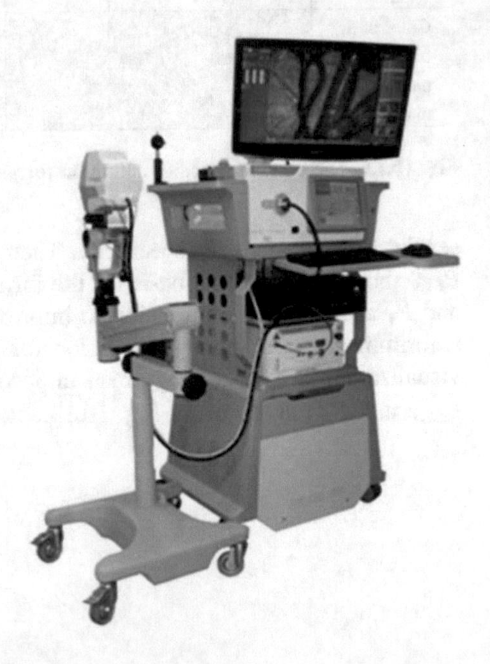

10.7.2 Clinical Data. Database of Verified Images and Records

For experimental testing of the proposed methods and algorithms, the real images obtained from three hospitals of South Korea were used. Eligible patients were at least 18 year old, not pregnant, were referred to the colposcopy clinic with an abnormal Pap-smear or with positive HPV. Written informed consent was obtained from each participant. For all patients we have results of Pap-smear, results of HPV test and HPV genotyping. For researches, two sets of images for each patient were obtained. The first set of images was acquired with the device LuxCol from each patient at baseline. The second set of images was acquired following the application of acetic acid 6% on the cervix for 2–3 min. Finally, the diagnose of colposcopist, cervix image with marked points, where biopsy was taken and results of biopsy, were obtained for each patient.

Thus, the input data for investigation were the following:

- The sets of images including the images obtained in white light and fluorescent images obtained with exciting wavelength 360 and 390 nm.
- The results of Pap-smear and HPV test.
- The diagnose of colposcopist.
- The cervix images with marked points of biopsy and result of biopsy for these points.

The short description of these images called the verified images is given in Table 10.1.

These verified images from hospitals were used for creating the special database of records, which is necessary for classifier training. According to verified images we put markers in the images received by LuxCol with white light illumination and with the assistant of special program RSS Colpo calculated the vector of color and brightness features for block of 20 $\times$ 20 elements around the marker. This process is illustrated in Fig. 10.15. Each such image block and its corresponding features vector we called as a record in database. In this way, we obtained database including records corresponded to the Norm, the CNI, and the CIN for further investigation.

Note that this special database of records was used in the algorithms of pre-processing checking and also for the most effective mode and features selection,

Table 10.1 The proposed training and testing image database parameters

Clinic	Image quantity	Histology						
		CNI	CIN					
			Total	CIN1	CIN2	CIN3	CIS
Clinic 1	48	31	17	8	3	4	2
Clinic 2	60	40	20	6	5	5	4
Clinic3	43	18	25	10	6	5	4

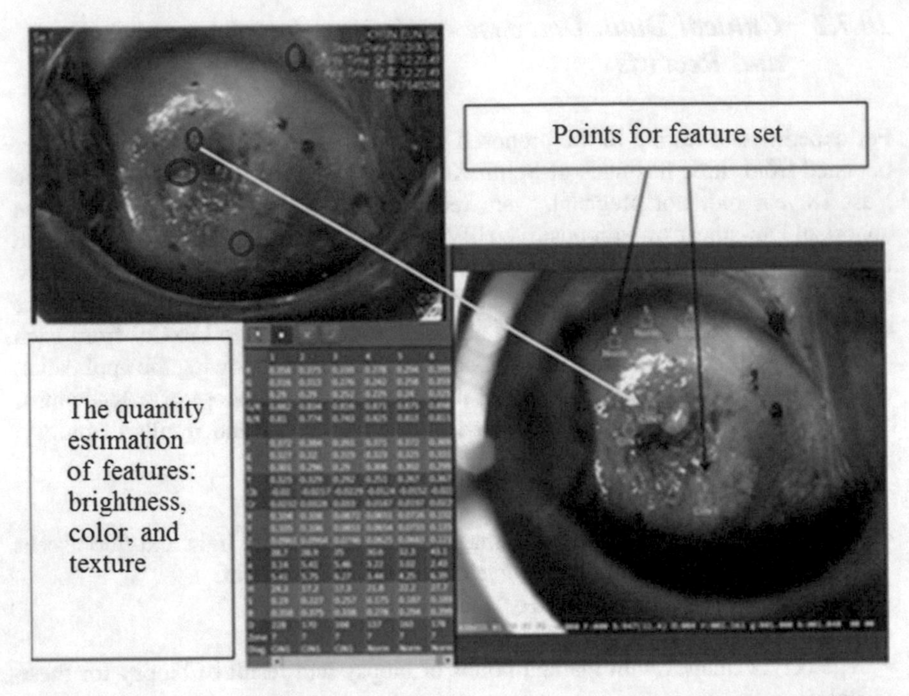

Fig. 10.15 The feature extraction and estimation (image in the *left top corner* is verified by biopsy results, image in *right bottom corner* is obtained with LuxCol). The points for classification training are marked

classifier training, estimation of main characteristics of the proposed algorithm, such as the specificity, sensitivity, and accuracy, and pathology map creation.

10.7.3 Experimental Research Methodology and Experimental Results

The experimental research includes two parts, such as the estimation of the specificity and sensitivity of proposed method for fluorescent image analysis and estimation of the correctness of the pathology map. Pathology map is an image shattered in the areas with the definite measure based on the fluorescents, acetowhite, and texture changes.

The results of fluorescent analysis for the border CNI/CIN obtained from the database of records with validation approach Leave–one-out-cross provide the sensitivity 0.85 and specificity 0.78. Additionally, we estimate the specificity and sensitivity on the border Norm/CNI. The results for the border Norm/CIN are the following: the sensitivity is 0.95 and specificity is 0.90. However, the main task of investigation is to estimate a quality of the obtained pathology map.

For estimation of pathology map correctness, the methodology mentioned below was used:

Step 1. According the database of records with Leave–one-out–cross approach, the classifier training was realized. In such way, we realized the training classifier in the fluorescent analysis, acetowhite analysis, and texture analysis. As one can see from Table 10.1, the quantity of images in group CIN I, group CIN II, and group CIN III is rather small and we had to join them in one common group CIN for classifier training. Thus, at current stage of investigation the classifier training was realized only for three classes Norm, CNI, and CIN.
Step 2. The pathology map indicates three classes Norm, CNI and CIN. In the pathology map, the cervix areas belonging to Norm have green color, the areas with CNI have yellow color, and the areas with CIN have red color. If area of cervix has intermediate state, for example between Norm and CNI, it color is mix of pure colors corresponded this diagnosis. So, in our example Norm/CNI area will have color including green and yellow. If state of tissue is intermediate between CNI and CIN, then a color of area in the pathology map will be mix of red and yellow (Fig. 10.16).
Step 3. The obtained pathology map was checked according the biopsy results for particular points. We matched the points of biopsy (obtained from clinics) with known results with areas in the obtained pathology map. This process is illustrated in Figs. 10.16 and 10.17. One can see from Figs. 10.16 and 10.17 that the arrows show the correspondences in the obtained pathology maps and images with biopsy points obtained from clinics. Analysis of the obtained pathology maps and clinics results gives a possibility to calculate the sensitivity and specificity.

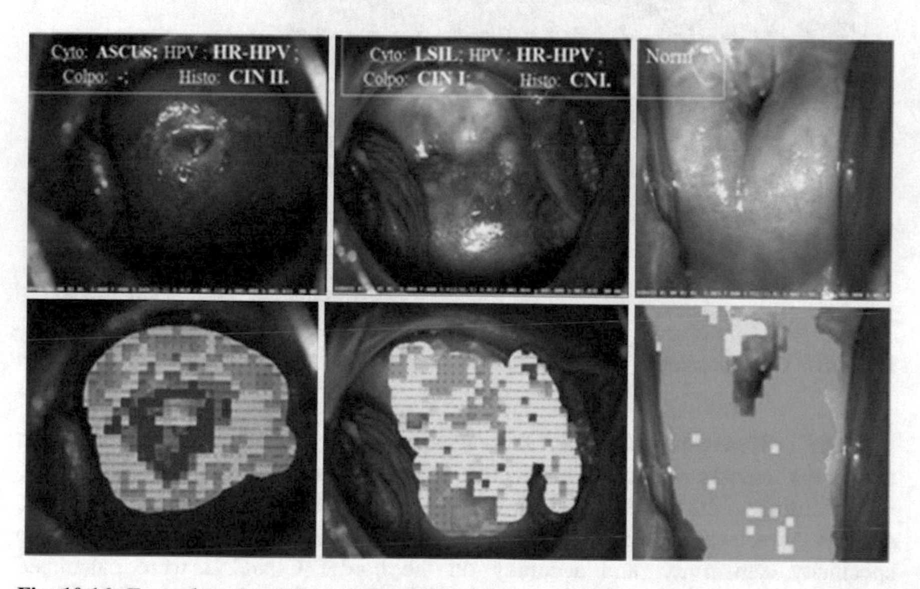

Fig. 10.16 Examples of pathology map with 3 classes Norm, CNI and CIN

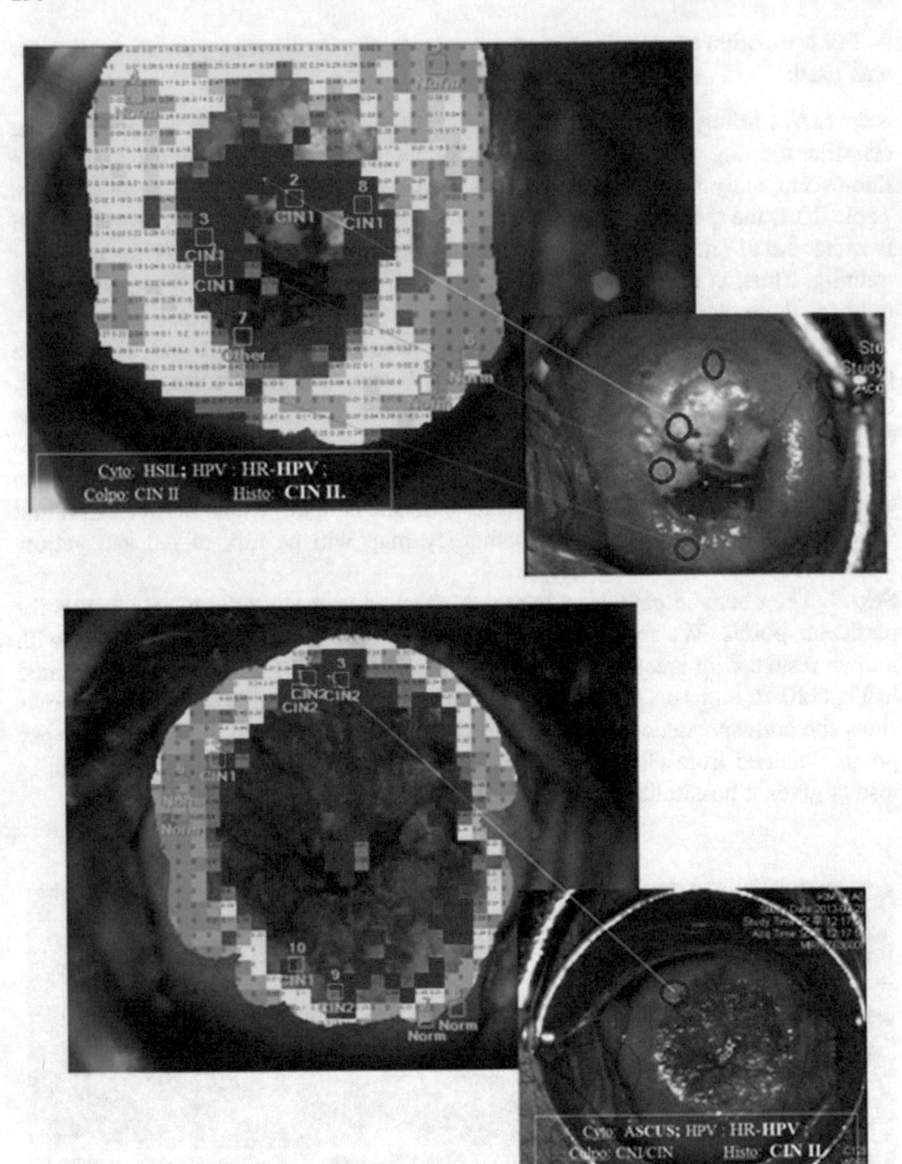

Fig. 10.17 Corresponding of biopsy points with diagnose CIN II to *red areas* of pathology map

Steps 1–3 were repeated for all images under investigation and according to results the main classifier characteristics were estimated.

According to the obtained values TP, TN, FP, and FN, the main characteristics, such as the accuracy, sensitivity, and specificity, were calculated. First of all, the specificity, sensitivity, and accuracy on the border CIN/CNI were calculated because the most interest and the most difficult for design task is the task of

Table 10.2 Results of pathology map investigation

Clinic	Image quantity	Histology						LuxCol				Characteristics of LuxCol		
		CNI	CIN					CNI		CIN		Border Norm+ CNI/CIN		
			Total	CIN1	CIN2	CIN3	CIS	True	Total	True	Total	Sen.	Sp.	Acc.
Clinic 1	48	31	17	8	3	4	2	21	31	15	17	0.88	0.67	0.75
Clinic 2	60	40	20	6	5	5	4	30	40	17	20	0.85	0.75	0.78
Clinic 3	43	18	25	10	6	5	4	13	18	22	25	0.88	0.72	0.81

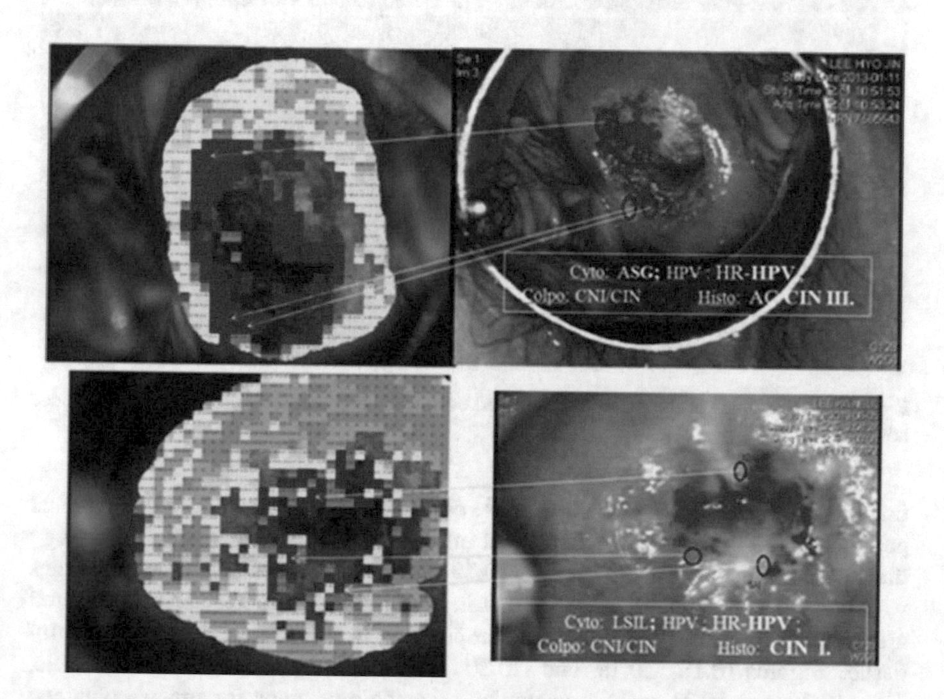

Fig. 10.18 Obtained pathology map and the biopsy result. The results of automatic analysis are matched with the result of histological analysis

classification the CNI and the CIN. This task is also the most actual in physician practice. The task of classification CIN/Norm is rather simple and it is not important for medical practice. The results of pathology map investigation are given in Table 10.2. From Table 10.2, one can see that the accuracy of proposed method on the border CNI/CIN is 0.8.

Multiple examples of the obtained pathology map are given in Figs. 10.17, 10.18, 10.19, 10.20, 10.21, 10.22 and 10.23. Figures 10.17 and 10.18 demonstrate

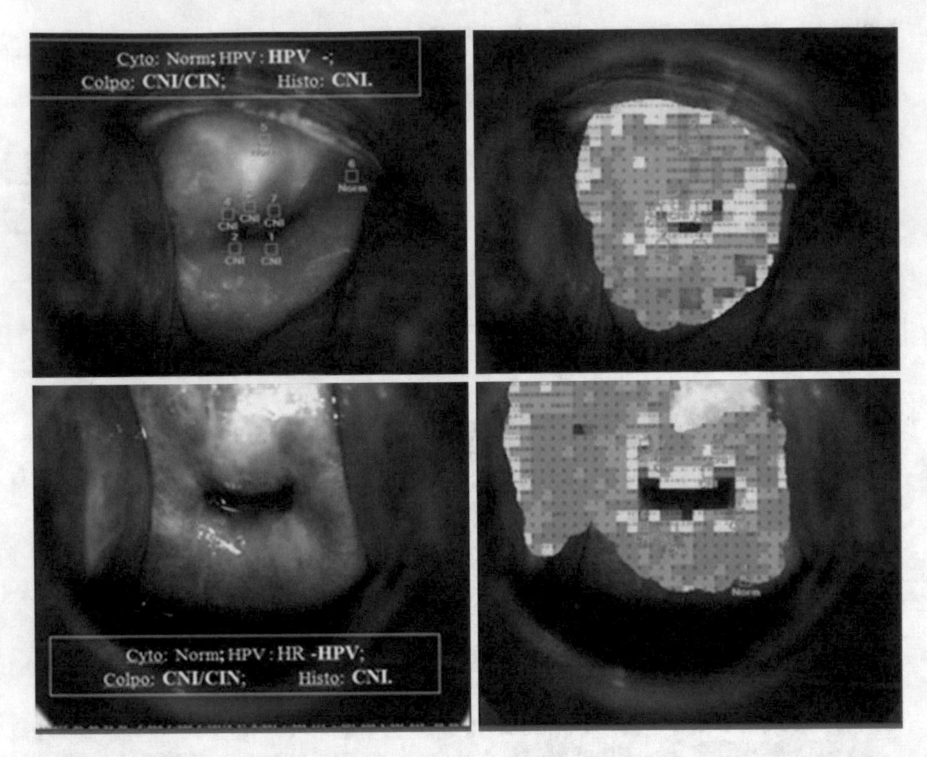

Fig. 10.19 Examples of the pathology map and corresponding results of biopsy CNI (Severance hospital)

that the points with biopsy result CIN are corresponded to red areas in the obtained pathology map. Also we gave all clinical information about these patients (cytology diagnosis, HPV, and so on). Figure 10.19 shows the analysis results for images with biopsy result CNI. One can see that in this case the pathology map has not red areas, but there are yellow blocks corresponding to the CNI and green for norm tissue. Figures 10.18, 10.19 and 10.20 are obtained from Severance Clinic, Figs. 10.20 and 10.21 include examples of pathology maps for images obtained

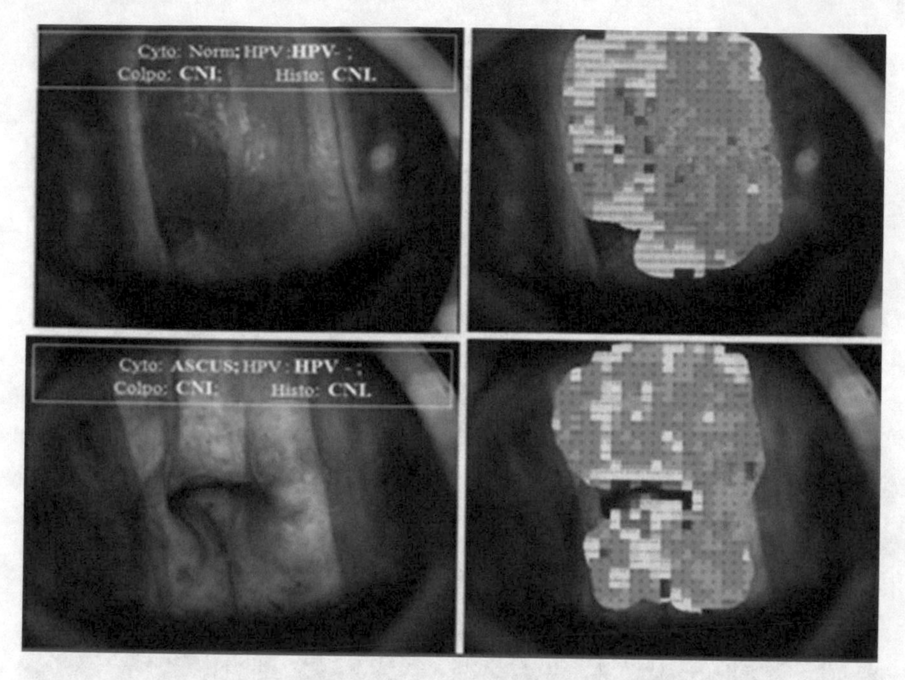

Fig. 10.20 Examples of pathology map corresponding results of biopsy CNI (Catholic Hospital)

from Catholic clinic, and Figs. 10.22 and 10.23 includes the examples of pathology maps using the images obtained from Asan clinic. One can see that the pathology maps have good correspondence with the biopsy results. Red and orange areas present only in images with CIN and in the case of CNI we have the pathology maps consisting from yellow and green colors. Thus, the pathology maps are well agreed with the biopsy results and can be useful in medical practice.

Additionally, we estimate the specificity and sensitivity on the border N/CNI. The algorithm of estimation was the same as the algorithm described above. The achieved results are the following: the sensitivity is 0.95 and specificity is 0.98.

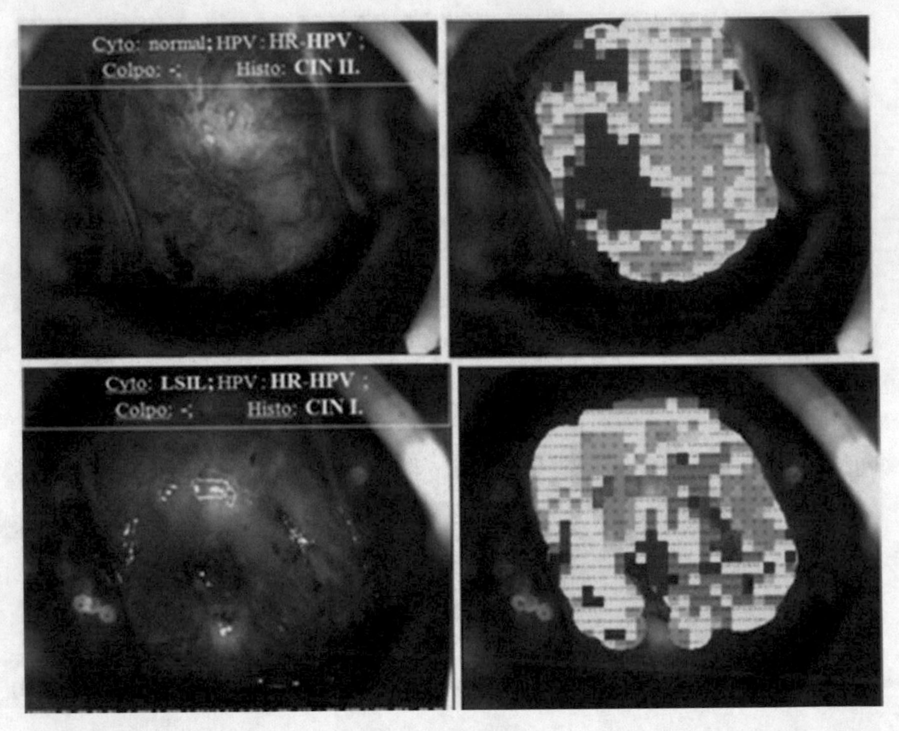

Fig. 10.21 Examples of pathology map corresponding results of biopsy CIN (Catholic Hospital)

10.8 Conclusions

The proposed method of image analysis allows to detect the following states of cervix tissues, such as the Norm, the CNI, and the CIN. For the border CIN/CNI, we got sensitivity 83% and specificity 72%, while for the border CIN/Norm we got sensitivity 95% and specificity 98%. These results correspond to the estimates of the sensitivity and specificity for colposcopic examination conducted by an experienced physician and exceed the characteristics of inexperienced physician, where the sensitivity and specificity have been reported to range from 87 to 96% and 48 to

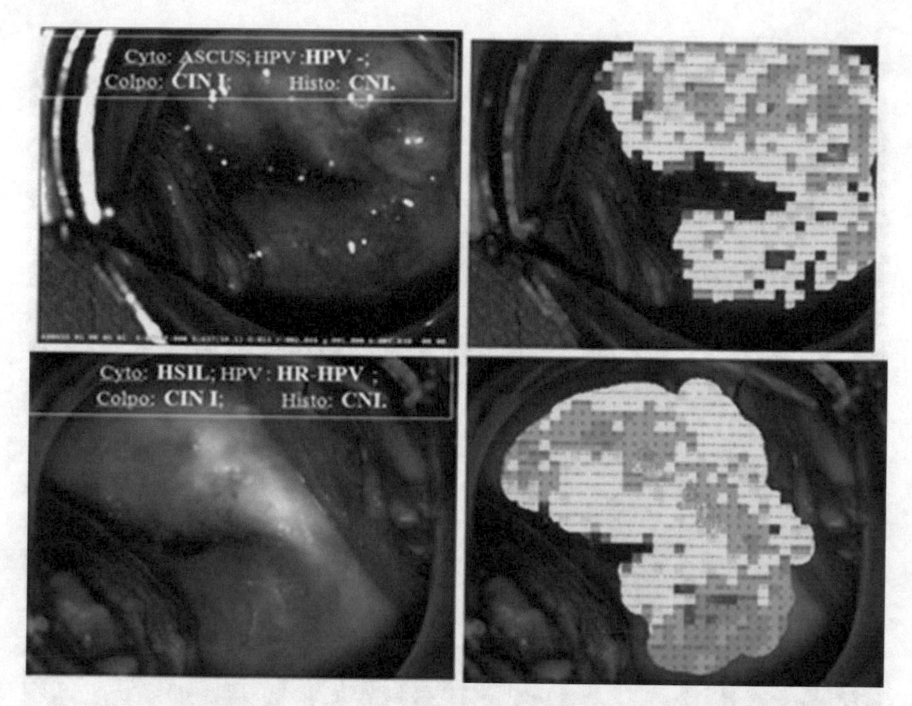

Fig. 10.22 Examples of pathology map corresponding results of biopsy CNI (Asan Hospital)

85%, respectively. The core of the developed algorithms is the pathology maps obtaining. In these maps, the degree of pathology confirming is obtained for each area of cervix image. Maps are based on three main classes, such as the Norm, the CNI, and the CIN. It is important that the classification border CIN/CNI is realized because it is the most difficult in the medical practice. Proposed algorithms provide a possibility to obtain the correct differential pathology map with probability 0.8. The obtained results demonstrated a possibility of practical application of the pathology maps for colposcopist examination directly.

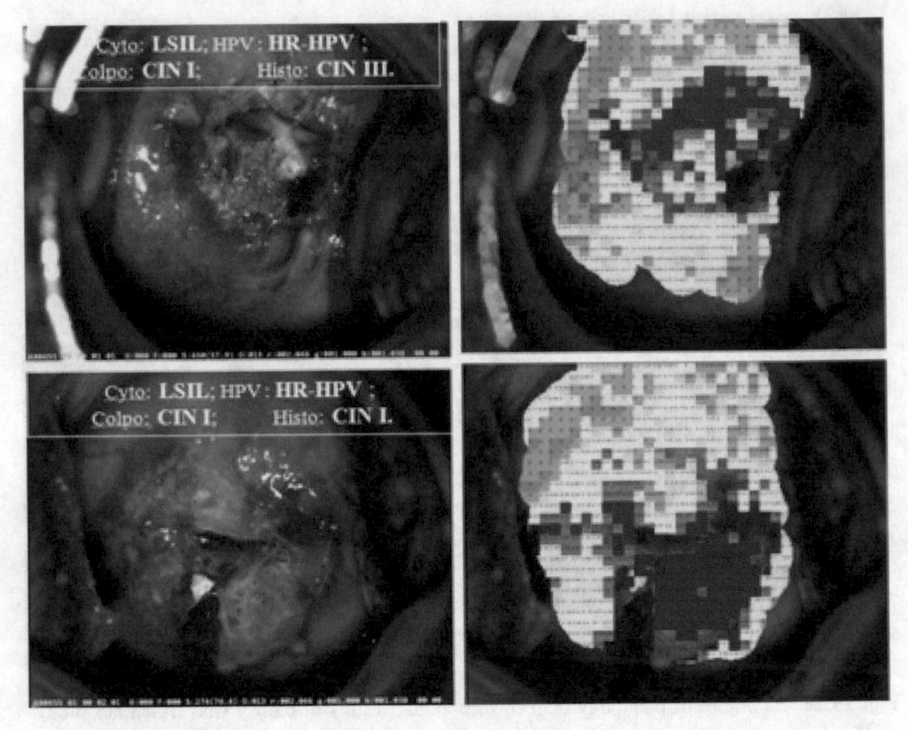

Fig. 10.23 Examples of pathology map corresponding results of biopsy CIN (Asan Hospital)

Acknowledgements Seoul Metropolitan Government and its Seoul Development Institute shall support and provide funds WR100001 for RSS in the frameworks of the Program for International Joint Research "Inviting & Supporting Project of Global Leading Institutions", a part of the Seoul Research & Business Development Support Program and Russian Foundation for Basic Research, grant № 17-07-00045.

References

1. Muhuri, S., Bhattacharjee, M.: Automated identification and analysis of cervical cancer. In: 3rd World Conference on Applied Sciences, Engineering & Technology, pp. 516–520 (2014)
2. Ramapraba, P.S., Ranganathan, H.: Automatic lesion detection in colposcopy cervix images based on statistical features. In: 4th International Conference on (ObCom' 2011), Part II, pp. 424–430 (2011)
3. Miranda, G.H.B., Barrera, J., Soares, E.G., Felipe, J.C.: Structural analysis of histological images to aid diagnosis of cervical cancer. In: 25th Conference on Graphics, Patterns and Images (SIBGRAPI'2012), pp. 316–323 (2012)
4. Liang, M., Zheng, G., Huang, X., Milledge, G., Tokuta, A.: Identification of abnormal cervical regions from colposcopy image sequences. In: 21st International Conference on Computer Graphics, Visualization and Computer Vision (WSCG'2013), pp. 130–136 (2013)

5. Srinivasan, Y., Hernes, D.L., Tulpule, B., Yang, S., Guo, J., Mitra, S., Yagneswaran, S., Nutter, B., Jeronimo, J., Phillips, B., Long, R., Ferris, D.G.: A probabilistic approach to segmentation and classification of neoplasia in uterine cervix images using color and geometric features. SPIE Med. Imaging **5748**, 995–1003 (2009)
6. Xiong, J., Wang, L., Gu, J.: Image segmentation of the acetowhite region in cervix images based on chromaticity. In: 9th International Conference on Information Technology an Applications in Biomedicine (ITAB'2009) (2009). doi:10.1109/ITAB.2009.5394329
7. Huang, X., Wang, W., Xue, Z., Antani, S.K., Long, L.R., Jeronimo, J.: Tissue classification using cluster features for lesion detection in digital cervigrams. SPIE Med. Images. **6914**, 69141Z-1–69141Z-8 (2008)
8. Alush, A., Greenspan, H., Goldberger, J.: Automated and interactive lesion detection anв segmentation in uterine cervix images. IEEE Trans. Med. Imaging **29**(2), 488–501 (2010)
9. Alush, A., Greenspan, H., Goldberger, J.: Lesion detection and segmentation in uterine cervix images using an arc-level MRF. IEEE International Symposium on Biomedical Imaging: From Nano to Macro (ISBI'2009), pp. 474–477 (2009)
10. Yang, S., Guo, J., King, P.S., Sriraja, Y., Mitra, S., Nutter, B., Ferris, D., Schiffman, M., Jeronimo, J., Long, R.: A multi-spectral digital cervigram analyzer in the wavelet domain for early detection of cervical cancer. SPIE Med. Imaging **5370**, 1833–1844 (2004)
11. Gordon, S., Zimmerman, G., Greenspan, H.: Image segmentation of uterine cervix images for indexing in PACS. 17th IEEE Symposium on Computer-Based Medical Systems (CBMS'2004), pp. 298–303 (2004)
12. Ji, Q., Engel, J., Craine, E.: Texture analysis for classification of cervix lesions. IEEE Trans. Med. Imaging **19**(11), 1144–1149 (2010)
13. Dattamajumdar, A.K., Wells, D., Parnell, J., Lewis, J.T., Ganguly, D., Wright Jr. T.C.: Preliminary experimental results from multi-center clinical trials for detection of cervical precancerous lesions using the @@cervicsantm system: a novel full field evoked tissue fluorescence based imaging instrument. In: IEEE 23rd Annual International Conference on Engineering in Medicine & Biology (EMBS'2001), pp. 3150–3152 (2001)
14. Benavides, J.M., Chang, S., Park, S.Y., Richards-Kortum, R., Mackinnon, N., MacAulay, C., Milbourne, A., Malpica, A., Follen, M.: Multispectral digital colposcopy for in vivo detection of cervical cancer. Opt. Express **11**(10), 1223–1236 (2003)
15. Park, S.Y., Collier, T., Aaron, J., Markey, M.K., Richards-Kortum, R., Sokolov, K., Mackinnon, N., MacAulay, C., Coghlan, L., Milbourne, A., Follen, M.: Multispectral digital microscopy for in vivo detection of oral neoplasia in the hamster cheek pouch model of carcinogenesis. Opt. Express **13**(3), 749–762 (2005)
16. Park, S.Y., Follen, M., Milbourne, A., Rhodes, H., Malpica, A., MacKinnon, N., MacAulay, C., Markey, M.K., Richards-Kortum, R.: Automated image analysis of digital colposcopy for the detection of cervical neoplasia. J. Biomed. Opt. **13**(1), 014029-1–014029-10 (2008)
17. Ghilani, C.D.: Adjustment Computations: Spatial Data Analysis, 5th edn. Wiley, Hoboken (2011)
18. Hastie, T., Tibshirani, R., Friedman, J.: Random Forests. In: Hastie, T., Tibshirani, R., Friedman J (auth) The Elements of Statistical Learning: Data Mining, Inference, and Prediction, 2nd edn., pp 587–604. Springer, Heidelberg (2009)

Chapter 11
A Novel Foot Progression Angle Detection Method

Jeffery Young, Milena Simic and Milan Simic

Abstract Foot Progression Angle (FPA) detection is an important measurement in clinical gait analysis. Currently, the FPA can only be computed, while walking in a laboratory with a marker-based or Initial Measure Unit (IMU) based motion capture systems. A novel Visual Feature Matching (VFM) method is presented here, measuring the FPA by comparing the shoe orientation with the progression, i.e. the walking direction. Both the foot orientation and progression direction are detected by image processing methods in rectified digital images. Differential FPA (DFPA) algorithm is developed to provide accurate FPA measurement. The hardware of this system combines only one wearable sensor, a chest or torso mounted smart phone camera, and a laptop on the same Wi-Fi network. There is no other prerequisite hardware installation or other specialized set up. This method is a solution for long-term gait self-monitoring in a home or community like environments. Our novel approach leads to simple and persistent, real time remote gait FPA monitoring, and it is a core of new bio-feedback medical procedure.

Keywords Gait monitoring · Foot progression angle · Differential foot progression angle · Progression direction · Monocular camera Image processing · Image rectification

J. Young (✉) · M. Simic
School of Engineering, RMIT University, GPO Box 2476, Melbourne, VIC 3001, Australia
e-mail: s3314581@student.rmit.edu.au

M. Simic
e-mail: milan.simic@rmit.edu.au

M. Simic
Faculty of Health Sciences, University of Sydney, PO Box 170 Lidcombe, Sydney, NSW 1825, Australia
e-mail: milena.simic@sydney.edu.au

M.N. Favorskaya and L.C. Jain (eds.), *Computer Vision in Control Systems-4*, Intelligent Systems Reference Library 136,
https://doi.org/10.1007/978-3-319-67994-5_11

11.1 Introduction

This chapter introduces a new implementation of feature detection in measuring foot progression angle. The FPA is an important parameter in the treatment of the people with medial knee osteoarthritis. Abnormal FPA is a clinical indicator of gait monitoring and retraining [1]. The FPA alternation is presented as a possible solution for reducing the knee loading and knee pain for individuals with knee osteoarthritis. In addition to that, changes in the FPA have been correlated with changes in the foot eversion moment [2], knee adduction moment [3], hip joint moment [4], foot pressure distribution [5], and foot medial loading [6, 7]. A comprehensive review of gait modification strategies for altering medial knee joint load is given in [8]. The importance of exercise, gait retraining, and also footwear and insoles for knee osteoarthritis treatment, is also highlighted in the literature [9, 10].

Currently, the FPA can only be computed, while walking in a laboratory equipped with marker-based or the IMU based motion capture systems. Specialized cameras or sensors are required to be installed by technicians in both methods. Those methods are too complex and not appropriate for long term non-clinical based FPA monitoring and walking habits retaining.

In our initial investigation we have considered applications of various wearable sensors, such as the accelerometers and gyroscopes. In addition to that, the touch sensors were considered for flat foot phase detection. All off those should be installed and used in non-clinical environments. As such, we present a different approach in this study.

First of all, let us explain what the positions are that our body takes, while we are walking naturally in a straight line. We go through the sequence of Gate Cycles. Each Gate Cycle has two phases: Stance Phase, which takes about 60% of the time and Swing Phase with 40% of the Gate Cycle time. During the Stance Phase one leg and foot is bearing the body weight, while during the Swing Phase the feet are not touching the ground. In this study, only Stance Phase is considered, since the FPA can be measured only when a foot is on the ground (see Fig. 11.1).

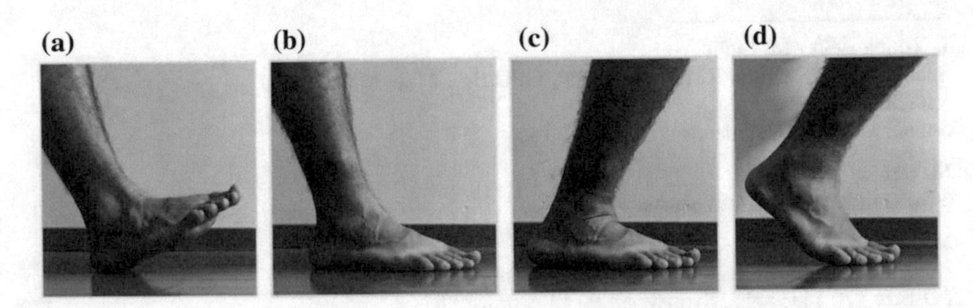

Fig. 11.1 Stance Phase is 60% of the gate cycle, which consists of four stages: **a** heel strike, **b** early flat foot, **c** late flat foot (mid stance), **d** toe off (push off)

Since in the Swing Phase a foot is off the ground, we cannot perform our measurements. We propose the VFM system to estimate the FPA using a monocular camera as a sole sensor in the model. The FPA is calculated as the angle between the foot vector, line joining the heel point center and the second metatarsal head, and the forward Progression Direction (PD) visualized as a pair of parallel lines in transverse plane. Measurements are conducted during foot flat phase of walking. The average time duration available for the FPA measurements is between 15 and 50% of stance [11]. The VFM method, for the first time is proposed for measuring the FPA via a chest mounted smart phone in non-clinical scenarios. The VFM model detects the FPA without installing any devices inside or outside of the shoes and keeps the intrusive and technician intervention to the minimum. In addition, the problem of matching the feature of shoe is addressed in this study. An image rectification is performed to remove the distortion caused by the monocular camera. The VFM modeling results are validated through comparison to the results from Digital Inclinometer Measurement (DIM).

This chapter includes following information. Experiment scenarios and system setup are described in Sect. 11.2. Image calibration and rectification are discussed in Sect. 11.3. Section 11.4 includes the VFM algorithm modules. Validation and results are represented in Sect. 11.5. Conclusions are given in Sect. 11.6.

11.2 Experiment Scenarios and Setup

The research presented here is approved by the Institutional Ethics Committee and all participants have provided written informed consent. In order to achieve Data AcQuisition (DAQ) goal, while still maintaining the low intrusion, low maintenance, high portable, and accessibility at this stage of research, the mobile phone is turned into a webcam via standard vision applications. Laptop is used for image processing. Mobile phone and laptop are on the same Wi-Fi network so that they can communicate. Currently, the VFM model of the FPA analysis is running on MATLAB 2014a environment using a laptop computer. Personal computer platform could, also, be used to analyze and present the gait data. The system is implemented entirely on mobile phone platform in the later stages of the research and biofeedback medical procedure is developed on mobile phone platform as well.

During the proceeding with computer vision and image processing, the optimal system setup became important. Mounting the camera downward on the torso (Fig. 11.2a, b) has a few advantages. First, the torso is just above the foot, which is the best angle to detect the FPA. Second, camera can monitor two feet with closed range, which is better than a vision range obtained from the head and ceiling mounting positions. Third, the position is close to the body weight center with high accessibility and stability. It is better than what can be obtained with other body mounting solutions. The chest mounted camera has similar property like torso one and is applied as an alternative solution.

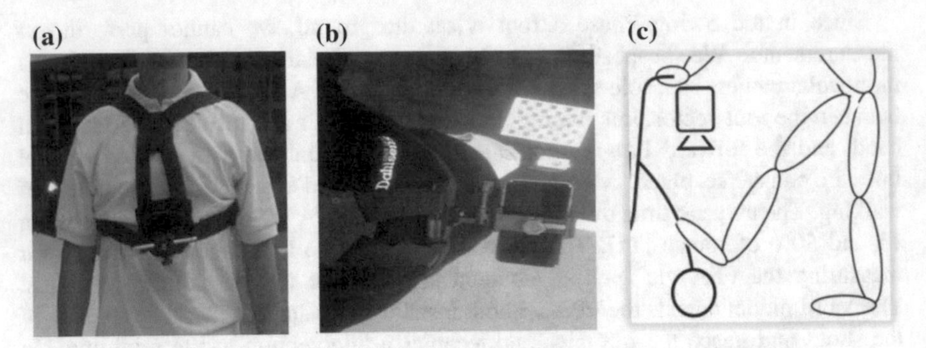

Fig. 11.2 Monitoring system: **a** front view, **b** top view of chest mounted camera and harness, **c** walking aid as platform

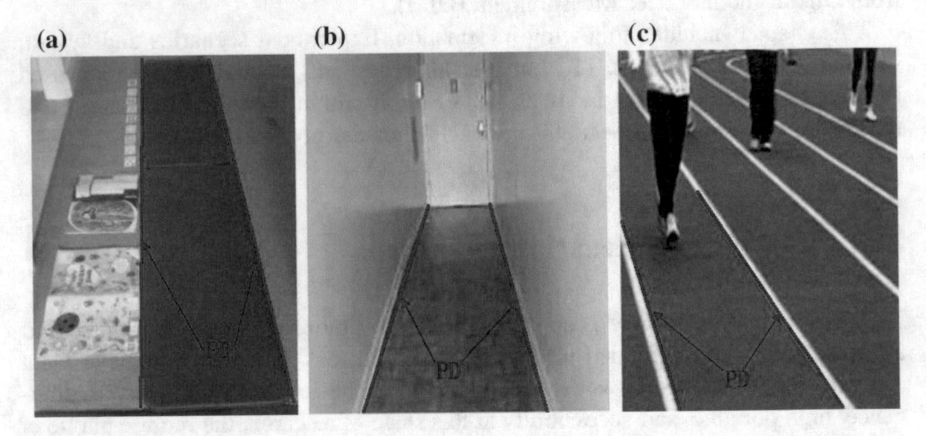

Fig. 11.3 Experimental environments with *straight lines* used for reference direction: **a** floor mat, **b** hall way, **c** tracks

Figure 11.2c shows a camera mounted on a walking aid. This approach is a subject of a separate investigation devoted to the people that need any kind of walking aid. The downward facing camera is mounted on torso via a harness shown in Fig. 11.2a, b. The camera axis does not have to be orthogonal to the plane of motion.

Experimental system is designed to be applied on the track, hallway, and tiles covered floor or any floors with visible straight lines or edges (see Fig. 11.3). Assume that the PD is the same as the direction defined by those parallel lines, which must be shown in camera's image acquired for the purpose of image rectification and lines detection. A smart phone camera is the only the DAQ hardware needed for the VFM model.

11.3 Image Calibration and Rectification

The VFM method of the FPA estimation is composed of seven modules, as shown in Fig. 11.4. The module 1 "Baseline data collection" is the starting module of the whole model. It is only executed once in each walking test. We will now analyze the initial data collection, while all other modules and the whole algorithm will be explained later. Onboard camera captures sequences of images that carry a large

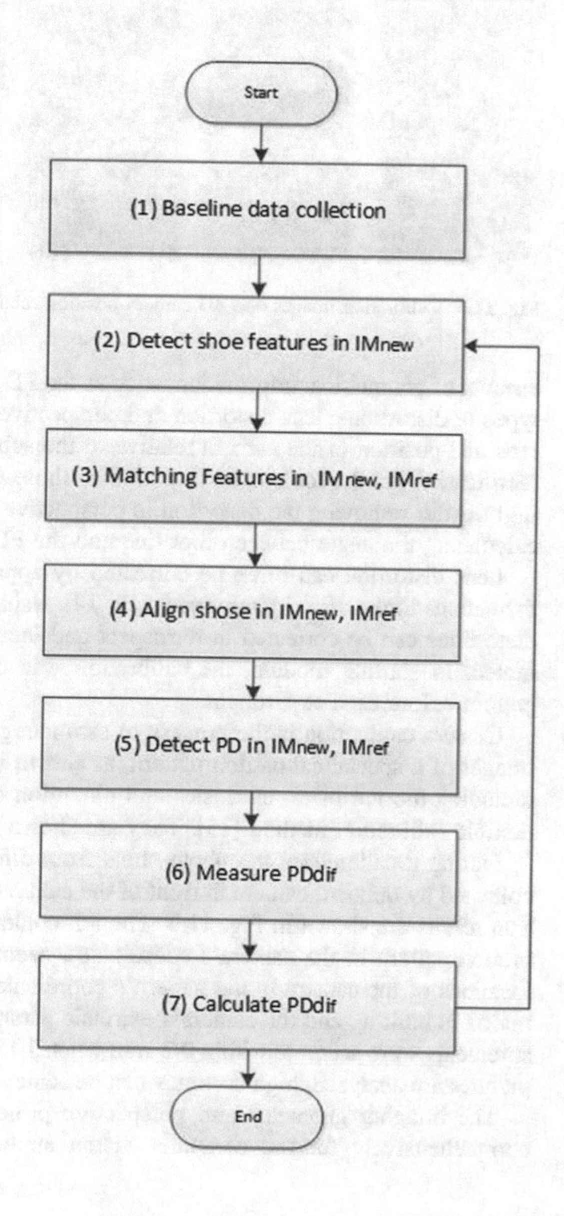

Fig. 11.4 The flow chart of VFM model. Module 1 collects baseline data likes FPA_{ref}, VF_{ref}. Module 2 extracts visual features from current frame of image. Module 3 matches VF_{new} and VF_{ref}. Module 4 aligns shoes of two successive images into same orientation. Module 5 detects the PD in Im_{new} and Im_{ref}. Module 6 measures PD_{dif} in aligned image. Module 7 calculate PD_{dif} and FPA_{new}

Fig. 11.5 Calibration images and 3D camera position rebuild

amount of geometrical information, such as the PD and the FPA. There are also two types of distortions: lens distortion and perspective distortion caused by the optical lens and position of the camera relative to the subject [12]. Image calibration and rectification are the tools used to eliminate those two types of distortion resoundingly. After removing the distortion in perspective image, the FPA is detectable by calculating the angle between foot line and the PD.

Lens distortion can often be corrected by applying suitable algorithmic transformations to the digital photograph [13, 14]. Using those transformations, the lens distortions can be corrected in retrospect and the measurement error can be eliminated. In starting module, the calibration was conducted above the calibration pattern before each experiment.

Camera calibration is the process of estimating parameters of the camera using images of a special calibration pattern, as shown in Fig. 11.5. Camera parameters, including the intrinsics, extrinsics and distortion coefficients, are extracted by the flexible calibration method [15]. They are shown in Fig. 11.5.

During a calibration, the photo shots from different angles and positions were collected by onboard camera in front of the calibration pattern board. The visualization results are shown in Fig. 11.6. The left column plots the locations of the calibration pattern in the camera's coordinate system and the right column plots the locations of the camera in the pattern's coordinate system. Based on that, 3D info matrix is built up and the camera's extrinsic parameters were calculated. The measurements were accurate within 0.2 mm when 10 images are collected. When more pictures are analyzed, high accuracy can be achieved, as given in literature [15].

The imaging geometry and perspective projection were already investigated comprehensively for the computer vision applications [16]. Under perspective

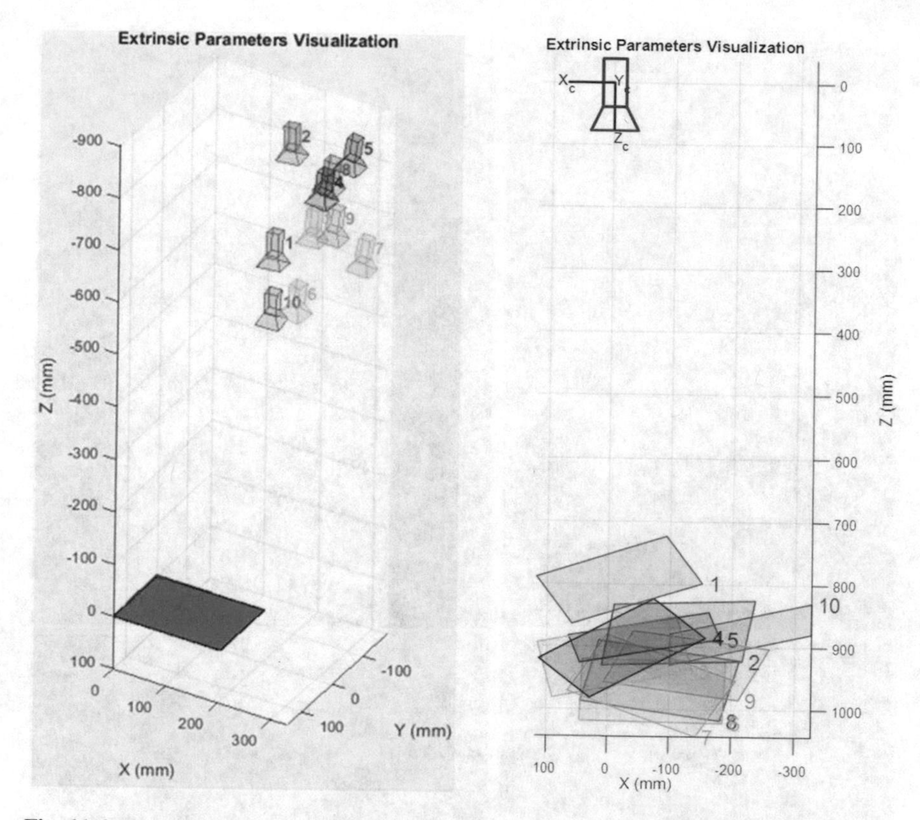

Fig. 11.6 Visualization results of camera calibration

image, a plane is mapped to the image by a plane projective transformation [12]. This transformation is used in many computer vision applications, including planar object recognition [17, 18]. The projective transformation is determined uniquely if Euclidean world coordinates of four or more image points are known [19].

Once the transformation is determined, the Euclidean measurements, such as lengths and angles, can be calculated on the world plane directly. In this research, only planar angle measurement is studied, which needs a pair of parallel lines on ground to rectify images (Fig. 11.7a, c). While walking along the parallel lines of track, hall way, or tile defined direction, the parallel lines are detectable by Hough transformation [20, 21]. The Hough transformation is applied in the module 5 "Detect PD line". The PD lines are shown in Fig. 11.7 before and after the Hough transformation. Those detected parallel lines are used as the reference lines to transfer the prospective image into the rectified image with metric angle information. The rectified image, as shown in Fig. 11.7b, d, allows the metric properties, such as planar angles to be measured on the world plane from those perspective images [19, 22].

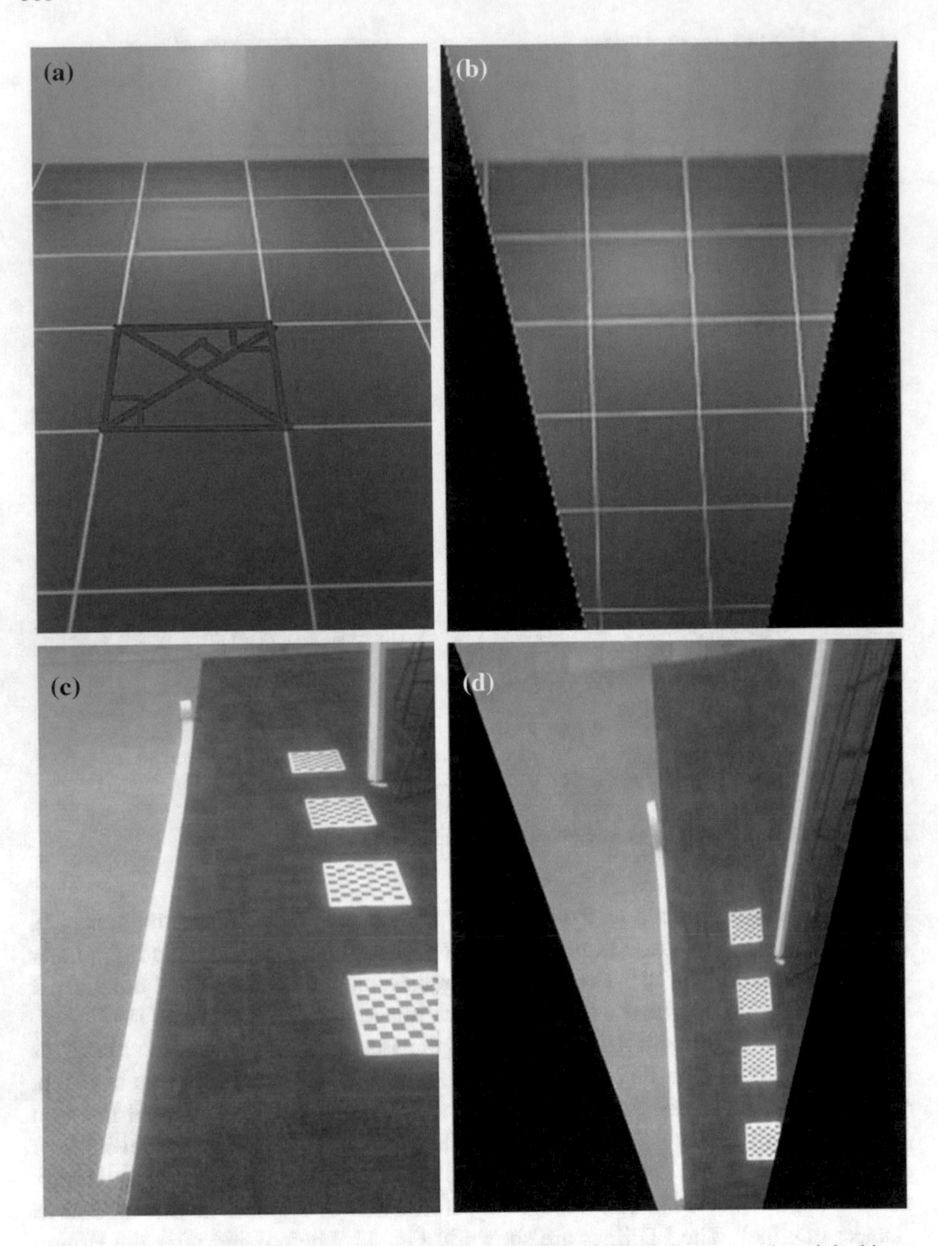

Fig. 11.7 Angle measurement in the rectified images of the planar surfaces: **a** and **c** original input image, tile, tape, **b** and **d** rectified image

In the Planar Angle Measurement (PAM) demonstration, two paper strips were placed upon a same plane and rotated constantly. The relative angle of two strips was calculated in the real time by the PAM method (Fig. 11.8a, b). In the demo, the

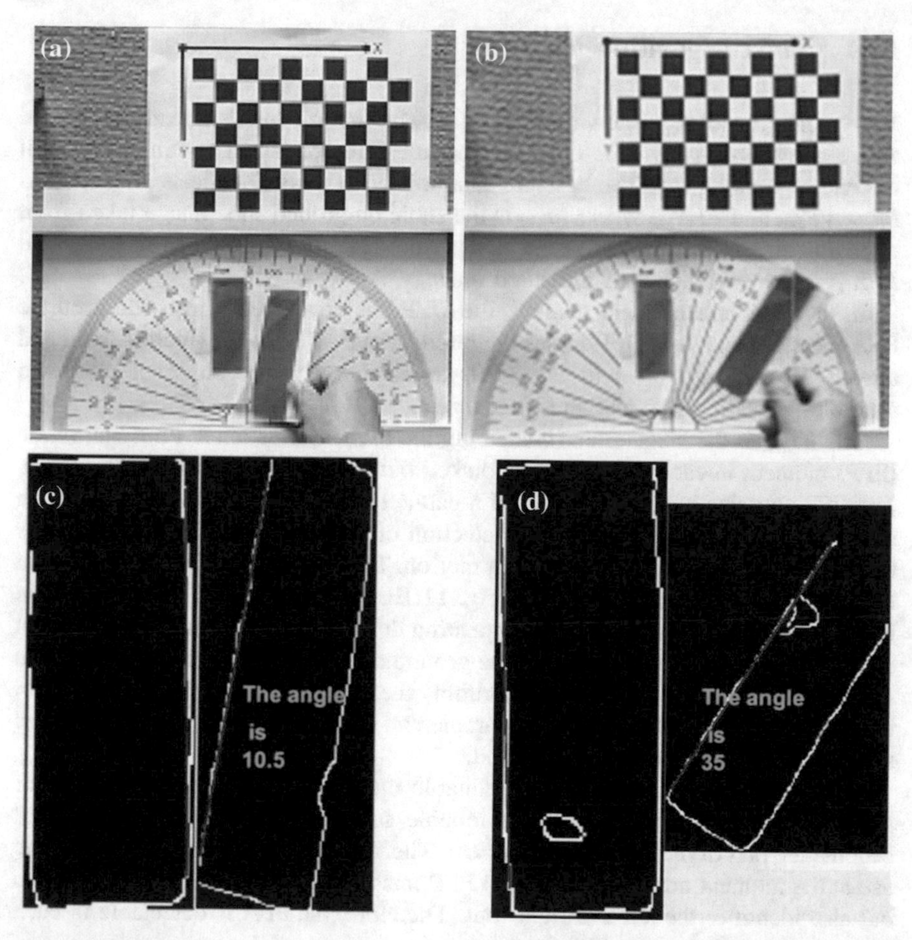

Fig. 11.8 Angle measurement in the rectified planar surface: **a** and **c** input images, **b** strip edges, **d** angle detection

paper strips were segmented out from each image frame in video stream, as shown in Fig. 11.8d [23, 24]. After calibration and rectification, the angles between two stripes were detected accurately with average error $E_{aver} \pm 0.05°$ shown in Fig. 11.8c, d. The PAM method is applied in the module 6 "Measure planar angle" to measure the PD_{dif}, as shown in Fig. 11.8. Our average error is well below the FPA differences that appear with every step, i.e. all step are different and accordingly all FPAs.

11.4 VFM Algorithm Modules

The whole procedure is shown in Fig. 11.4. The first module is already explained above comprehensively. The other six modules are conducted continuously until the end of walking test. The baseline data collection module collects data such as Im_{ref}, VF_{ref}, and FPA_{ref}, where Im_{ref} is the static on-ground shoe image taken from onboard camera, VF_{ref} refers to the shoe visual features extracted from Im_{ref}, and FPA_{ref} is the reference FPA measured in Im_{ref}.

In the FPA enhancement method, the DFPA is developed to provide accurate FPA measurement. The DFPA uses the reference FPA (FPA_{ref}), which is fixed and already known, as shown in Fig. 11.9, to obtain a precise angle of unknown rotation by relating it to known object Im_{ref} via VF_{ref}.

The Im_{new} is the image frame containing the FPA information at Foot Flat Phase (FFP) moment in each step, which is picked out by the FFP estimation algorithm. The FFP optimization is the subject of a parallel investigation. The main approach in that complementary research is a detection of the sequence of still foot images that implies stationary phase in the foot motion. The sequence is appearing from the Early Flat Foot stage, as shown in Fig. 11.1b, to the Late Flat Foot, shown in Fig. 11.1c. In that foot flat position, appearing during the stance phase after the heel contact, a foot stays in contact with the ground and the location of the foot should not be changing. For real time algorithm, the time duration of the FFP is an important factor. Some of the other parameters, such as the FPA, step length, and step width, are detectable in this period.

Another approach that is also a valuable option would be the application of additional tactile sensors. The most reliable solution will be selected and will become the part of the biofeedback system. The vision based FFP algorithm that we use at the moment not only detects the FFP image frame in real-time video stream but also identifies the left and right foot. Therefore, the FPA is detectable in each image that is indicated by FFP moment.

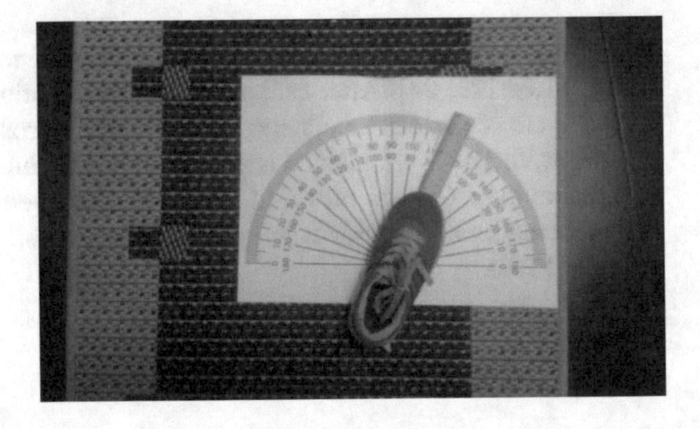

Fig. 11.9 The known FPA measured applying a protractor is used in defining reference angle FPA_{ref}

Hereinafter, consider the foot feature extraction and matching, alternative feature extraction method, and the FPA measurement in Sects. 11.4.1–11.4.3, respectively.

11.4.1 Foot Feature Extraction and Matching

The Speeded-Up Robust Feature (SURF) algorithm is based on extraction of feature points from image [25]. This method is also used for description of feature points and comparison of these points in the module 2 "Detect shoe features" and module 3 "Matching features", as shown in Fig. 11.10. The main advantage of the SURF is its speed. Higher speed processing is achieved using the integral convolution method and approximation of Gaussian function [15]. The feature points of shoes

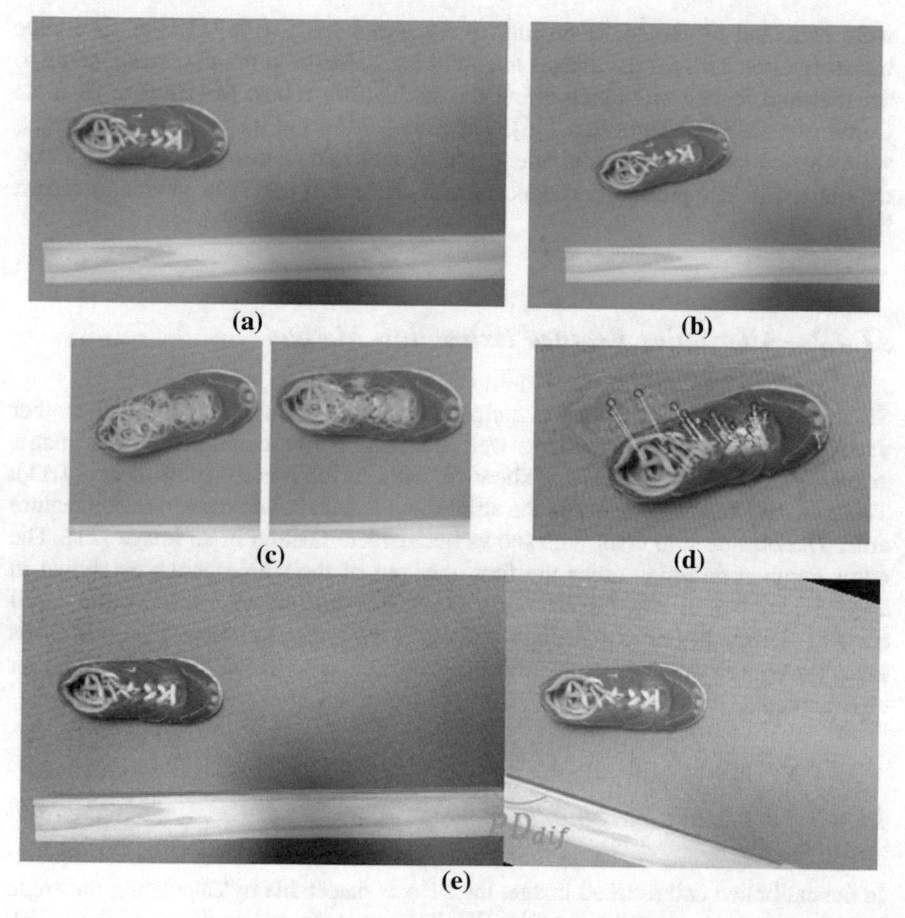

Fig. 11.10 Shoes alignment: **a** initial image Im_{ref}, **b** obtained image Im_{new}, **c** feature point extraction, **d** matched features, **e** shoes alignment and PD_{dif} measurement

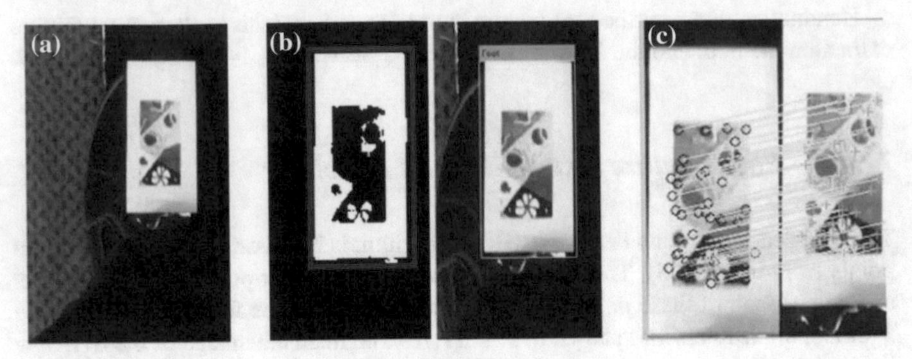

Fig. 11.11 Shoe indicator (an alternative feature extraction method): **a** foot indicator strip, **b** segmented foot area, **c** matched features

were extracted by the SURF in images Im_{ref} and Im_{new} (Fig. 11.10c). The shoe transformation between the images Im_{ref} and Im_{new} has been found corresponding to the matched feature point pairs using the statistically robust M-estimator SAmple Consensus (MSAC) algorithm [26]. The shoes from the images Im_{ref} and Im_{new} were aligned into same size and orientation by matched feature point pairs for PD_{dif} measurement. The process is conducted in module 2 of the VFM model, as shown in Fig. 11.4.

11.4.2 Alternative Feature Extraction Method

Since the color and textile of participant's shoes are unexpected and the other variable environment conditions as well, the shoe is not identifiable in the image. A pair of paper strips, instead of shoes, is used in the investigations (Fig. 11.11). There are two functional areas in the strip: outside solo color area and inside feature area. The outside solo color area shows better identification in an image [11]. The color segment detection crops the foot area out of the whole photo, as shown in Fig. 11.11a, b [23, 24]. Furthermore, the inside feature area reduces the VFM model's detection time cost without affecting the angle accuracy. This method is ideal for participants to monitor the FPA in their own shoes and in the familiar environments.

11.4.3 FPA Measurement

In the calibrated and rectified image, the FPA is detectable by calculating the angle between the foot orientation and the PD. The numbers and positions of the SURF feature points in each image are variable. Accordingly, there is no a certain

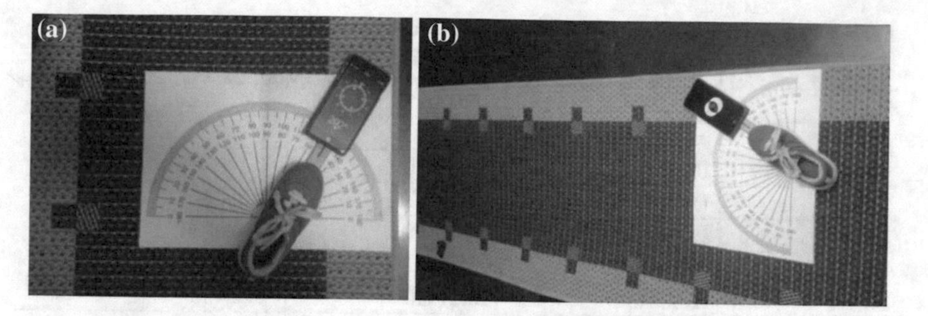

Fig. 11.12 Validation methods: **a** the DIM, **b** the PM

geometry shape of a group of feature points to describe the same shoe in each image. Instead of measuring angle between the feature points and the PD, angles of VF_{dif} are measured in aligned Im_{ref} and Im_{new} via VF_{ref} and VF_{new}. VF_{dif} is the relative position between VF_{ref} and VF_{new} in Im_{ref} and Im_{new}. Due to the fact that the human body is not a fixture, the positions between the onboard camera and shoes are variable in Im_{ref} and Im_{new}. For example, when the camera rotates with body and the foot keeps still on ground, FPA_{dif} is zero degree theoretically, while VF_{dif} reports a rotation value due to the body sway. Therefore, FPA_{dif} is defined the same as PD_{dif} (the rotation offset of the PD measured in Im_{ref} and Im_{new} in the shoe aligned picture). It is different to the scenario that can be seen in Fig. 11.10e.

Thus, FPA_{new} is calculated by using PD_{dif} and FPA_{ref}, as given by Eq. 11.1.

$$FPA_{new} = PD_{dif} + FPA_{ref} \tag{11.1}$$

11.5 Validation and Results

To test the performance of the proposed VFM model of the FPA estimation, the validation tests were conducted both on static protractor patterns and in process of walking along a straight track. The accuracy of the VFM model was evaluated based on the average errors with respect to the FPA computed from the DIM and Protractor Measurement (PM) methods. One-way ANalysis of VAriance (ANOVA) was used to determine if there was any difference in errors of FPA estimation based on the DIM and the VFM model.

In the static validation, the shoe was placed upon a large protractor laying on the track and rotated transversely from 0° to 40°, as shown in Fig. 11.12. The FPA results estimated by the PM, the DIM and the VFM are shown in Table 11.1. The average FPA errors of the DIM and the VFM were $0.0063 \pm 0.000009°$ and $0.312578 \pm 0.0494°$. As digital inclinometer and protractor are in the same plane, the DIM and the PM recorded the FPA errors within 0.64%, which is negligible.

Table 11.1 Validation results of the PM, the DIM, and the VFM

PM (°)	DIM (°)	VFM (°)	Errors of VFM (°)
0	0.007953	0.019978	0.012025
5	5.004283	4.898605	–0.10568
10	10.009435	10.19978	0.190345
15	15.004838	14.70017	–0.30467
20	19.998932	20.41958	0.420648
25	25.008534	24.47842	–0.53011
30	30.008432	29.37410	–0.63433
35	35.0052843	35.69447	0.689186
40	40.0069274	40.74925	0.742323

Table 11.2 Walking validation results of the DIM and the VFM estimations

Left foot			Right foot		
#	DIM (°)	VFM (°)	#	DIM (°)	VFM (°)
1	4.39	4.56	2	5.53	5.56
3	5.29	5.46	4	5.71	6.02
5	4.41	4.62	6	4.44	4.68
7	5.50	5.59	8	5.23	5.55
9	5.47	5.61	10	5.37	5.70
11	5.33	5.39	12	4.63	4.90
13	5.22	5.44	14	4.62	4.87
15	4.47	4.47	16	5.65	5.89
17	5.04	5.24	18	4.55	4.55
19	5.52	5.86	20	4.88	5.10

Furthermore, it is not feasible to place protractor under each step, therefore only the DIM is applied to validate the VFM model in the walking test.

From the column of errors of the VFM in Table 11.1, we can see that the error of the VFM estimation increases, as foot rotates away from the original orientation (0°). This error can be minimized by setting original reference of shoes orientation into the middle of the subjects' FPA range. For example, if the subject's FPA range is observed from 20° to 40°, then the reference shoes orientation should be set to 30°, which minimizes the error ranges.

In the walking test, the subject's nature FPA was observed to be around 4°, which was set as original reference of shoes orientation. The DIM and the VFM estimations of the FPA during a trial are shown in Table 11.2. In general, the FPA estimations from the VFM model closely followed the DIM estimations under the walking test. The average FPA errors were $-0.15 \pm 0.13°$ for normal gait.

For comparison, consider that a recently published result in a foot wore Magneto-Inertial Sensing system yielded corresponding error measures of $-0.15 \pm 0.24°$ [2]. Our proposed FPA estimation uses only a smart phone as input

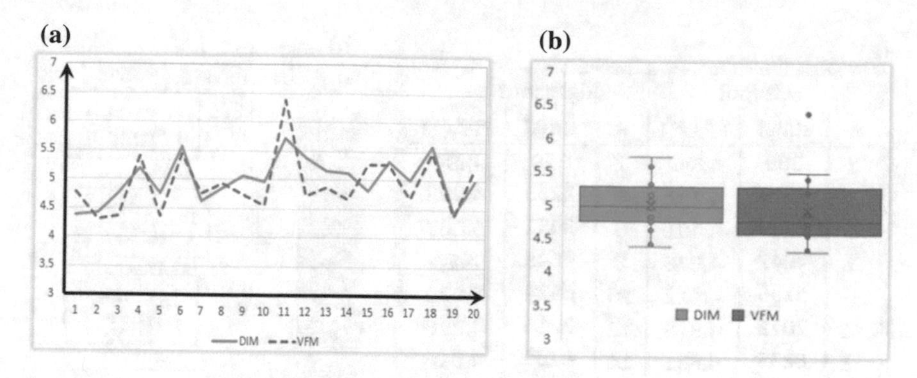

Fig. 11.13 FPA validation results and error analysis: **a** plots, **b** variance

sensor and achieves similar error rates, while being substantially less complex to implement.

Validation results are shown in Fig. 11.13. Solid and dashed lines represent the DIM and the VFM estimations over 20 steps in the trial respectively, see Fig. 11.13a. There are no significant differences between the DIM and the VFM methods of FPA estimates for the nature walking condition. The average errors of the VFM are slightly larger than the DIM method errors, and one outliner (6.4°) is found at 11th step, as shown in Fig. 11.13a. This is likely due the curly edges of the carpet, which could affect the accuracy of the PD and the FPA measurements.

The VFM model was tested to analysis the FPA with respect to three variables: Step Length, Step Width, and Body Swing Angle (BSA) [27]. The 10-step free-cadence walking at the self-selected velocity was recorded firstly as baseline. Cadence is the walking rate expressed in steps per minute. Everyone can calculate his/her free-cadence walking rate. Average is in the range of 100–130 steps/min calculated based on the walking speed in km per hour and the step length. There are huge differences for each person.

Ten real-time numeric outputs of gait information for free-cadence walking were collected and are presented in Fig. 11.14a, while their corresponding graphic patterns are displayed in Fig. 11.14b. The graphic patterns, one per step, are instantly updated on the computer screen. Gait information like the foot progression angle, body swing angle, and flat-foot stance time of each step are displayed in text box of Fig. 11.14b. The gait information of two successive steps, such as the step length and step width, is illustrated between two frames.

The step length, step width and the BSA variability are modified to present the effect to the baseline FPA through this method, as shown in Fig. 11.15. It can be seen that three groups of the FPA values are computed in relevant to the variability of participant's step length. The first increment of step length is 100 mm, and the second is 200 mm. The blue dots present results for walking using nature step length that have the largest FPA. As the step length increases, the FPA is observed to be smaller, i.e. foots ate more aligned to the progression direction. The regression function between the FPA variability and step length variables is given by Eq. 11.2,

(a)

	Left foot			Right foot	
#	t(ms)	FPA(°)	#	t(ms)	FPA(°)
1	389	4.542	2	979	4.34
3	1573	4.132	4	2265	4.095
5	2859	4.401	6	3452	5.267
7	4047	4.876	8	4645	4.74
9	5236	4.412	10	6424	5.026
11	7022	4.378	12	7616	4.218
13	8208	4.881	14	8799	4.616
15	9392	5.019	16	9991	4.213
17	10587	4.47	18	11182	5.086
19	11776	4.943	20	12358	4.497

(b)

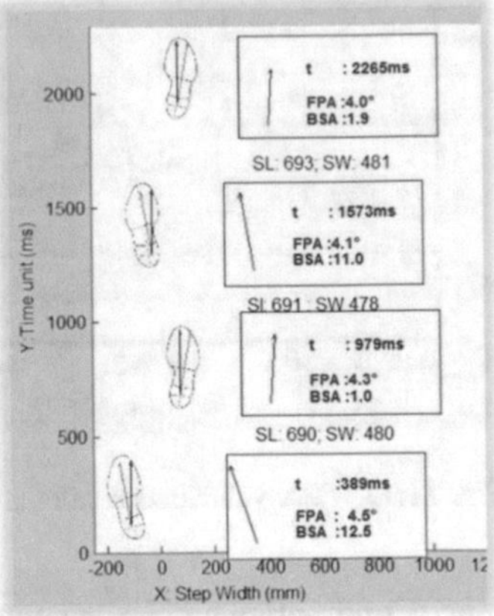

Fig. 11.14 Real time numeric outputs: **a** numerical outputs, **b** graphical outputs collected during the walk. Axis OY corresponds to the PD. Motion time is expressed in ms. Axes OX and OY together show 2D layout of the experimental place. Time *t* is the elapsed time from the beginning of the video, while the FPA and the BSA, as all other quantities, are shown for each step (# is a step number)

where 1 degree increase in the FPA variability is associated with 129.29 mm reduction in *step_length* variability.

$$FPA = 13.203 - 1.2929 * step\_lenth \tag{11.2}$$

Figure 11.15b reflects the participant's FPA in respect to step widths, in his free-cadence, and in targeting 10° and 20° larger, respectively. Changing step width affects the FPA dramatically as the FPA ranges from 2° to 12° in Fig. 11.15b. The FPA is affected but not on same direction. Mean FPA is 1.5° less than the baseline FPA, when the participant targeting 10 mm larger step width and 6° larger in targeting 20 mm larger step width. It depends on how the participant deals with the gait modification in his/her own most comfort way.

Figure 11.15c plots the effect of the BSA in respect to the FPA on three walks: free-cadence walking manner, increasing the BSA by 10° and 20° on that. The FPA stays at a stable value, while the participant can keep the balance by swinging other part of body. The chart does not address the relative importance related to the FPA contributed by the BSA. The FPA shows low correlation with the body sway compared to the step length and step width. The differences between the FPA values measured for various BSA are not significant. Thus, all three groups of data represent the same mean FPA value at increasing BSA.

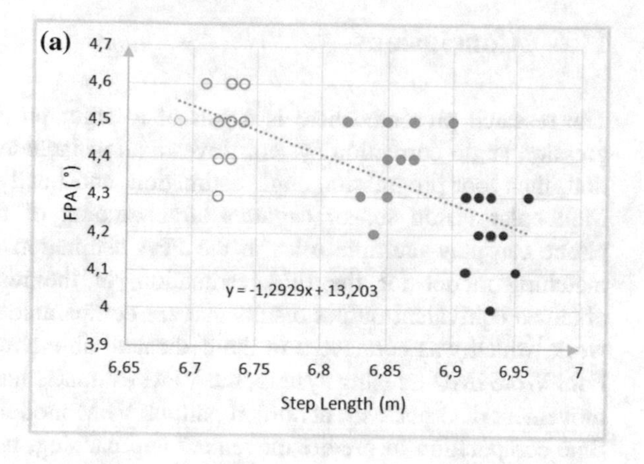

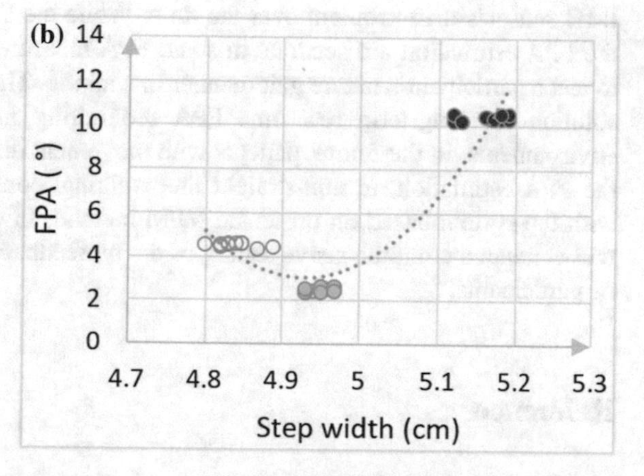

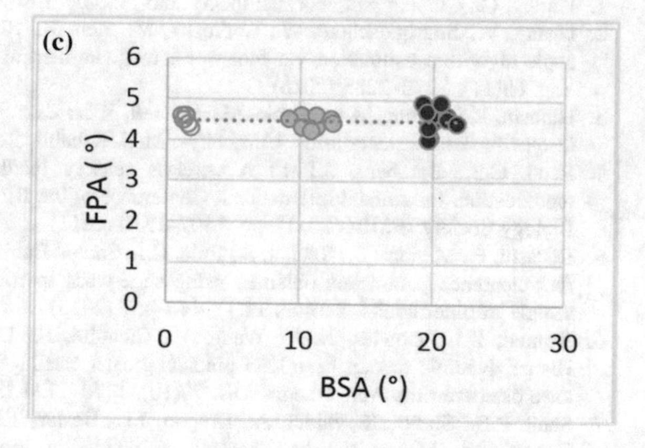

Fig. 11.15 The statistic charts of the FPAs in respect to: **a** step length in, **b** step width in, **c** the BSA in. *White dots* with *blue circle* represent the gait info of free-cadence as a baseline, *green* and *red dots* represent data collected on first and second gait modifications

11.6 Conclusions

The research presented here is a part of a larger project called mobile foot progression angle correction system. Investigation includes the flat foot phase detection, then foot progression angle estimation, and finally, correction system design. Monocular vision sensor captures large amount of data simultaneously. Smart phone can play multiple roles in the FPA estimation process. The visual feature matching model for the FPA estimation, in the research presented here, has obtained equivalent output results that are comparable with the recently published work, which was conducted in the dedicated lab environments. Comparing to the Foot Wore IMU Sensing system, the VFM estimates has the advantage in detecting movement disorder with abnormal gait; as VFM model, it does not need large real time computation to predict movement approaching, nor need the coping with the IMU sensor's drift problem over the time. While the PM and the DIM methods of the FPA estimation are accurate in static FPA measurements, they are not feasible to test a participant's nature gait manner in walking. Therefore, the VFM model is a solution for long-term real time FPA monitoring in home or community like environments. In the future, patients with movement disorders or abnormal gait and the FPA estimation, in non-straight line walking, could be diagnosed and treated with the system based on presented VFM method. In order to prepare for that, we will conduct a comprehensive testing of the biofeedback method with large number of participants.

References

1. Kirtley, C.: Clinical gait analysis: theory and practice. Elsevier Health Sciences (2006)
2. Huang, Y., Jirattigalachote, W., Cutkosky, M., Zhu, X., Shull, P.: Novel foot progression angle algorithm estimation via foot-worn, magneto-inertial sensing. IEEE Trans. Biomed. Eng. **63**(11), 2278–2285 (2016)
3. Hinman, R.S., Hunt, M.A., Simic, M., Bennell, K.L.: Exercise, gait retraining, footwear and insoles for knee osteoarthritis. Curr. Phys. Med. Rehabil. Rep. **1**, 21–28 (2013)
4. Redd, C.B., Bamberg, S.J.M.: A wireless sensory feedback system for real-time gait modification. In: Annual International Conference of the IEEE Engineering in Medicine and Biology Society (EMBC'2011), pp. 1507–1510 (2011)
5. Dadashi, F., Mariani, B., Rochat, S., Büla, C.J., Santos-Eggimann, B., Aminian, K.: Gait and foot clearance parameters obtained using shoe-worn inertial sensors in a large-population sample of older adults. Sensors **14**(1), 443–457 (2013)
6. Bennell, K.L., Bowles, K.-A., Wang, Y., Cicuttini, F., Davies-Tuck, M., Hinman, R.S.: Higher dynamic medial knee load predicts greater cartilage loss over 12 months in medial knee osteoarthritis. Ann. Rheum. Dis. **70**(10), 1770–1774 (2011)
7. Shull, P.B., Shultz, R., Silder, A., Dragoo, J.L., Besier, T.F., Cutkosky, M.R., Delp, S.L.: Toe-in gait reduces the first peak knee adduction moment in patients with medial compartment knee osteoarthritis. J. Biomech. **46**(1), 122–128 (2013)
8. Simic, M., Hinman, R.S., Wrigley, T.V., Bennell, K.L., Hunt, M.A.: Gait modification strategies for altering medial knee joint load: a systematic review. Arthritis Care Res. **63**(3), 405–426 (2011)

9. Ferrigno, C., Stoller, I.S., Shakoor, N., Thorp, L.E., Wimmer, M.A.: (2016) The feasibility of using augmented auditory feedback from a pressure detecting insole to reduce the knee adduction moment: a proof of concept study. J. Biomech. Eng. 138(2):Article 021014
10. Xu, W., Huang, M.C., Amini, N., Liu, J.J., He, L., Sarrafzadeh M.: Smart insole: a wearable system for gait analysis. In: 5th International Conference on Pervasive Technologies Related to Assistive Environments (PETRA'2012), Article No. 18 (2012)
11. Simic, M., Wrigley, T., Hinman, R.S., Hunt, M., Bennell, K.: Altering foot progression angle in people with medial knee osteoarthritis: the effects of varying toe-in and toe-out angles are mediated by pain and malalignment. Osteoarthr. Cartil. **21**(9), 1272–1280 (2013)
12. Semple, J., Kneebone, G.: Algebraic Projective Geometry. Oxford University Press, Oxford (1979)
13. Duane, C.B.: Close-range camera calibration. Photogram. Eng. **37**(8), 855–866 (1971)
14. Faig, W.: Calibration of close-range photogrammetric systems: mathematical formulation. Photogram. Eng. Remote Sens. **41**(12), 1479–1486 (1975)
15. Zhang, Z.: A flexible new technique for camera calibration. Pattern analysis and machine intelligence. IEEE Trans. Pattern Anal. Mach. Intell. **22**(11), 1330–1334 (2000)
16. Ma, Y., Soatto, S., Kosecka, J., Sastry, S.S.: An invitation to 3-D vision: from images to geometric models. Springer Science & Business Media (2012)
17. Kitt, B.M., Rehder, J., Chambers, A.D., Schonbein, M., Lategahn, H., Singh, S.: Monocular visual odometry using a planar road model to solve scale ambiguity. In: 5th European Conference on Mobile Robots (ECMR'2011), 43–48 (2011)
18. Zienkiewicz, J., Davison, A.: Extrinsics autocalibration for dense planar visual odometry. J. Field. Robot. **32**(5), 803–825 (2015)
19. Liebowitz, D., Zisserman, A.: Metric rectification for perspective images of planes. In: IEEE Computer Society Conference on Computer Vision and Pattern Recognition (CVPR'1998), 482–488 (1998)
20. Young, J., Elbanhawi, M., Simic, M.: Developing a navigation system for mobile robots. In: Damiani E., Howlett R.J., Jain L.C., Gallo L., De Pietro G. (eds.) Intelligent Interactive Multimedia Systems and Services, SIST, vol. 40, pp. 289–298. Springer, Berlin (2015)
21. Ballard, D.H.: Generalizing the hough transform to detect arbitrary shapes. Pattern Recogn. **13**(2), 111–122 (1981)
22. Johnson, M.K., Farid H.: Metric measurements on a plane from a single image. Department of Computer Science, Dartmouth College, Tech. Rep. TR2006-579 (2006)
23. Cheng, H.-D., Jiang, X., Sun, Y., Wang, J.: Color image segmentation: advances and prospects. Pattern Recogn. **34**(12), 2259–2281 (2001)
24. Sural, S., Qian, G., Pramanik S.: Segmentation and histogram generation using the HSV color space for image retrieval. In: IEEE International Conference on Image Process (ICIP'2002), vol. 2, pp. 1–4. (2002)
25. Juan, L., Gwun, O.: A comparison of SIFT, PCA-SIFT and SURF. Int. J. Image Process **3**(4), 143–152 (2009)
26. Jacobs, L., Weiss, J., Dolan, D.: Object tracking in noisy radar data: comparison of Hough transform and RANSAC. In: IEEE International Conference on Electro/Information Technology (EIT'2013), (2013). doi:10.1109/EIT.2013.6632715
27. Owings, T.M., Grabiner, M.D.: Step width variability, but not step length variability or step time variability, discriminates gait of healthy young and older adults during treadmill locomotion. J. Biomech. **37**(6), 935–938 (2004)

Printed in the United States
By Bookmasters